ELECTRICAL DESIGNS

COMPRISING INSTRUCTIONS FOR CONSTRUCTING SMALL MOTORS, TESTING INSTRUMENTS AND OTHER APPARATUS; WITH WORKING DRAWINGS FOR EACH DESIGN.

REPRINTED FROM

THE AMERICAN ELECTRICIAN

Electrical Designs

by American Electrician Magazine

Originally published by
American Electrician Company
New York

Reprinted by
Lindsay Publications Inc
Bradley IL 60915

ISBN 0-917914-22-8

3 4 5 6 7 8 9 0

1998

WARNING

Remember that the materials and methods described here are from another era. Workers were less safety conscious then, and some methods may be downright dangerous. Be careful! Use good solid judgement in your work, and think ahead. Lindsay Publications Inc. has not tested these methods and materials and does not endorse them. Our job is merely to pass along to you information from another era. Safety is your responsibility.

PREFACE.

The chapters of this book originally formed articles written for the AMERICAN ELECTRICIAN by the designers of the apparatus described, which, in many cases, had been actually built and used prior to the publication of the description. The designs were all prepared with a view to reducing to the simplest degree the tools and facilities necessary for the construction of the apparatus. The designs for such of the small motors as have not been built have been calculated with ample allowance for variations in the quality of iron and steel, so that any discrepancy between the performance predicted for these machines and that actually realized will be in favor of the motor, if the material is of a grade at all allowable in good foundry practice.

WARNING

Remember that the materials and methods described here are from another era. Workers were less safety conscious then, and some methods may be downright dangerous. Be careful! Use good solid judgement in your work, and think ahead. Lindsay Publications has not tested these methods and materials and does not endorse them. Our job is merely to pass along to you information from another era. Safety is your responsibility.

Write for a catalog or other unusual books available from:

Lindsay Publications
PO Box 12
Bradley, IL 60915-0012

TABLE OF CONTENTS

CHAPTER I.

ONE-SIXTH HORSE-POWER MOTOR WITH DRUM ARMATURE.

In preparing this design and those which follow, it has been assumed that any one who is sufficiently interested in the subject to undertake the construction of a motor or dynamo will be sufficiently familiar with electro-mechanics to exercise individual judgment in the matter of fitting the various parts, and also in the design and construction of journal boxes, brush holders, terminal blocks and such other parts as are not of vital importance in the electrical design of the machines. Detailed descriptions of these parts will, therefore, not be given; the reader may easily

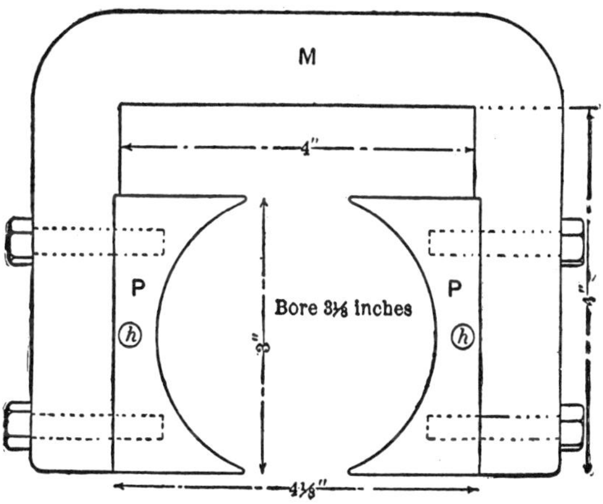

FIG. I.—ELEVATION OF FIELD MAGNET.

inform himself concerning these, if necessary, by inspecting a finished machine of almost any type, or by reference to any good text-book.

The accompanying sketches are intended to serve as working drawings in the construction of a 1-6 horse-power motor, for operation upon a 110-volt direct-current circuit. In Figs. 1 and 2, *M* is the field magnet, consisting of a bar of wrought iron three inches wide and one inch thick, bent into the shape shown; the inner surface of each limb is machined smooth a distance of three inches, forming shallow mortises to receive the pole-pieces, *P P*, which are secured by ¼-in. cap screws passing through the magnet limbs. The pole-pieces, *P P*, are of gray cast iron, and should be finished on all sides to remove the scale as well as to improve the appearance of the completed machine. The magnet, *M*, might be made to look neater by touching up its sides on a coarse emery wheel; it should be well annealed after bending and finishing.

Two holes, *h*, *h*, are bored through the pole-pieces, after these are fitted to the magnet, but before the armature chamber is bored out. These holes are 17-64 in. diameter, and they must be 3¾ ins. apart, center to center, and equ distant from the center of the armature chamber; if the magnet limbs conform strictly to the measurement given from face to face of the finished part of the limbs, the centers of the holes, *h*, *h*, will each be 3-16 in. from the joint between the magnet and the pole pieces. In these holes are to be inserted ¼-in. iron or steel rods 7¼ ins. long, threaded at each end a distance of ⅜ in.

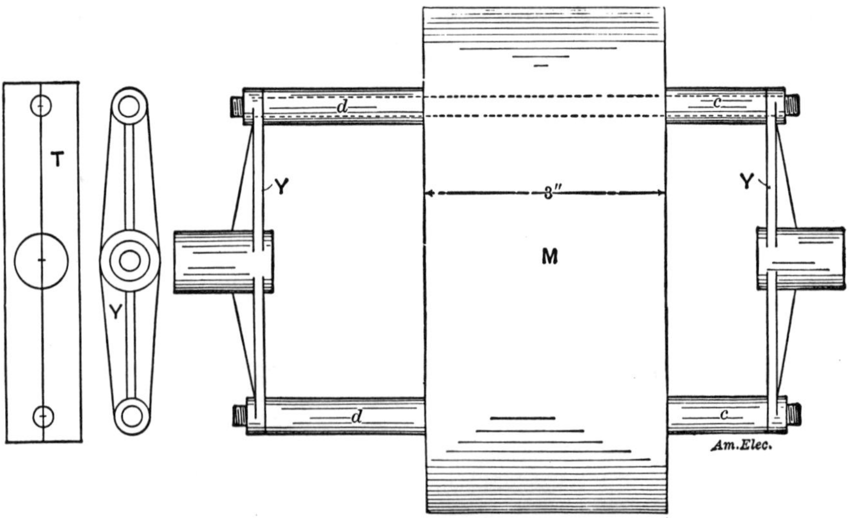

FIG. 2.—PLAN OF FIELD MAGNET AND JOURNAL YOKES.

Fig. 2, which gives a plan view of the magnet and the journal yokes, *Y*, *Y*, shows the function of these rods; they support the yokes and carry distance-pieces, *c*, *c*, *d*, *d*, made of brass tubing just large enough to slip over the rods, and having ⅛-in. walls. The pieces, *c*, *c*, are 1⅜ ins. long, and *d*, *d*, are 2⅛ ins. long. The yokes are held in place by brass nuts, not shown in Fig. 2.

The journal yokes, *Y, Y*, are alike. They are of cast brass, 1/8 in. thick, with a stiffening rib 1/8 in. thick, on each side of the journal box. The inner end of one box should be trued up to receive the brush arm or quadrant. The yokes may be much more easily and accurately fitted if a steel template is used. This may be cheaply provided by taking a piece of flat steel, 1 in. wide and 4 1/2 ins. long, scribing a straight line approximately down the center, and drilling three holes as shown by *T*, Fig. 2, the center one 11-16 in. and the others 1/4 in. in diameter. After the box is bored, mount it on a mandrel and turn down the inner end to fit the center hole in the template *T*, and at the same time face up the ends of the yoke where they are to touch the distance-pieces; put the template on the end of the box and scribe the positions of the 1/4 in. holes on the ends of the yoke. This template should also be used to fix the distance apart of the holes, *h, h* (Fig. 1). The boxes are bored out 9-16 in. in diameter and fitted with bushings of 1/4 in. bore and 1 in. long; oil grooves should be cut at each end of the box and provision made for taking out the oil. Oil cups may be used to feed the bearings.

After the yokes are fitted the frame may be centered in a lathe as follows, for boring out the pole-pieces. Take a piece of 1/2-in. steel rod 11 ins. long, and make the shaft *S* (Fig. 3); the distance from *e* to *g* is 3 1/4 ins., and the diameter there is 1/4 in.; from *g* to *i* is 3 11-16 ins. and the diameter 3/8 in.; from *i* to *j* is 1 1-16 ins. and the diameter 1/2 in.; from *j* to *k* is 3 ins., and the diameter is 1/4 in. Turn the ends of the shaft down to a point, like that of a lathe center; put it in the boxes, bolt the yokes in place, and then put the frame on the lathe carriage, adjusting it until the sharp ends of the shaft are in exact line with the lathe centers. Bolt the motor down in this position, remove the yokes and shaft, and bore out the pole-pieces. The ends of the shaft should afterwards be squared off, care being taken to cut exactly 1/2 in. off each end, leaving the shaft 10 ins. long.

The armature (Fig. 3) is built up of iron discs 3 ins. in diameter and not more than 1-32 in. thick; there are twelve slots, each 5-16 in. wide and 7-16 in. deep. These may be punched in each disc separately, if a stamping press is available, or they may be milled after the discs are assembled on the shaft. If the slots are milled the discs should be taken off the shaft afterwards and the burrs dressed off, care being taken to reassemble them exactly as they were when the slots were milled; this may be accomplished by taking a very slight cut with a metal saw along the top of one tooth, using the mark as a guide to get the proper slots together. In order to get them in exact alignment, a rectangular bar of metal should be made to fit snugly in one slot before taking the discs off;

when they are put back this bar is inserted in the slot to which it was fitted and the nut is set up hard. End plates, *WW*, of brass, 2 ins. in diameter and 3-16 in. thick, serve to prevent the end discs from buckling when they are compressed. A nut (not shown) fitted to the thread which begins at *g* on the shaft, serves to clamp the discs, which are held at the other end by the shoulder, *i*; no key is necessary to prevent the discs from turning on the shaft in so small a machine, but it is essential that they should be clamped as tightly as a fairly strong man can clamp them, using a six-inch wrench on the nut. The shaft may be held in a pipe vise between *i* and *j* when setting up the nut; the nut should be made of very hard bronze metal in preference to steel, as the latter attracts magnetic lines of force and is liable to heat.

The commutator may be made as shown in the sketch, or according to any other modern plan, a number of which were described in the

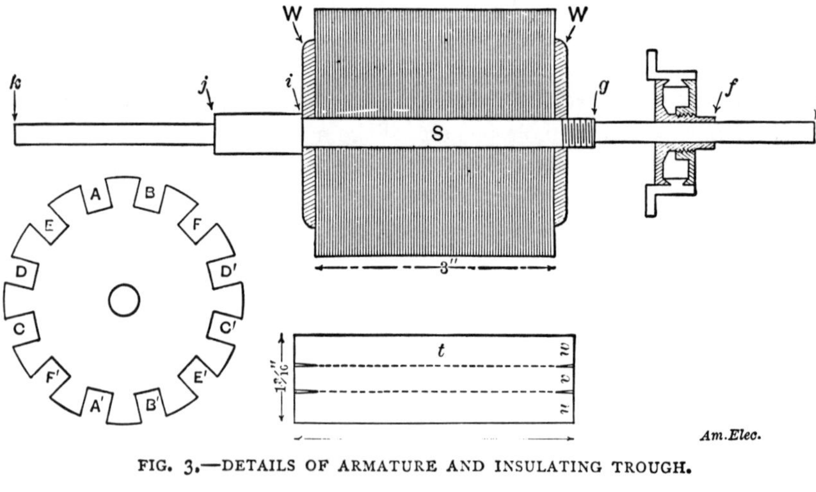

FIG. 3.—DETAILS OF ARMATURE AND INSULATING TROUGH.

"American Electrician" for July, 1896. The only essential features are the space along the shaft which must not exceed ¾ in., the width of face, which should not be less than ½ in., and the number of segments, which must be 12. The commutator here shown is intended to be secured to the shaft by a small steel set-screw through the hub or boss at the front; the end of this hub, *f*, must be 1¼ ins. from the end of the shaft. Extreme care must be taken to insulate the segments from the shell as well as from each other; mica is the only reliable material for this purpose. Carbon brushes ½ in. wide and ¼ in. thick should be used.

The armature core is next prepared for winding. Cut four discs of heavy drilling (so-called twilled muslin), 2¼ ins. in diameter, with a ⅜ in. hole in the center; varnish the ends of the armature core with shellac and varnish two of the cloth discs, each on one side; thread them on the shaft, one at each end, with the varnished sides next to the core, and press them tightly on the core. While the varnish is hardening cut 24 pieces of drilling the shape of *t* (Fig. 3); cut two slits ¼ in long in each end, 7-16 in. from each side and 5-16 in. from each other; varnish the strips on one side, and when nearly dry bend them along the dotted lines so as to form troughs, with the varnish inside the trough. Varnish the outside of each trough and the walls of the slots in the core; put two troughs in each slot and turn the flaps, *u*, *v*, *w*, flat against the end of the core, applying enough fresh shellac to hold them down. Then put on the two remaining end discs of cloth, first varnishing the sides next to the armature; after they are in place varnish the outsides and put the core in an oven to bake, being careful that the oven is not hot enough to scorch the cloth. A temperature of 130 degs. Fah. is sufficient. After baking, tape the shaft

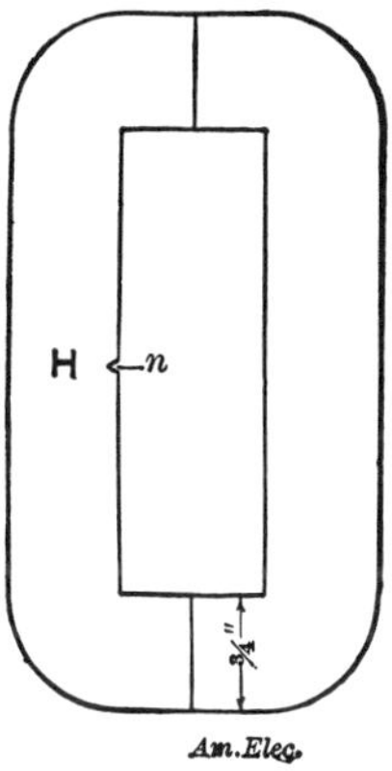

FIG. 4.—MAGNET-COIL HEAD.

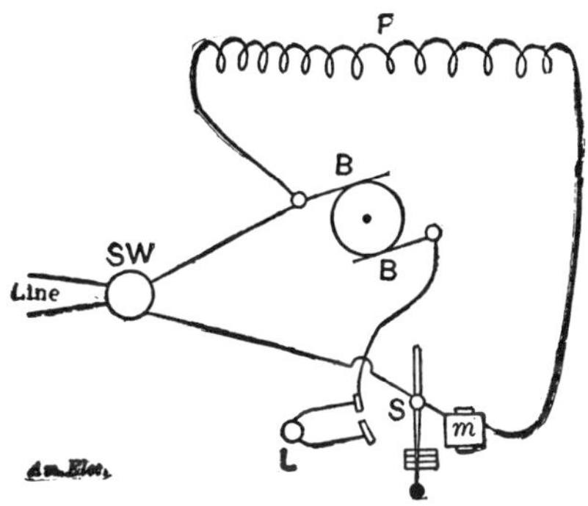

FIG. 5.—DIAGRAM OF STARTING SWITCH.

thoroughly from *i* to *j*, and from the other end of the core to where the commutator will come.

The coils consist of 48 turns of No. 24 double cotton-covered wire each, wound 8 turns wide and 6 deep in the slots, but spread out as flat as possible across the heads. Wind coil No. 1 in slots, *A A'*; coil No. 2 in *B B'*; No. 3 in *C' C*; No. 4 in *D D'*; No. 5 in *E E'*, and No. 6 in *F F'*. Coil No. 7 goes in *A' A*, on top of coil No. 1, but beginning on the opposite side of the core, as indicated by the lettering; No. 8 in *B' B*; No. 9 in *C' C*; No. 10 in *D' D*; No. 11 in *E' E*, and No. 12 in *F' F*. After winding each coil bring the finishing end across to the slot where the starting end enters and twist the two lightly together. When all the coils are on untwist the coil ends and twist the last end of each coil to the starting end of the coil in the slot next to it on the right; these twisted ends go each to a commutator segment, in regular order.

The field magnet is easily made ready to wind by taping the horizontal part of the magnet, two layers deep, with varnished muslin and putting on two fibre heads. One of these heads is shown by *H* (Fig. 4). It is in two pieces, the seams being at the ends, and is cut from ⅛-in. sheet fibre. The two halves may be clamped together on the core by means of a small brass wire drawn around the outer edge, laying in a shallow groove, the ends being twisted and cut close. The pole pieces should be removed before taping and putting on the heads, to facilitate these opera-

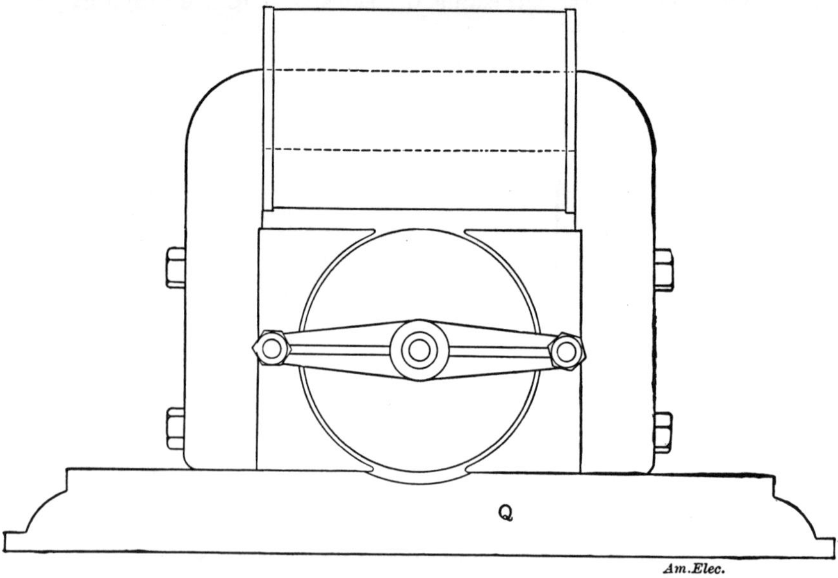

FIG. 6.—COMPLETED MOTOR FROM THE PULLEY END.

tions as well as the winding of the coil. One fibre head has a notch, *n*, half way of its inner long side, to enter the field wire. The coil consists of No. 28 wire, B. & S. gauge, 34 layers deep and 170 turns long, making 5,780 turns in all. The field winding is connected in shunt to the brushes, and it would be a good plan to provide a starting switch and resistance lamp connected up as shown diagrammatically by Fig. 5, where *F* is the field coil, *B B* the brushes, *L* a 32-candle power, 100-volt incandescent lamp, *S* the starting switch, *m* a magnet, and *S W* a double-pole snap switch. This arrangement could be mounted on the base of the motor. Fig. 6 shows the complete motor on a wooden base, *Q*, without the pulley; the latter may be any diameter between 1½ ins. and 2½ ins., with a 1-in. crown face or ½-in. grooved face. The motor is secured to the base by flat head machine screws from below, entering the ends of the wrought iron and countersunk in the under side of the wood. This machine will stand a momentary overload of 100 per cent., and will work up to ¼ horse-power for half an hour at a time.

WINDINGS FOR BATTERY SERVICE.

In order to adapt this motor for use in connection with a battery the following windings, etc., must be substituted for those specified above: The armature to be wound with six coils of No. 12 wire, each having twelve turns (three wide and four deep in a slot). The field wire will be No. 19, wound 17 layers deep and 83 turns in length. The commutator will have six segments, and should have a brush surface ¾ in. wide; copper brushes ⅜ × ½ in. should be used, the contact faces being cut to such a bevel as to present an area of ½ in. square at least. Connect the field winding in shunt with the armature, instead of in series as is usually done. This winding is for 6 volts. The machine thus wound will stand an armature current of 25 to 30 amperes.

CHAPTER II.

ONE-SIXTH HORSE-POWER MOTOR WITH RING ARMATURE.

In Figs 7 and 8 *M* is a wrought iron magnet core, *P P* cast-iron pole-pieces, *C* the armature core, and *Y* the journal yoke. The magnet core, *M*, is made from a ¾ in. × 4½ in. bar of commercial wrought iron bent to the shape shown. The faces of the arms are machined to a depth of 1-16 in., where the pole-pieces, *P P*, are attached, so as to form a magnetic joint of as low reluctance as possible. The pole-pieces are secured to the magnet arms by ¼-in. cap screws passing through smooth holes in the arms and tapped into the pole-pieces; the latter are of grey cast-iron, and should be finished on all sides sufficiently to remove the scale. The magnet, *M*, might be improved in appearance by touching up its sides with a coarse emery wheel; it should be thoroughly annealed after bending and finishing. It will be noticed by reference to Fig. 8 that the ends of the magnet arms project slightly beyond the outer faces of the pole-pieces; this is done in order to furnish a guide for the flanges of the journal yoke arms. After fitting the pole-pieces to the magnet arms the complete magnet frame is bolted to the lathe carriage in position for boring out the pole-pieces; before this is done it is necessary to drill a hole through the back of the magnet to allow the boring bar to pass through, and also to form a seat for the rear bearing. This hole is ¾ in. in diameter, and the magnet frame must not be allowed to move from its original position on the lathe carriage from the time the hole is drilled until all the circular tooling on it is accomplished.

After drilling the hole in the back of the magnet adjust the boring bar and bore the armature chamber out, 4 11-16 ins. in diameter. Next adjust the boring tool so that it will scribe on the ends of the magnet arms arcs of a circle 6 ins. in diameter; then cut away the wrought iron inside the scribed marks, down flush with the pole-pieces, as shown in Fig. 7, forming recesses for the flanges of the journal yoke. The yoke and box are cast in one piece of brass or other non-magnetic composition; the shell of the box is 1¼ ins. long, and projects ¼ in. beyond the inner face

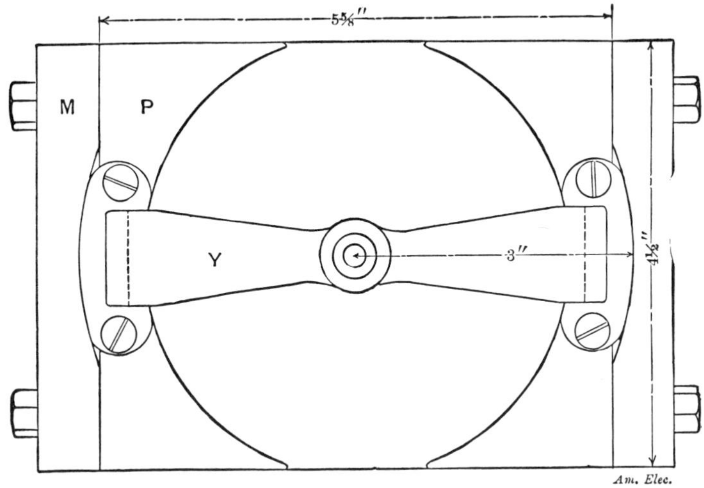

FIG. 7.—ELEVATION OF FIELD MAGNET AT THE COMMUTATOR END.

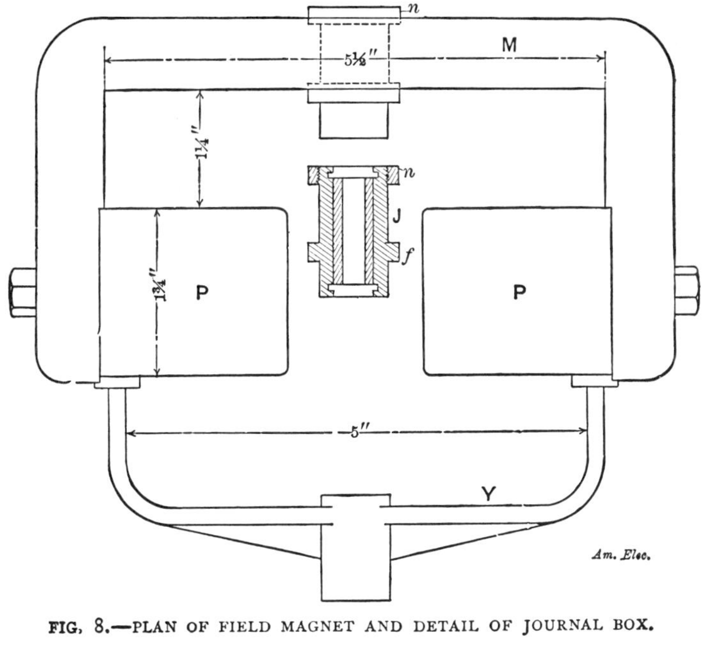

FIG. 8.—PLAN OF FIELD MAGNET AND DETAIL OF JOURNAL BOX.

of the yoke; the outer diameter of the shell is ¾ in., and it is bored out to 7-16 in. inner diameter and bushed to ¼ in. The yoke and arm portions are 3-16 in. thick, with a ⅛ in. stiffening rib on each side of the box, and the arms taper from 1 in. wide at the flanges to about ½ in. near the box. The flanges are 2 ins. long, ¾ in. wide and ⅛ in. thick after facing; the arms, beyond the bends, are sufficiently long to make the distance from the face of the pole-piece to the inner face of the yoke 2 ins. After boring the box it is mounted on a stiff mandrel and the surfaces of the flanges that go next the magnet are faced up true; next, the outer edges of the flanges are skimmed off until the yoke fits snugly between the curved edges of the recesses previously cut in the ends of the wrought iron magnet. Care must be taken in making the pattern for the yoke that the inner edges will not project inward beyond the bore of the pole-pieces. The yoke is fastened to the pole-pieces by screws, as indicated in Fig. 7.

The rear bearing, *J*, is a little peculiar in construction. The box portion is similar to that part of the yoke, but it is cast with a flange, *f*, 1 in. from the farthest end of the shell, which is 1½ ins. long. A collar, *n*, is fitted to screw onto the outer end of the shell, which is threaded for that purpose. The shell is turned down outside to fit snugly in the hole drilled in the back of the magnet, and when it is inserted in the hole the collar, *n*, is put on and screwed up tight. This box, like the front one, is bushed to ¼ in. bore. The drawing shows the flange, *f*, and collar, *n*, countersunk in the metal of the magnet; this will not be necessary if the magnet is smoothed up with an emery wheel, as above suggested, the object in countersinking being to provide smooth, true bearing surfaces for the flange and collar.

The armature core, spider and shaft are shown, partly in cross-section, by Fig. 9. The core is built up of charcoal iron (not steel) rings, 4½ ins. outside diameter and 2½ ins. inside, not more than 1-32 in. thick; these are assembled on a brass drum, shown by Fig. 11, which should be 2⅝ ins. outside diameter before finishing, so that it may be turned down to exactly fit the inner circle of the armature rings; the wall of the drum is ⅛ in. thick after finishing, and there are four equidistant projecting lugs, *l*, ½ in. long, on each end by which the drum is secured to the spider (see Figs. 9 and 10). The rings forming the core, *C* (Fig. 9), are compressed and held on the drum, *r*, by two brass washers, *w*, *w*, 3-16 in. thick and 3⅝ ins. outer diameter, which screw onto the ends of the drum. The core, when compressed, is 1¾ ins. long, and has 20 slots ¼ in. wide and 7-16 in. deep; the washers, *w*, *w*, must be set up as tight as the threads will stand.

The spider, *s* (Figs. 9 and 10), is made of brass, and consists of a hub

($\frac{3}{4}$ in. diameter, 2 ins. long and $\frac{1}{2}$ in. bore) and four arms having T-shaped ends, the wide part or heads of which project beyond the arms and hub at each end, the length of these heads being $2\frac{7}{8}$ ins. and their width $\frac{3}{8}$ ins. The heads of the spider arms are turned off to fit very closely inside the drum, *r*, which is mounted on the spider in such a position as to bring the spider arms in alignment with the lugs, *l*, of the drum;

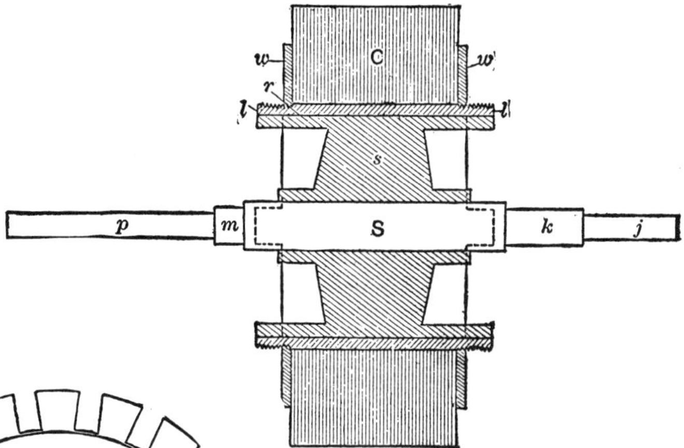

FIG. 9.—AXIAL SECTION OF ARMATURE CORE.

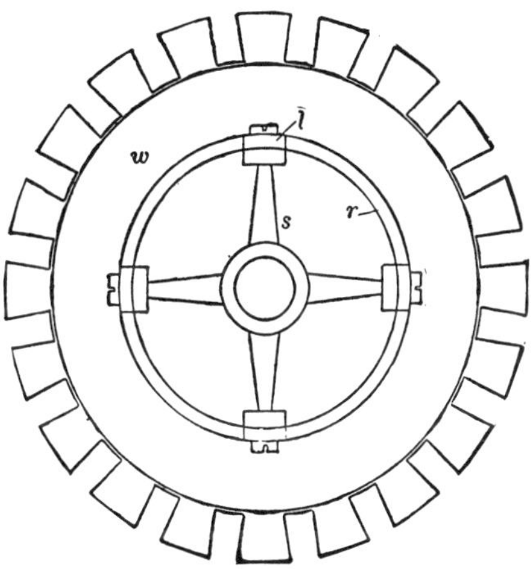

FIG. 10.—END OF ARMATURE CORE.

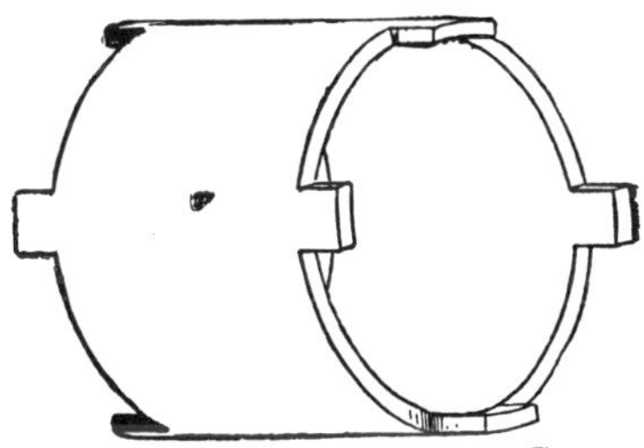

FIG. 11.—ARMATURE CORE DRUM.

screws through the spider arms into the lugs hold the drum and spider together.

The shaft, *S*, is $8\frac{1}{2}$ ins. long; the portion *j* is $1\frac{1}{8}$ ins. long and $\frac{1}{4}$ in diameter; *k* is 1 in. long and $\frac{3}{8}$ in. diameter; the part passing through the core is 3 ins. long and $\frac{1}{2}$ in. in diameter; *m* is $\frac{3}{8}$ in. long and $\frac{3}{8}$ in. diameter; and *p* is 3 in. long and $\frac{1}{4}$ in. diameter. The spider, *s*, may be secured to the shaft by a key or a set-screw; the set-screw is sufficient in so small a machine. The commutator (not shown) must not be more than ¾ in. over all, along the shaft; it must have ½ in. brush surface and 20 segments; other details may be made to suit the will of the builder. The front end of the commutator must be not less than 3-16 in. from the shoulder where *j* and *k* join.

The armature is next prepared for winding by removing the drum and core from the spider and insulating the ends and interior of the core and the walls of the slots. Cut four rings of heavy drilling of a size to cover the washers, *w w*, and the ends of the drum, *r*; varnish two of them on one side with shellac, and apply them to the ends of the armature body. While these are hardening cut forty strips of drilling 1⅛ ins. wide and 2¼ ins. long; in each end of each of these cut two slits ¼ in. long parallel with the sides, and located 7-16 in. from each side of the strip. Varnish these on one side, and when nearly dry fold them into troughs to fit the slots, two troughs to a slot, one within the other; fold them so that the varnish will be on the inside of the trough.

When these are dry varnish the slots and the outsides of the troughs and put the latter in the slots, bending the ends flat against the core and securing them there with a little fresh varnish. Then varnish the ends of the core (two cloth rings being on them), and one side of the two remaining rings of drilling; put these rings on top of the first ones, varnish them on the outside, and put the core in an oven to bake. The armature coils consist of No. 24 double cotton-covered wire, wound six turns wide and twelve layers deep. Before winding them four strips of wood 3 ins. long. ⅜ in. wide and ½ in. thick should be screwed to the inner wall of the brass drum, in line with the lugs, *l*, so as to preserve the spaces for the four arms of the spider. A double thickness of drilling should also be applied to the interior of the drum to insulate the coils from it. The connections are the simple Gramme ring arrangement.

The field winding is necessarily divided into two coils, on account of the rear bearing passing through the magnet. Each coil consists of No. 28 double cotton-covered wire, wound 17 layers deep and 181 turns or more long; the two coils are connected in series with each other and in shunt to the brushes. Heads of hard fibre ⅛ in. thick should be used to protect the ends of each coil; one of these is shown by *H* (Fig. 12), but the width should be 7-16 in. instead of ¾ in. as marked.

It is in two pieces, the seams being at the ends, and is cut from ⅛ in. sheet fibre. The two halves may be clamped together on the core by means of a small brass wire drawn around the outer edge, laying in a shallow groove, the ends being twisted and cut close. The pole-pieces should be removed before taping and putting on the heads, to facilitate these operations as well as the winding of the coil. One fibre head has a notch, *n*, half way of its inner long side to enter the field wire. The pole-pieces should, of course, be removed before winding the field coil, and the magnet core should be wrapped with two layers of varnished drilling where the coils are to go. The entering end of each coil should be remote from the journal, and this means that the magnet must be turned end for end after one coil is wound, or else the two coils must be wound in opposite directions in order that the free ends at the center of the magnet may be connected together. It is advisable to provide a starting switch similar to the one shown diagrammatically by Fig. 13, where *F* is the field coil; *B B* the brushes; *S* the starting switch lever: *L* a 32-candle-power 110-volt lamp; *M* a magnet, and *SW* a double-pole snap switch.

The motor is intended to be mounted on a wooden base-board 8 ins. × 8 ins., a cleat 3 ins. wide and 7-16 in. thick being put under the pole-

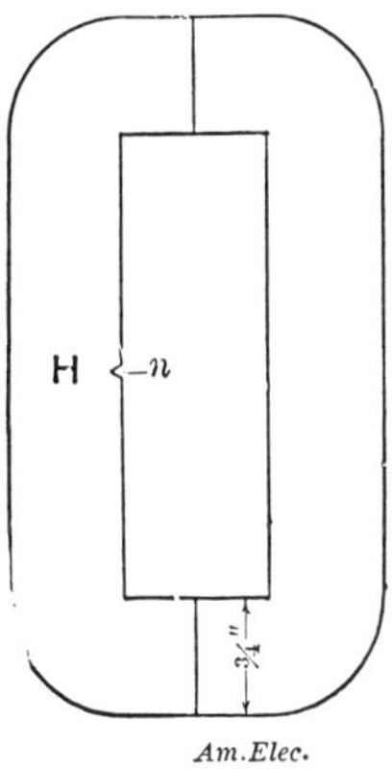

FIG. 12.—MAGNET COIL HEAD.

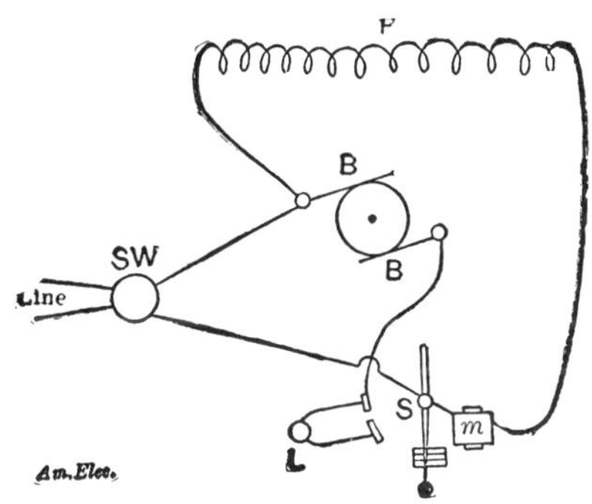

FIG. 13.—DIAGRAM OF STARTING SWITCH.

pieces so as to clear the field coil. Bolts from beneath, tapped into the magnet and countersunk in the under side of the base, should be used to hold the motor on the base. The pulley may be any diameter from 1 in. to 3 ins. by 1 in. face, if crowned, or ½ in. if grooved.

WINDINGS FOR BATTERY CURRENT.

The armature will be wound with 10 coils of No. 12 wire, each coil having 16 turns and occupying two (adjacent) slots. The field wire will be No. 18, wound 8 layers deep and 40 turns long in each coil; the two coils containing 640 turns in all. The commutator must have 10 segments and a brush surface ¾ in. wide; copper brushes ⅜ × ½ in. should be used, the contact faces being cut to such a bevel as to present a surface at least ½ in. square each. The field winding is to be connected in shunt to the brushes, instead of in series as is usually the practice in battery motor construction. This winding is for 6 volts at the terminals; the current required will depend upon the work done; the machine is capable of standing an armature current of 25 to 30 amperes.

CHAPTER III.

ONE-FOURTH HORSE-POWER MOTOR WITH DRUM ARMATURE.

Fig. 14 represents the field magnet, and Fig. 15 one of the journal yokes. The magnet is of the familiar single-coil type employed by Westinghouse, Jenney and others. The core is of round Norway iron, 2 ins. in diameter and 9 ins. long over all. The ends are turned tapering, as indicated by dotted lines, to insure intimate contact with the yokes; the taper is from the full diameter to 1¾ ins., and begins 2 ins. from each end. The pole-pieces are of cast-iron. Fig. 16 gives a plan view and a face view of one pole-piece, from which all the essential dimensions may be obtained. The arms which support the journal yokes are cast solid with the pole-pieces, and their horizontal thickness tapers from ½ in. at the pole-piece to ¼ in. where the yoke is bolted on.

In fitting the magnet frame together the best procedure is to bore the tapered holes in the lower part of each pole-piece and turn the ends of the magnet core to the same taper, but just a trifle large; then dress each taper down very gradually with a fine file (the core being run in a lathe) until the pole-piece can be pushed on by hand far enough to bring the end of the core within 1-32 in. of the back surface of the cast-iron. The pole-pieces and ends of the core should be punch-marked, so as to insure finally mounting each pole-piece on the end to which it was fitted. After dressing down the ends of the core as above described, drill and tap in each end a hole for a ¼-in. machine screw, the purpose of which will be apparent by glancing at the right-hand end of the magnet in Fig. 14, where *C* is a four-armed claw or spider with a hole through the center where the arms intersect. The arms are 3-16 in. thick, measured at right angles to the bolt, and taper from 3-16 in. to ⅜ in. thick measured parallel with it. One of these spiders is used at each end, though the drawing shows it at only one end of the machine.

After drawing one pole-piece home solid by means of its spider and bolt, slip the other pole-piece on loosely and clamp the pole-pieces lightly

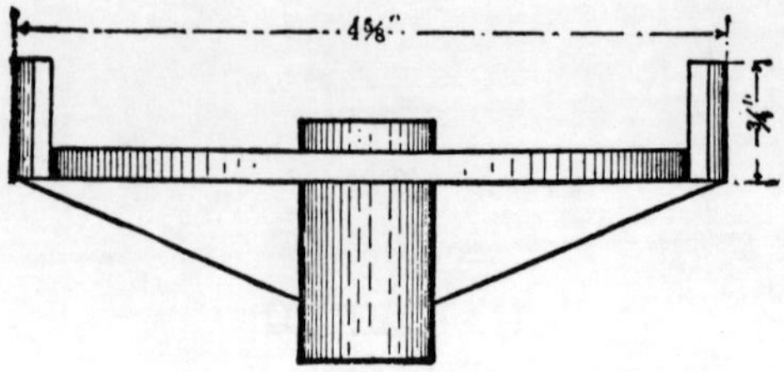

FIG. 15.—PLAN OF JOURNAL YOKE.

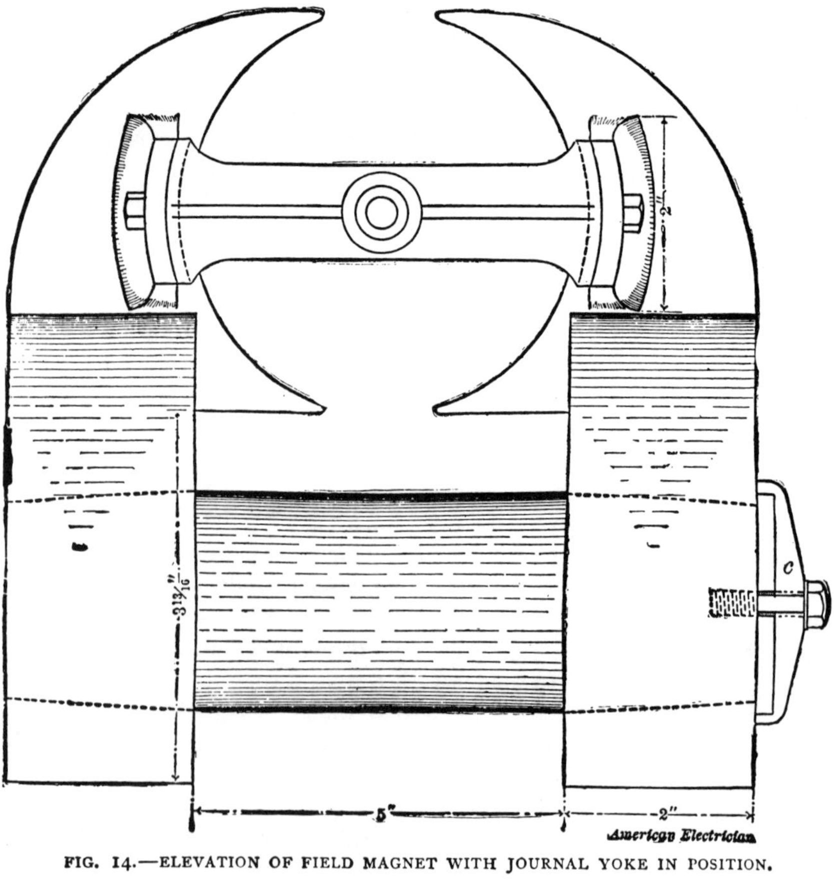

American Electrician

FIG. 14.—ELEVATION OF FIELD MAGNET WITH JOURNAL YOKE IN POSITION.

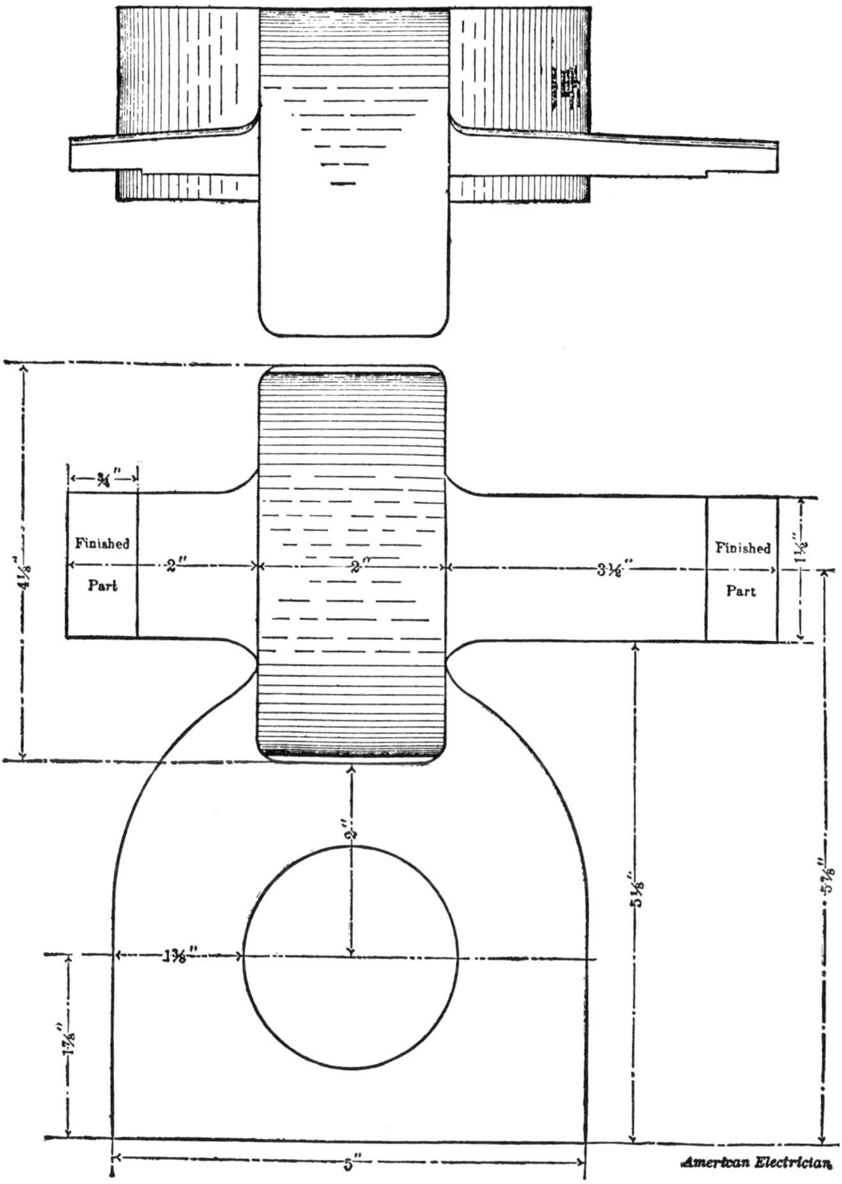

FIG. 16.—PLAN AND FACE VIEWS OF ONE FIELD-MAGNET POLE-PIECE.

between two iron plates with planed surfaces, applied between the journal arms, so as to keep the four horns of the pole-pieces in alignment; then force the second pole-piece home by means of its bolt and spider, and clamp the horns hard between the iron plates. The bottom surfaces of the cast-iron pieces should then be trued up on a planer or shaper and the clamps taken off the pole-piece horns.

The next operation is boring the armature chamber and the seats for the journal yokes. The armature chamber bore is 4 3-16 ins; the seats for the journal yokes, marked "finished part" in Fig. 16, are bored or cut to 4⅝ ins. diameter, and this must be done before the position of the machine is disturbed after boring the armature chamber. This completes the machine work on the magnet, except the bolt holes.

The journal yoke may be made of brass or any composition metal. The bar is 3-16 in. thick and 1 in. wide, except near the ends, where it

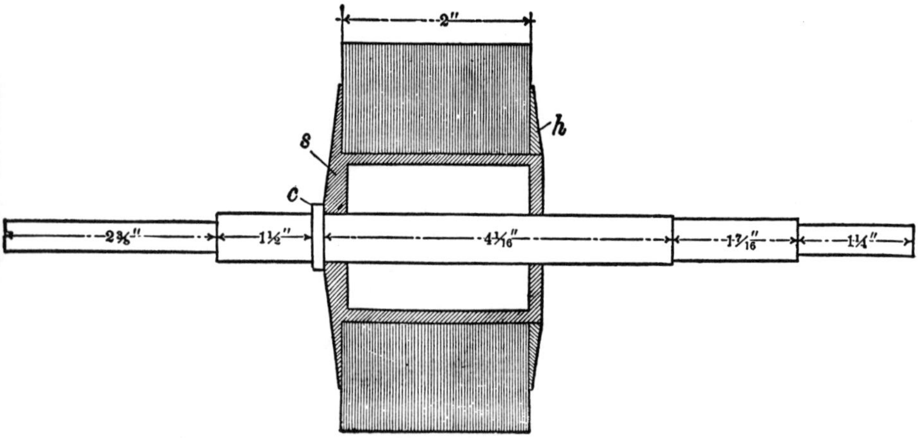

FIG. 17.—SHAFT AND CROSS SECTION OF ARMATURE CORE.

flares to correspond with the width of the arms. At each end is a right-angled lug, ⅛ in. thick after machining; these lugs fit the seats in the ends of the iron arms, and the yokes should be fitted to the magnet immediately after finishing the machine work on the latter, and before it is taken apart to put on the coil. The box portion is 1½ ins. long over all, 3-16 in. of its length being on the inside of the yoke and 1⅛ ins. on the outside. As shown by the plan view of the yoke in Fig. 15, there are stiffening webs starting flush near the ends of the yoke and attaining a width of ¾ in. at the box; these are ⅛ in. thick. The box is ⅞ in. in outer diameter, and bored to 17-32 in. inside; it is bushed to ⅜ in. diameter. These latter dimensions, excepting the final inside diameter of the bushing, may be varied to suit individual ideas, as may also the design of the box. The only essential measurements are those of the yoke-bar, the length of the box and the bore of the journal bushing. The journal yokes are held in place by ¼ in. cap screws passing through the iron arms and tapping into the lugs of the yokes.

Figs. 17, 18 and 19 show the shaft and armature core (the latter in cross-section), an armature disc, and the shell and head. The discs are of charcoal iron, 4 ins. outside diameter with a 1¼-in. hole in the center and a ⅛-in. key-seat, annealed after punching and key-seating; there are eighteen slots ⅜ in wide and ⅜ in. deep. The shell and one head are cast in one piece (of brass), and consist of a barrel 1¼ ins. outside diameter (when finished) and 2 ins. long, with a head, *s*, at one end, 3⅛ ins. in diameter and tapered in thickness from ¼ in. near the center to 1-16 in. at the periphery; at the opposite end of the barrel is a cross-bar ⅓ in. thick, cast with the barrel and of the shape shown, being ⅜ in. wide where it joins the barrel and ¾ in. at the center. A ½-in. hole is drilled in the center of this cross-bar and another in the center of the head, *s*, at the other end of the barrel; the shell is mounted on a mandrel, the barrel turned down to fit the hole in the armature discs, and both sides of the head faced off smooth. A ⅛-in key-seat 3-16 ins. deep is cut in the barrel so as to come in the center of one end of the cross-bar, as shown; a ⅛-in. × ¼-in. feather, or parallel key, is laid in the key-seat, and the discs threaded on the barrel and compressed against the head by the collar, *h*, and two bolts (not shown) passing through the collar and inside the barrel, and tapping into the head at the other end. This collar, *h*, is of brass, 3⅛ ins. in diameter and tapering from 3-16 to 1-16 in. in thickness when finished. The opening in the center should fit the outline of the cross-bar on the end of the barrel at least closely enough to prevent the collar from shifting under stress of centrifugal force; the collar must be finished up smooth on both sides. A disc of insulation should be put on next to the brass head before the iron discs are put on, and another insulating disc should go between the last iron disc and the clamping collar, *h*.

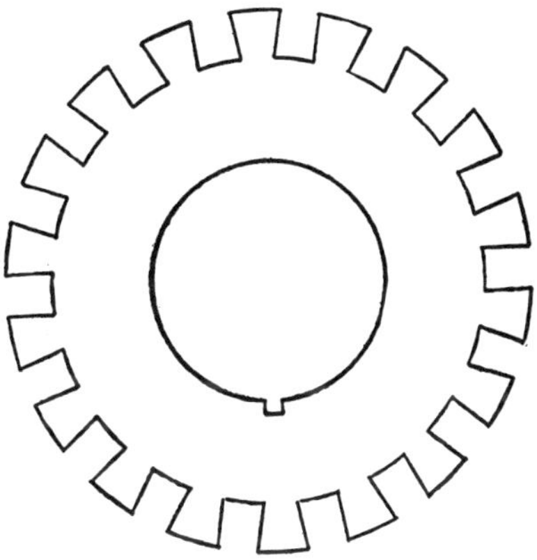

(*American Electrician*)

FIG. 18.—AN ARMATURE DISC.

If the slots are cut in the core with a milling machine the discs must all come off the barrel to have the burrs removed, and also be reannealed; the key-seat will insure their returning in the original angular position. It is much better to have discs with the slots punched before the first annealing. The shaft is 10¾ ins. long over all; ½ in. in diameter in the largest part, 7-16 in. where the commutator goes and ⅜ in. in the journals. A 1-16-in × ⅛-in. collar, *c*, is shown back of the armature, the purpose of which is merely to "locate" the armature shell; it is not absolutely necessary, however, and may be left off if desired. The easiest way to provide for it is to make the shaft of ⅝-in. stock, leaving the original metal to form the collar when turning the shaft to proper diameter. The armature shell may be keyed to the shaft or pinned obliquely through the thick part of the head; it must be positively secured by some such means.

The commutator shell must be bored to fit the 7-16-in. portion of the shaft, and must not exceed 1¼ ins. along the shaft. The lugs where the wires are attached to the segments may project toward the armature ⅛

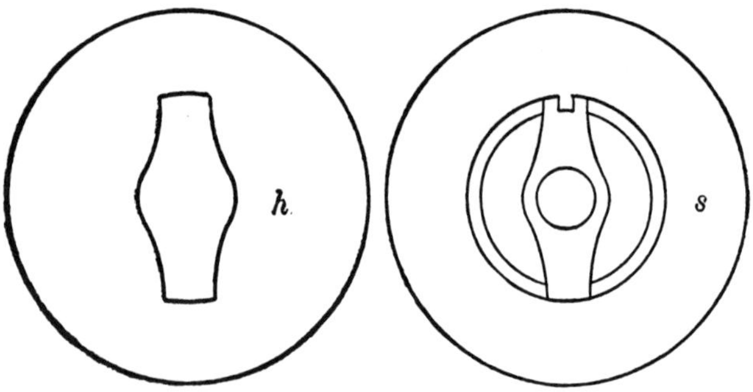

FIG. 19.—ARMATURE CORE DRUM AND HEAD.

in. or so. There must be eighteen segments, and a diameter of 2 ins. is recommended. The quadrant carrying the brush-holders should be fitted to the inner end of the journal box, and carbon brushes not smaller than ¼ in. × ½ in. (one on each side) on the contact surface should be used. If the machine be used as a dynamo (it will maintain five or six 110-volt lamps) metal brushes of the same surface should be used to reduce the resistance of the brush contact.

The field coil contains 37 layers of No. 28 double cotton-covered wire. After the magnet is fitted as described in the beginning of the article it is taken apart and two circular magnet heads of fibre ⅛ in. thick and 3¾ in. outer diameter are put on with a driving fit, care being taken that the distance along the core from outside to outside of the heads corresponds with the distance between the pole-pieces when the whole is assembled. A groove must be cut on the inner face of one head from the center to the outer edge in order to lead out the starting end of the field wire, and this must be covered with two layers of oil paper to prevent short-circuiting the successive layers of the coil. The core must be insulated with three layers of shellaced muslin between the heads and the field wire put on evenly, care being taken not to "spread" the heads; if the winding is carefully done the coil will be 216 turns in length. The number of turns in length is not a vital matter, but the depth must be 37 layers. The ampere turns are the same no matter what the length of the coil, but it should be as long as possible in order to reduce the heat loss.

After winding the coil and securing the ends one pole-piece is put on solid and the other one slipped on until it begins to bind, when the journal yokes must be inserted between their arms and the bolts put in as far as possible without jamming. Then by tightening up the journal yoke bolts and the pole-piece bolt together, being particular never to draw the yoke bolt hard against the arm, the frame will come together in its original position. As an additional precaution it may be set on a true plane surface, and if the base of the loose pole-piece gets out of alignment tap the horn lightly until the frame is true on the bottom. The magnet frame must be provided with a non-magnetic base; hard wood is as good as anything, the frame being secured by flat-head brass machine screws from below, two in each casting, countersunk in the wood.

The armature winding is divided into eighteen coils, each having 45 turns of No. 22 double cotton-covered wire, 9 turns wide and 5 turns deep in the slot. The slots must be insulated with troughs of muslin and mica, or preferably flexible micanite, 0.03 in. thick. The troughs are easily made by cutting the material into strips 2½ ins. long by 1⅛ ins. wide, and slitting the ends so as to permit the projecting portion of the trough to be folded back flat against the core. Before putting in the troughs a disc of heavy drilling 3¼ ins. in diameter should be secured to each end of the core by means of varnish, and the outer faces varnished and allowed to nearly dry. Then put in the troughs and put on two more muslin discs, varnishing the whole, and bake until thoroughly dry. Instead of winding each coil in diametrically opposite slots, take slots lacking one of being precisely opposite.

A good plan is to make a sketch of an armature disc and number the slots from left to right successively around the periphery. Then wind the coils as follows, the coil numbers indicating the order in which the coils are put on, not the order in which they are connected to the commutator.

COIL NO.—	1	2	3	4	5	6	7	8	9	10	11	12	13	14	15	16	17	18
STARTS IN SLOT NO.—	1	10	13	4	7	16	2	11	15	6	14	5	18	9	3	12	8	17
ENDS IN SLOT NO.—	9	18	3	12	15	6	10	1	5	14	5	13	8	17	11	2	16	17

Each pair of coils must be covered with muslin where they cross the heads before the next pair is put on, and before coil No. 8 is wound on top of coil No. 1 in slot No. 1 the bottom coil must be insulated by a strip of micanite laid in the slot; this is true of every bottom coil.

After the winding is on, and before connecting up to the commutator, the band wires should be put on. Use No. 19 B. W. G. soft tinned-iron wire, known by hardware dealers as "white stove-pipe wire," for the bands, and put them on under as heavy pressure as possible without endangering the armature shaft. Two bands of eight turns each, ½ in. from each end of the core, will suffice. A strip of mica between two strips of fullerboard must go under each band, and the bands should be soldered at intervals, not all the way around. Four tin clips located equidistantly, with a dab of solder at each, will give ample security.

The technical data for this machine are as follows:

TERMINAL E. M. F., 110 VOLTS.

Armature current, normal	1.9 amps.
" " maximum	2.3 "
" resistance, warm	3.33 ohms
Field current at 110 volts	.25 amp.
" resistance, warm	440 ohms
C^2R loss in field	27½ watts
C^2R loss in armature	12+ "
Hysteresis loss in armature	20— "
Magnetic flux per square inch:	
In field core	90,000 lines
In pole-pieces	39,000 "
In air-gap	25,250 "
In armature teeth	68,000 "
In armature core	56,000 "
Co-efficient of leakage	1.4
Electrical efficiency	84 per cent.
Commercial efficiency (friction 10 p. c. estimated)	65 "
Revolutions per minute	2,000

If it is desired to build a smooth-core machine the armature core must be made 3⅝ ins. in diameter, and two grooves ⅛ in. wide and 5-16 in. deep must be cut in the face of the core at opposite points for the reception of driving teeth. These are two pieces of fibre ⅛ in. thick, ½ in. wide and 2 ins. long, set on edge in the grooves, and projecting 3-16 in. above the surface of the core. The core must be thoroughly covered with two layers of micanite cloth. The number of coils is the same as before, but the coils will be 18 turns wide and 5 deep; and in this case they are not superposed, the depth of a coil (5 layers) being the total depth of the winding. The guiding diagram, therefore, must divide the periphery of the armature into 36 spaces instead of 18, because each space now contains one side of only one coil. The smoothest winding will be as follows:

COIL NO.—	1	2	3	4	5	6	7	8	9	10	11	12	13	14	15	16	17	18
STARTS IN SPACE NO.—	1	19	7	25	31	13	3	21	27	9	33	15	23	5	11	29	35	17
ENDS IN SPACE NO.—	18	36	24	6	12	30	20	2	8	26	14	32	4	22	28	10	16	34

Care must be taken in connecting up either of the armature windings to take the starting ends of the coils in proper succession to the commutator segments; the outer end of each coil goes to the segment on the right of the one to which the starting end is led. The smooth-core armature is banded just as the slotted one is, except that soft brass wire must be used instead of tinned iron.

CHAPTER IV.

ONE-FOURTH HORSE-POWER MOTOR WITH RING ARMATURE.

This machine has a field magnet of exactly the same design as the one last described, the only difference being in the dimensions. The instructions for fitting up the magnet shown by Figs. 14 and 15, therefore, apply to this one. The size of the magnet core and yokes shown by Fig. 14 also apply to this magnet. Figs. 21 and 22 give all of the dimensions for this magnet frame that differ from those of the previous one, excepting the bore of the armature chamber, which is 5 3-16 ins. instead of 4 3-16 ins. The lugs that support the journal yokes are set one inch wider apart than in the drum armature motor, and the seats for the ends of the journal yokes are bored or cut to 5⅝ ins. diameter. As in the former case, this boring must be done before the frame is moved from the position it occupied during the boring of the armature chamber.

The journal yokes may be made of anything except iron and steel. The bar is 3-16 in. thick and 1 in. wide., except near the ends, where it flares to correspond with the width of the arms. At each end is a right-angled lug, ⅛ in. thick after machining; these lugs fit the seats in the ends of the iron arms, and the yokes should be fitted to the magnet immediately after finishing the machine work on the latter, and before it is taken apart to put on the coil. The box portion is 1½ ins. long over all, 3-16 ins. of its length being on the inside of the yoke and 1⅛ ins. on the outside. As shown by the plan view of the yoke, Fig. 22, there are stiffening webs starting flush near the ends of the yoke and attaining a width of ¾ in. at the box; these are ⅛ in thick. The box is ⅞ in. in outer diameter, and bored to 17-32 in. inside; it is bushed to ⅜ in. diameter. Most of the dimensions of the yoke and box may be varied to suit individual ideas, as may also the design of the box. The only essential measurements are the length of the yoke-bar, the length of the box and the bore

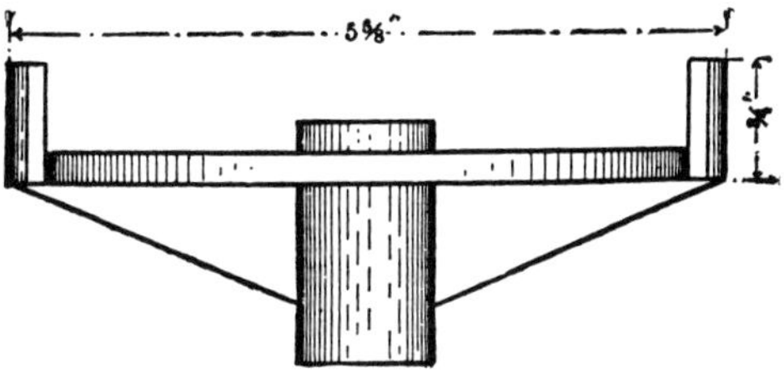

FIG. 22.—PLAN OF JOURNAL YOKE.

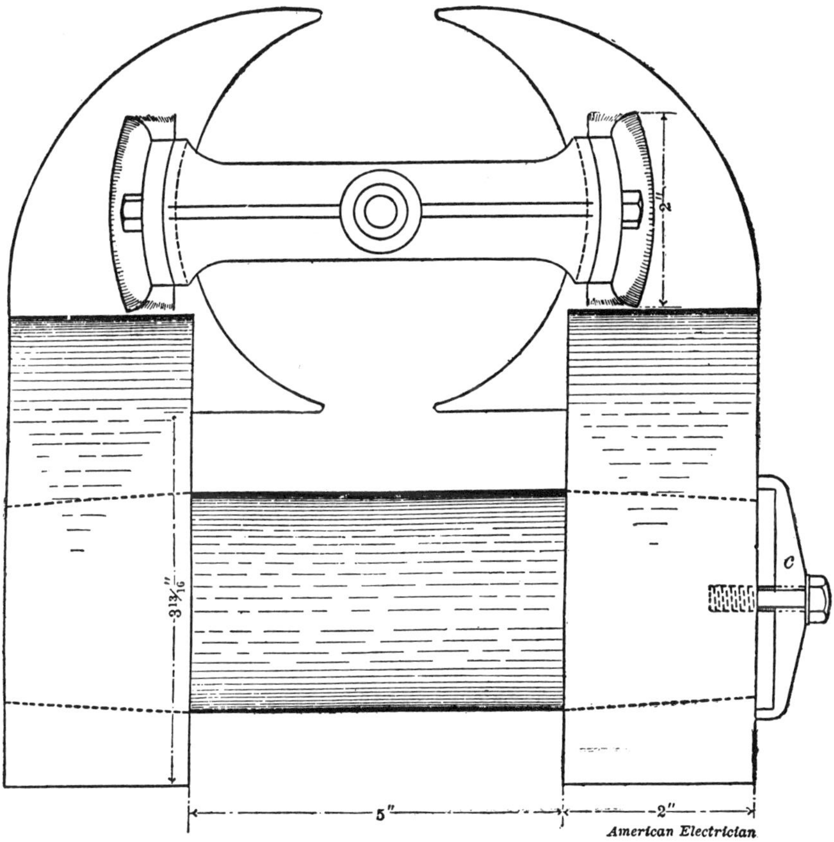

FIG. 20.—ELEVATION OF FIELD MAGNET WITH JOURNAL YOKE IN POSITION.

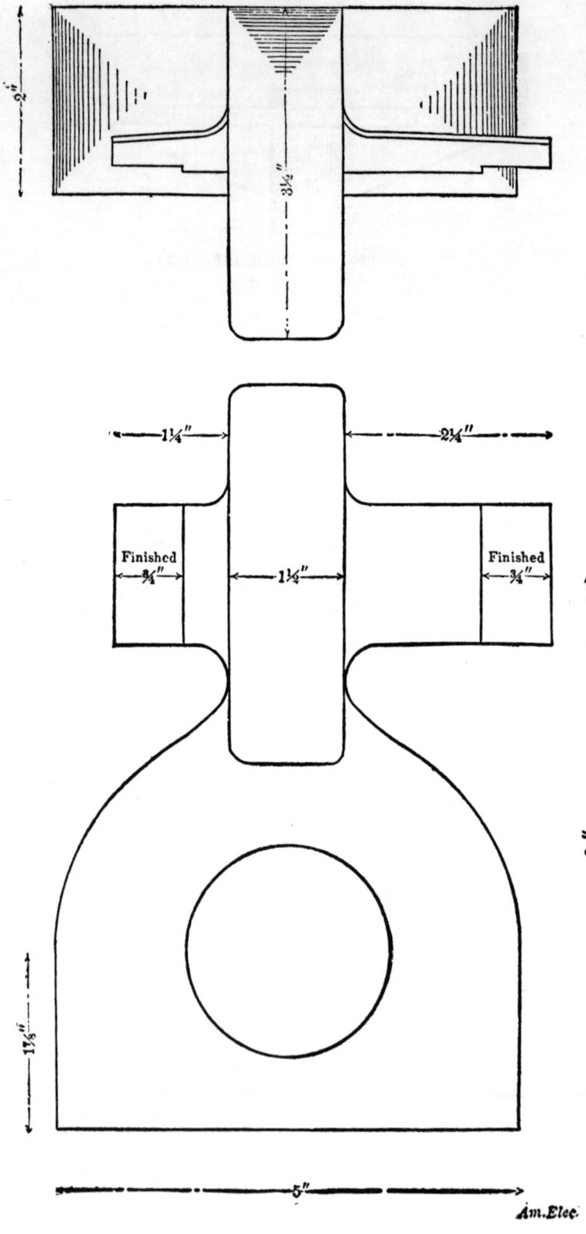

FIG. 21.—PLAN AND FACE VIEWS OF ONE FIELD-MAGNET POLE-PIECE.

of the journal bushing. The journal yokes are held in place by 1/4 in. cap-screws passing through the iron arms and tapping into the lugs of the yokes.

The armature core, spider and shaft are shown, partly in cross-section, by Figs. 23 and 24. The core is built up of charcoal iron (not steel) discs 5 ins. outside diameter and 2 5/8 ins. inside, not more than 1-32 in. thick; these are assembled on a brass drum 1 5/8 ins. long (Fig. 25). which

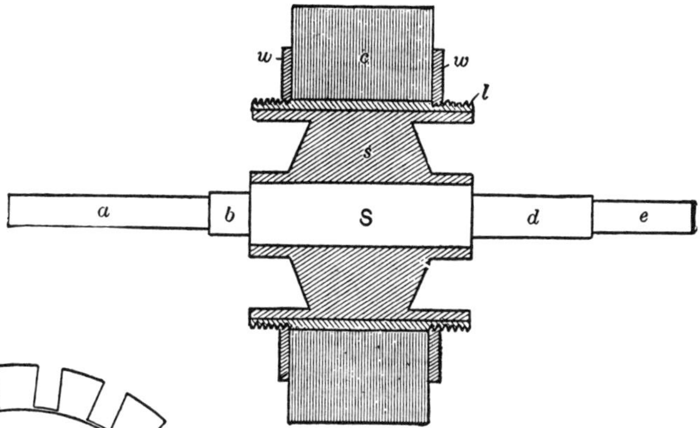

FIG. 23.—ARMATURE CORE AND SHAFT.

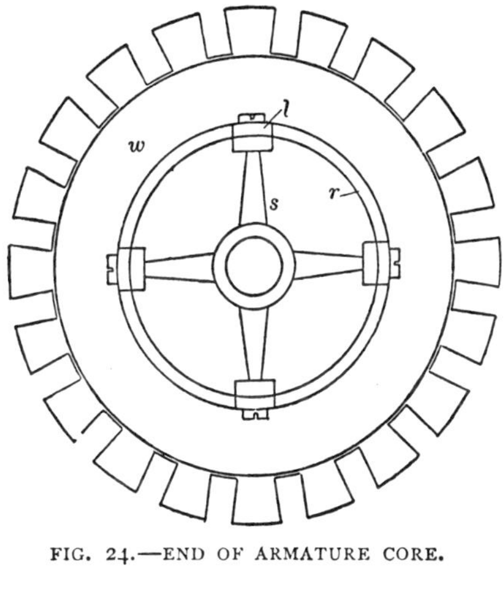

FIG. 24.—END OF ARMATURE CORE.

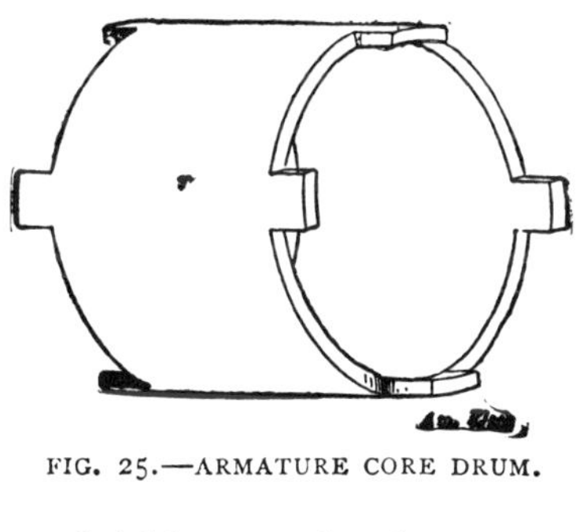

FIG. 25.—ARMATURE CORE DRUM.

should be 2 3/4 ins. outside diameter before finishing, so that it may be turned down to exactly fit the inner circle of the armature rings; the wall of the drum is 1/8 in. thick after finishing, and there are four equidistant projecting lugs, *l*, 3/8 in. wide and 1/2 in. long, on each end, by which the drum is secured to the spider (see Figs. 22 and 23). The rings forming the core, *C* (Fig. 22), are compressed and held on the drum, *r*, by two brass washers, *w*, *w*, 3-16 in. thick and 3 7/8 ins. outer diameter, which screw onto the lugs and ends of the drum. The core when compressed is 1½ ins. long, and has 20 slots 3-10 in. wide and ½ in. deep; the washers, *w w*, must be set up as tight as the threads will stand.

The spider, *s* (Figs. 23 and 24), is made of brass, and consists of a hub (⅞ in. in diameter, 2½ ins. long and ⅝ in. bore) and four arms having T-shaped ends, the wide part or heads of which project beyond the arms at each end, the length of these heads being 2⅝ ins. and their width ⅜ in. The heads of the spider arms are turned off to fit very closely inside the drum, *r*, which is mounted on the spider in such a position as to bring the spider arms in alignment with the lugs, *l*, of the drum; screws through the spider arms into the lugs hold the drum and spider together.

The shaft, *S*, is 8¼ ins. long; the portion, *a*, is 2½ ins. long and ⅜ in. in diameter; *b* is ½ in. long and ½ in. in diameter; the part passing through the core is 2½ ins. long and ⅝ in. in diameter; *d* is 1½ ins. long and ½ in. in diameter, and *c* is 1¼ ins. long and ⅜ in. in diameter. The spider, *s*, should be secured to the shaft by a key, the key-seat being located at the base of one of the arms. The front end of the commutator must be located not less than 3-16 in. from the shoulder where *d* and *c* join.

The armature is next prepared for winding by removing the drum and core from the spider and insulating the ends and interior of the core and the walls of the slots. Cut four rings of heavy drilling of a size to cover the washers, *w w*, and the ends of the drum, *r*; varnish two of them on one side with shellac, and apply them to the ends of the armature body. While these are hardening cut twenty strips of micanite cloth, 25-1000 in. thick, 1⅜ ins. wide and 2 ins. long; in each end of each of these cut two slits, ¼ in. long, parallel with the sides and located 17-32 in. from each side of the strip. Varnish these on one side, and when nearly dry fold them into troughs to fit the slots; fold them so that the varnish will be on the inside of the trough.

When these are dry varnish the slots and the outsides of the troughs and put the latter in the slots, bending the ends flat against the core and securing them there with a little fresh varnish. Then varnish the ends of the core (two cloth rings being on them), and one side of the two remaining rings of drilling; put these rings on top of the first ones, varnish them on the outside and put the core in an oven to bake. The armature coils consist of No. 22 double cotton-covered wire, wound seven turns wide and thirteen layers deep. Before winding them four strips of wood 3 ins. long, ⅜ in. wide and ½ in. thick should be screwed to the inner wall of the brass drum, in line with the lugs, *l*, so as to preserve spaces for the four arms of the spider. A double thickness of drilling should also be applied to the interior of the drum to insulate the coils from it. The connections are the simple Gramme ring arrangement. Before connecting up to the commutator the band wires should be put on. Use No. 19 B. W. G. soft tinned-iron wire, known by hardware dealers as "white stove-pipe wire," for the bands, and put them on under as heavy pressure as possible without endangering the armature shaft. Two bands of eight turns each, ½ in. from each end of the core, will suffice. A strip of mica between two strips of fullerboard must go under each band, and the bands should be soldered at intervals, not all the way around. Four tin clips located equidistantly, with a dab of solder at each, will give ample security.

The commutator (not shown) must be bored to fit the ½ in. portion, *d*, of the shaft, and must not exceed 1¼ in. along the shaft; it must have a brush tread 1 in. wide. The lugs where the wires are attached to the segments may project toward the armature ⅛ in. or so. There must be 20 segments, and a diameter of 2½ ins. is recommended. The quadrant carrying the brush-holders should be fitted to the inner end of the journal box, and carbon brushes not smaller than ¼ in. × ⅝ in. (one on each side) on the contact surface should be used. If the machine be used as a dynamo (it will maintain five or six 110-volt lamps) metal brushes of the same surface should be used to reduce the resistance of the brush contact.

The field coil contains 37 layers of No. 28 double cotton-covered wire. After the magnet is fitted as described in the beginning of the article it is taken apart and two circular magnet heads of fibre ⅛ in. thick and 3¾ ins. outer diameter are put on with a driving fit, care being taken that the distance along the core from outside to outside of the heads corresponds with the distance between the pole-pieces (5 ins.) when the whole is assembled. A groove must be cut on the inner face of one head from the center to the outer edge, in order to lead out the starting end of the field wire, and this must be covered with two layers of oil paper to prevent short-circuiting the successive layers of the coil. The core must be insulated with three layers of shellac muslin between the heads, and the field wire put on evenly, care being taken not to "spread" the heads; if the winding is carefully done the coil will be 216 turns in length. The number of turns in length is not a vital matter, but the depth must be 37 layers. The ampere turns are the same no matter what the length of the coil, but it should be as long as practicable to reduce the heat loss.

After winding the coil and securing the ends one pole-piece is put on solid and the other one slipped on until it begins to bind, when the journal yokes must be inserted between their arms, and the bolts put in as far as possible without jamming. Then by tightening up the journal-yoke bolts and the pole-piece bolt together, being particular never to draw the yoke bolt hard against the arm, the frame will come together in its original position. As an additional precaution it may be set on a true plane surface, and if the base of the loose pole-piece gets out of alignment tap the horn lightly until the frame is true on the bottom. The magnet frame must be provided with a non-magnetic base; hardwood is as good as anything, the frame being secured by flat-head brass machine screws from below, two in each casting, countersunk in the wood.

The technical data for the above machine are as follows:

TERMINAL E. M. F., 110 VOLTS.

Armature current, normal......	1.9 amps.
" " maximum....	2.3 "
" resistance, warm....	4.15 ohms
Field current at 110 volts......	.25 amp.
" resistance, warm.........	440 ohms
C^2R loss in field...............	27½ watts
C^2R loss in armature...........	15 "
Hysteresis loss in armature....	48 "
Magnetic flux per square inch.	
In field core.............	76,000 lines
In pole-pieces...........	36,000 "
In air-gap...............	23,000 "
In armature teeth........	65,000 "
In armature core.........	85,000 "
Co-efficient of leakage.......	1.4
Electrical efficiency..........	82 per cent.
Commercial efficiency (windage and friction losses 10 p. c., estimated)...............	52 "
Revolutions per minute......	2,000

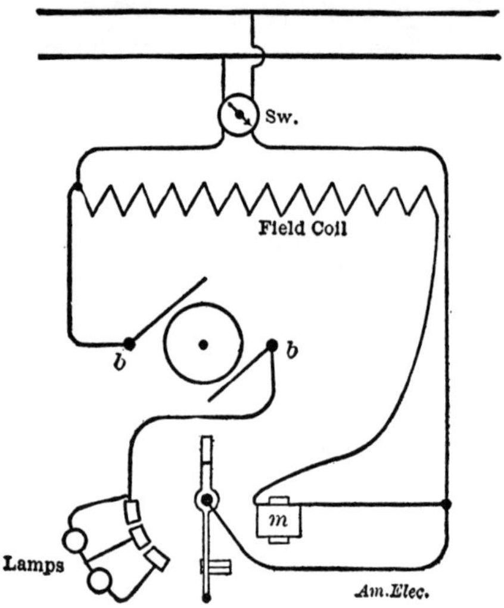

FIG. 26.—DIAGRAM OF STARTING SWITCH.

It is advisable to provide a starting switch similar to the one shown diagrammatically by Fig. 26, where *b b* are the brushes; *S* the starting switch lever; *m* a magnet, and *Sw* a double-pole snap switch. The lamps shown are 50-volt, 32-candle-power lamps. The handle of the starting switch is provided with a spring tending to keep it in the position shown by the sketch. This starting switch is also suitable for use with the motor described in Chapter III.

If it is desired to build a smooth-core machine the armature core must be made 4½ ins. in diameter, and two grooves ⅛ in. wide and 5-16 in. deep must be cut in the face of the core at opposite points for the reception of driving teeth. These are two pieces of fibre ⅛ in. thick, ½ in. wide and 2 ins. long, set on edge in the grooves and projecting 3-16 in.

above the surface of the core. The core must be thoroughly covered with two layers of micanite cloth. The number of coils is the same as before, but the coils will be twenty turns wide and five deep on the outside; on the inside of the ring the coils must lap, making ten layers of wire. The smooth core is banded just as the slotted one is, except that soft brass wire must be used instead of tinned iron.

CHAPTER V.

ONE-HALF HORSE-POWER MOTOR, WITH DRUM ARMATURE.

For this size of motor three types of field magnet are described, the single-coil Jenny, like those previously described, a bipolar one-piece magnet of the so-called iron-clad type, and a similer form with four poles

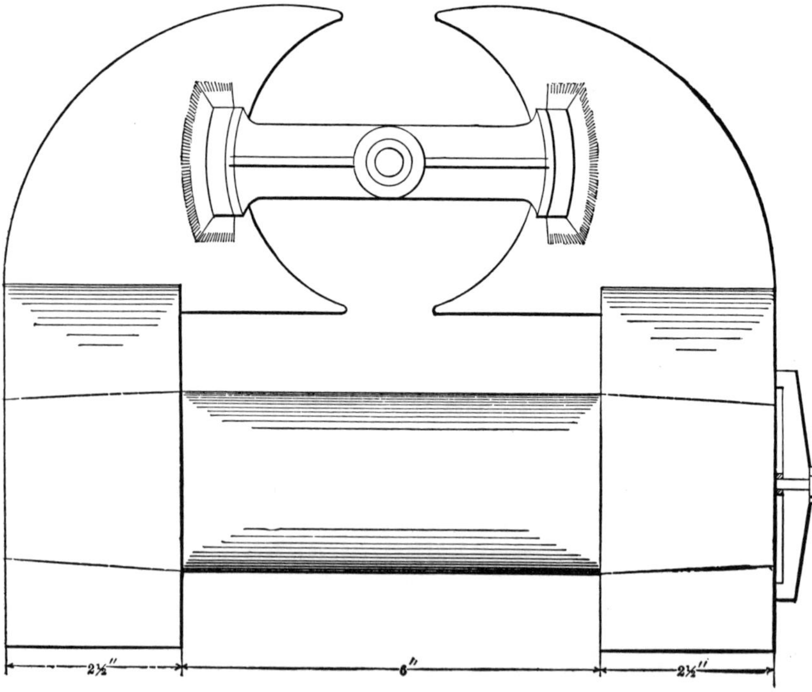

FIG. 27.—END ELEVATION OF FIELD MAGNET.

(Kapp type). The armature core and shaft are the same in each case, excepting the number of slots in the four-pole machine. The machine is a ½ horse-power motor to operate on a 110-volt constant potential circuit at a speed of 2,000 revolutions per minute. The single-coil magnet

(Figs. 27 and 28) has a round core of commercial wrought iron 2½ ins. in diameter and 11 ins. long over all. The ends are turned tapering, as indicated by the dotted lines, to insure intimate contact with the yokes; the taper is from the full diameter to 1¾ ins., and begins with 2½ ins. from each end. The pole-pieces are of cast-iron. The arms which support the journal yokes are cast solid with the pole-pieces, and their horizontal

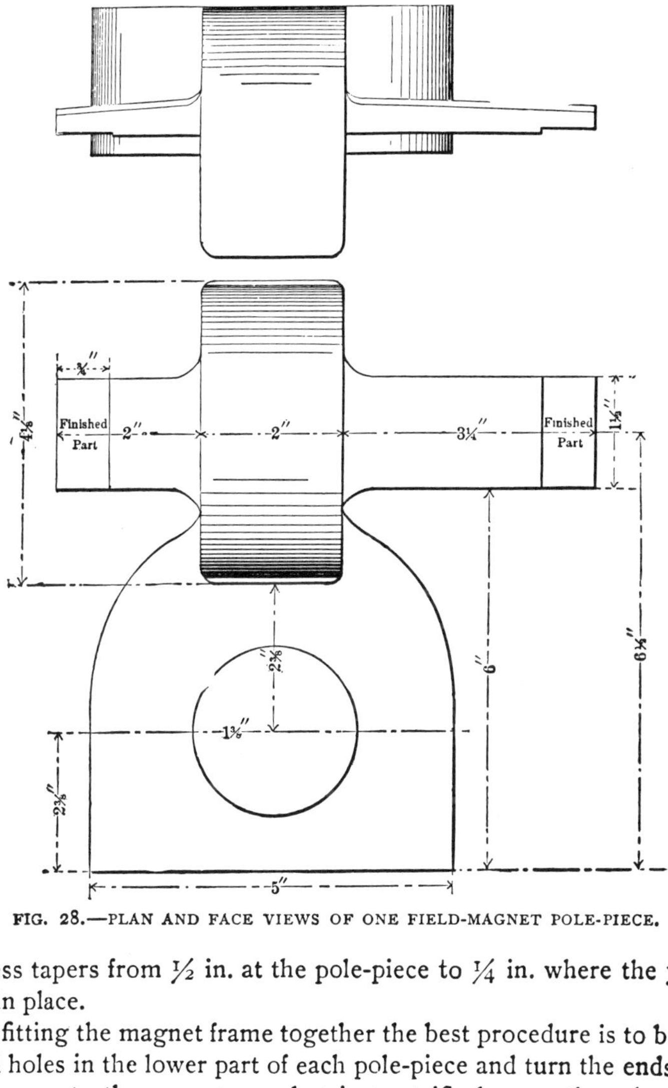

FIG. 28.—PLAN AND FACE VIEWS OF ONE FIELD-MAGNET POLE-PIECE.

thickness tapers from ½ in. at the pole-piece to ¼ in. where the yoke is bolted in place.

In fitting the magnet frame together the best procedure is to bore the tapered holes in the lower part of each pole-piece and turn the ends at the magnet core to the same taper, but just a trifle larger; then dress each tapered end of the core down very gradually with a fine file (the core being run on a lathe) until the pole-piece can be pushed on by hand far enough to bring the end of the core within 3-64 in. of the surface of the cast-iron. The pole-pieces and ends of the core should be punch-marked so as to insure finally mounting each pole-piece on the end which was fitted to it. After dressing down the ends of the core as above described drill and tap in each end a hole for a ¼-in. machine screw, the purpose of which will be made apparent by a glance at the right-hand end of the complete magnet in Fig. 27, where *C* is a four-armed claw or spider, with a hole through the center where the arms intersect. The arms are 3-16 in. thick, measured at right angles to the bolt, and taper from 3-16 to ⅜ in. thick, measured parallel with it. One of these claws or spiders is used at each end of the core, though the drawing shows it at one end only.

After drawing one pole-piece home solid by means of the spider and bolt, slip the other on the other end of the core loosely and clamp the pole-pieces lightly between two iron plates with planed surfaces, applied between the journal arms, so as to keep the four horns of the pole-pieces in alignment; then force the second pole-piece home and clamp the horns hard between the iron plates. The bottom surface of the pole-pieces are then to be turned up on a shaper or planer and the iron clamping plates removed from the horns.

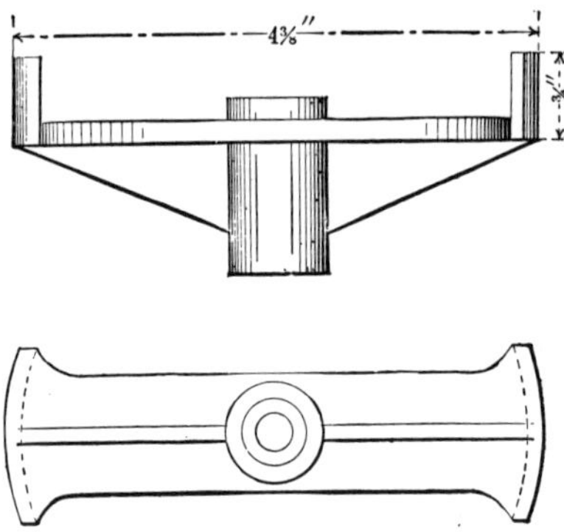

FIG. 29.—JOURNAL YOKE.

The next operation is boring the armature chamber and the seats for the ends of the journal yokes. The bore of the armature chamber is 4 3-16 ins.; the seats for the journal yokes are machined to a 4⅝-in. circle for ¾ in. from the outer ends. These operations must be completed before the original position of the frame on the lathe or boring machine is altered. This completes the machine work on the magnet, with the exception of the holes through the ends of the supporting arms and holes in the bottom surfaces of the pole-pieces for bolting to the base.

The journal yoke must be made of brass or some similar composition. The bar is 3-16 in. thick and 1 in. wide, except near the ends, where it flares to correspond with the width of the supporting arms. At each end is a right-angle lug, ⅛ in. thick after machining; these lugs fit the machined seats in the ends of the iron arms, and the yokes should be fitted to these arms before the frame is taken apart to put on the magnet coil. The journal box is 1½ ins. long over all, 3-16 in. of its length projecting on the inside of the yoke bar, and 1⅛ ins. on the outside. As shown by the plan view of the yoke in Fig. 29, there are stiffening webs starting flush near the ends of the yoke and attaining a height of ¾ in., where they join the box; these ribs are ⅛ in. thick. The box is 1 in. in outside diameter and bored out 9-16 in.; the bore is bushed to ⅜ in. No particular form of oiling device is specified, as any amateur of sufficient ability to build such a motor will be fully competent to decide this detail for himself. The journal yokes are held in place by ¼-in. cap-screws passing through the ends of the supporting arms and tapping into the lugs on the yokes.

The field coil contains 35 layers of No. 26 double cotton-covered magnet wire. After the magnet is fitted as above described it is taken apart and two circular fibre heads 4½ ins. in diameter and ⅛ in. thick are put on the core with a driving fit, care being taken that the distance from outside to outside of the heads corresponds with the space between the perpendicular faces of the pole-pieces when the frame is assembled; this measurement should be taken prior to dismantling the frame. A groove must be cut on the outer face of one head, from the center to the outer edge, in order to form a channel for leading out the starting ends of the coil when the frame is re-assembled, at which time two discs of oil paper with one of mica between them must be threaded on the core outside of this head to insulate the leading-out wire from the pole-piece. Before winding the coil insulate the core with a strip of muslin just wide enough to go between the heads, and long enough to wrap around the core three times; this should be heavily shellacked before it goes on. If the coil is carefully wound it will be 210 turns in length along the core; the umber of turns in this direction is not particularly essential, but as many should be put on as possible without jamming the insulation, in order to reduce the heat loss. The depth of the winding must be 35 layers.

After winding the coil and securing the ends, put one pole-piece on solid and slip the other on loosely. When it begins to bind bolt the journal yoke to the lugs on the pole-piece first put on, and insert the bolts through the lugs of the one that is loose. Then tighten up the spider bolt at the end of the core and force it into place, the bolts through the lugs serving as guides to keep the pole-piece from twisting on the core. These bolts should be set up little by little with the spider bolt, so as to keep the bolt heads within 1-16 in. of the surface of the lugs. As an additional precaution the frame may be set on a true surface and tried at intervals to see if it gets out of alignment; if it does, tap the horn of the loose pole-piece until the bottom surface agrees with the guide. The magnet frame must be provided with a non-magnetic base, preferably composition metal, but allowably of wood.

Figs. 30, 31 and 32 show an armature disc, the shaft and armature core (the latter in cross-section), and the shell and head. The discs are of charcoal iron, 4 ins. outside diameter, with a 1 in. hole in the center and an 1/8 in. key-seat, annealed after punching and key-seating; there are 18 slots 3/8 in. wide and 3/8 in. deep. The shell and one head are cast in one piece (of brass), and consist of a barrel 1 in. outside diameter (when finished), and 2 ins. long, with a head, *s*, at one end, 3 1/8 ins. in diameter and tapered in thickness from 1/4 in. near the center to 1-16 in. at the periphery; at the opposite end of the barrel is a cross-bar 1/8 in. thick, cast

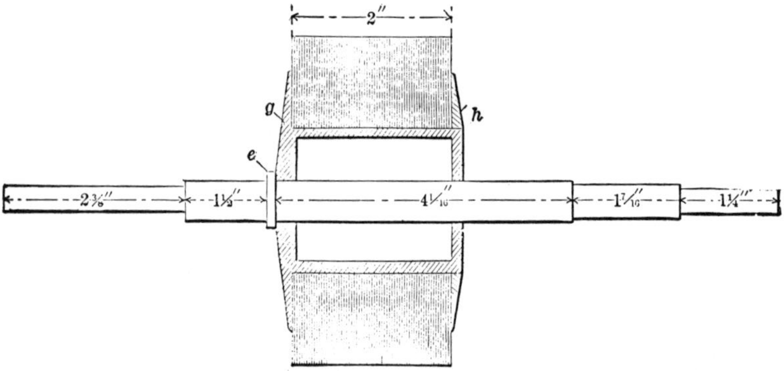

FIG. 31.—ARMATURE SHAFT AND AXIAL SECTION OF ARMATURE CORE.

with the barrel and of the shape shown, being 3/8 in. wide where it joins the barrel and 3/4 in. at the center. A 1/2-in. hole is drilled in the center of this cross-bar and another in the center of the head, *s*, at the other end of the barrel; the shell is mounted on a mandrel, the barrel is turned down to fit the hole in the armature discs, and both sides of the head are faced off smooth. A 1/8 in. key-seat 3-16 in. deep is cut in the barrel, so as to come in the center of one end of the cross-bar, as shown; a 1/8 in. × 1/4 in. feather, or parallel key is laid in the key-seat, and the discs are threaded on the barrel and compressed against the head by the collar, *h*, drawn down by two bolts (not shown) passing through the collar and inside the barrel, and tapping into the head at the other end. This collar, *h*, is of brass, 3 1/8 ins. in diameter and tapering from 3-16 to 1-16 in. in thickness when finished. The opening in the center should fit the outline of the cross-bar on the end of the barrel at least closely enough to prevent the collar from shifting under stress of centrifugal force; the collar must be finished up smooth on both sides. A disc of insulation should be put on next to the brass head before the iron discs are put on, and another insulating disc should go between the last iron disc and the clamping collar, *h*.

If the slots are cut in the core with a milling machine the discs must all come off the barrel to have the burrs removed, and also be re-annealed; the key-seat will insure their returning in the original angular position. It is much better to have discs with the slots punched before the first annealing. The shaft is 10¾ ins. long over all; ½ in. in diameter in the largest part; 7-16 in. where the commutator goes, and ⅜ in. in the journals. A 1-16 in. × ⅛ in. collar, *e*, is shown back of the armature, the purpose of which is merely to "locate" the armature shell; it is not absolutely necessary, however, and may be left off if desired. The

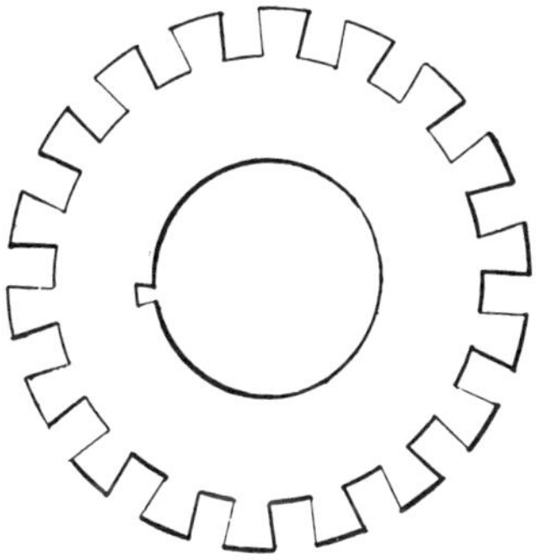

FIG. 30.—ARMATURE DISC.

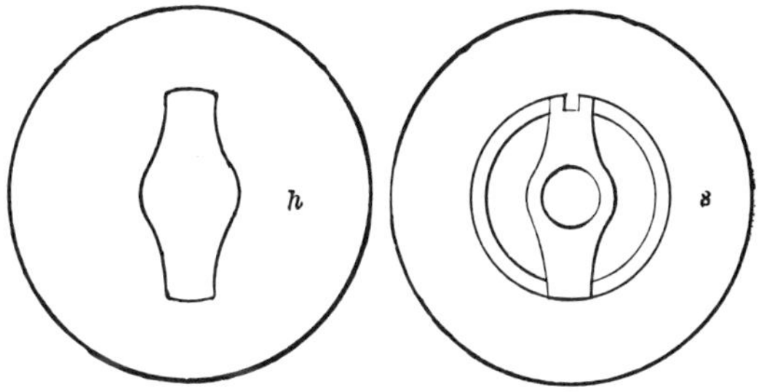

FIG. 32.—ARMATURE CORE DRUM AND HEAD.

easiest way to provide for it is to make the shaft of ⅝ in. stock, leaving the original metal to form the collar when turning the shell to the proper diameter. The armature shell may be keyed to the shaft or pinned obliquely through the thick part of the head; it must be positively secured by some such means.

The commutator shell must be bored to fit the 7-16 in. portion of the shaft, and must not exceed 1¼ ins. along the shaft. The lugs where the wires are attached to the segments may project toward the armature ⅛ in. or so. There must be 18 segments, and a diameter of 2 ins. is recommended. The quadrant carrying the brushholders should be fitted to the inner end of the journal box, and carbon brushes (one on each side) not smaller than ¼ in. × ½ in. on the contact surface should be used. If the machine be used as a dynamo (it will maintain about ten 110-volt lamps) metal brushes of the same surface should be used to reduce the resistance of the brush contact. The armature winding is divided into 18 coils, each having 32 turns of No. 20 double cotton-covered wire, eight turns wide and four turns deep in the slot. The slots must be insulated with troughs of muslin and mica, or preferably flexible micanite, 0.03 in. thick. The troughs are easily made by cutting the material into strips 2½ ins. long by 1⅛ ins. wide, and slitting the ends so as to permit the projecting portion of the trough to be folded back flat against the core. Before putting in the troughs a disc of heavy drilling 3¼ ins. in diameter should be secured to each end of the core by means of varnish, and the outer faces varnished and allowed to nearly dry. Then put in the troughs and put on two more muslin discs, varnishing the whole, and bake until thoroughly dry. Instead of winding each coil in diametrically opposite slots, take slots lacking one of being precisely opposite.

A good plan is to make a sketch of an armature disc and number the slots from left to right successively around the periphery. Then wind the coils as follows, the coil numbers indicating the order in which the coils are put on, not the order in which they are connected to the commutator:

COIL NO.—	1	2	3	4	5	6	7	8	9	10	11	12	13	14	15	16	17	18
STARTS IN SLOT NO.—	1	10	13	4	7	16	2	11	15	6	14	5	18	9	3	12	8	17
ENDS IN SLOT NO.—	9	18	3	12	15	6	10	1	5	14	1	13	8	17	11	2	16	7

Each pair of coils must be covered with muslin where they cross the heads before the next pair is put on, and before coil No. 8 is wound on top of coil No. 1 in slot No. 1 the bottom coil must be insulated by a strip of micanite laid in the slot; this is true of every bottom coil.

After the winding is on, and before connecting up to the commutator, the band wires should be put on. Use No. 19 B. W. G. soft tinned-iron wire, known by hardware dealers as "white stove-pipe wire," for the bands, and put them on under as heavy pressure as possible without endangering the armature shaft. Two bands of eight turns each, ½ in. from each end of core will suffice. A strip of mica between two strips of fullerboard must go under each band, and the bands should be soldered at intervals, not all the way around. Four tin clips located equidistantly, with a dab of solder at each, will give ample security.

If cast steel be available, one of the iron-clad types of magnet, shown by Figs. 33 and 34, is somewhat preferable because of the small amount of machine work required. Of these two the four-polar type is considered preferable by the writer, being much lighter in weight and having an "open-head" armature winding. Each of the iron-clad magnets is a single casting; the essential dimensions are shown in the sketches, with

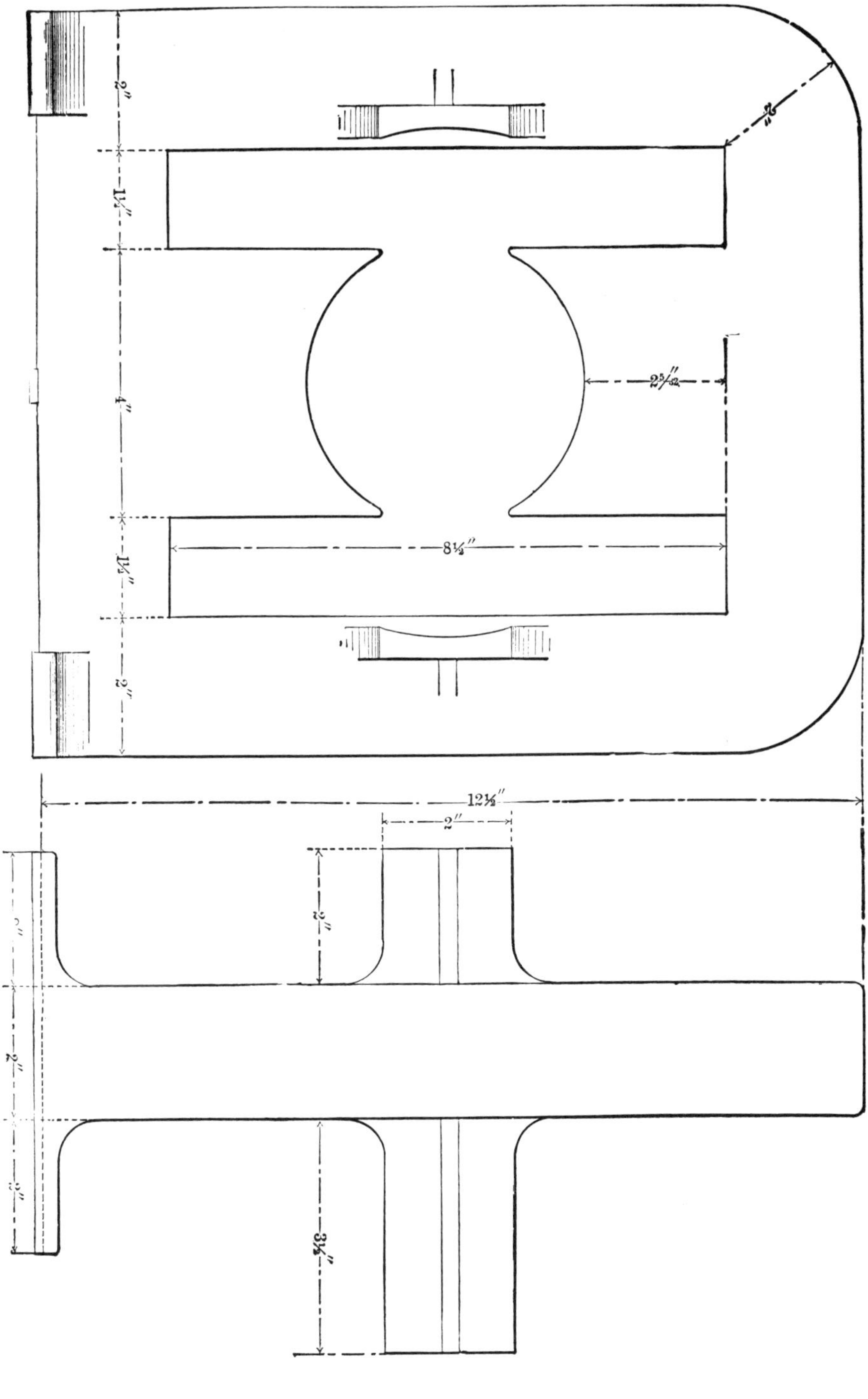

FIG. 33.—SIDE AND END ELEVATIONS OF BIPOLAR IRON-CLAD FIELD MAGNET.

the exception of the bore of the armature chamber, which is, of course, the same as for the single-coil magnet—4 3-16 ins. As the two magnets require the same treatment, varying only in dimensions, the following remarks apply to both:

It will be noticed that the feet of the machine project ⅛ in. below the body and that there is a transverse rib under the center of similar depth. These are to give the machine a floor bearing which may be trued up on a shaper or planer without finishing the whole bottom of the machine. The first operation on the casting is chipping off the numerous fins and lumps with which steel castings are invariably afflicted. An emery wheel may be used for this purpose around the outside of the frame,

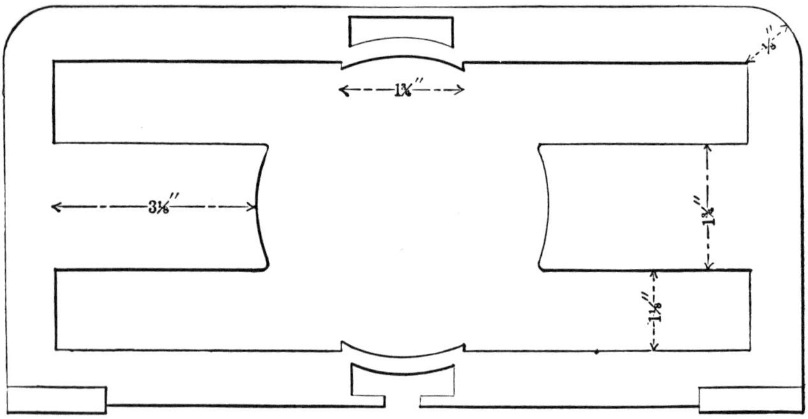

FIG. 34.—SIDE ELEVATION OF FOUR-POLE IRON-CLAD FIELD MAGNET.

but in the corners of the coil spaces a cape chisel and lots of muscular exertion will be required.

Next, the bearing surfaces are trued up, and ½-in. holes drilled in the feet; then the magnet is mounted for boring out the armature chamber and the seats for the journal yokes, all of which must be done with one mounting. This finishes the magnet frame, unless it is desired to put a terminal block on the machine instead of on the base and do away with the latter. In this event four ¼-in. holes are to be drilled in the top surface of the frame and tapped for machine screws to hold the block, which may be 2 by 6 ins. and 1½ ins. thick. The journal yoke and journal are the same as shown in Fig. 29, except that for the bipolar magnet the yoke is 7½ ins. long instead of 4⅝ ins.

The field coils for the bipolar machine consist of No. 25 double cotton-covered wire wound 45 layers deep, and each coil is 80 turns long.

The coils are to be wound in fibre bobbins, as shown by Fig. 35. The heads of the bobbin must be 2⅛ ins. apart, and the body must be ⅛ in. wider and longer than the magnet core, actual measurement. Before winding the coil the bobbin must be mounted on a wooden core of proper size to fit the opening through the center, and having flanges or heads at each end to "back up" the heads of the bobbins; one of these heads is put on permanently and the other is secured by two screws so as to be removable. A spindle of 1-in. iron goes through the center of the wooden core upon which to mount it in the lathe for winding.

When a coil is completed bend the wire back upon itself near the end, tie a linen thread in the loop formed, and secure the end of the coil by passing the thread several times around the coil and tying its ends together. Then varnish the outside heavily and bake the coil at a low temperature—100 to 125 deg. Fah.—until the varnish is hard.

The coils for the four-polar machine are 35 layers deep, and 110 long,

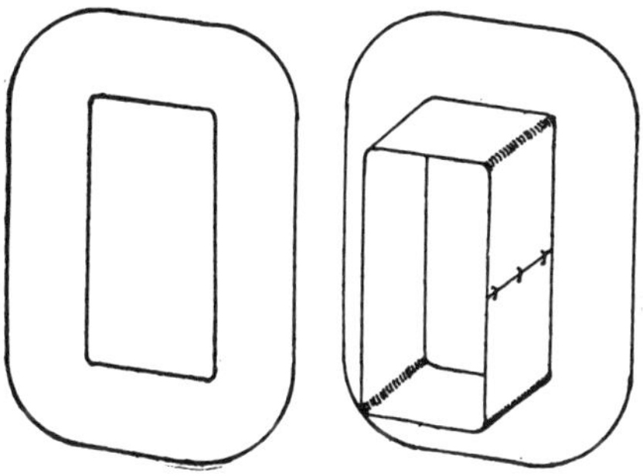

FIG. 35.—MAGNET COIL BOBBIN.

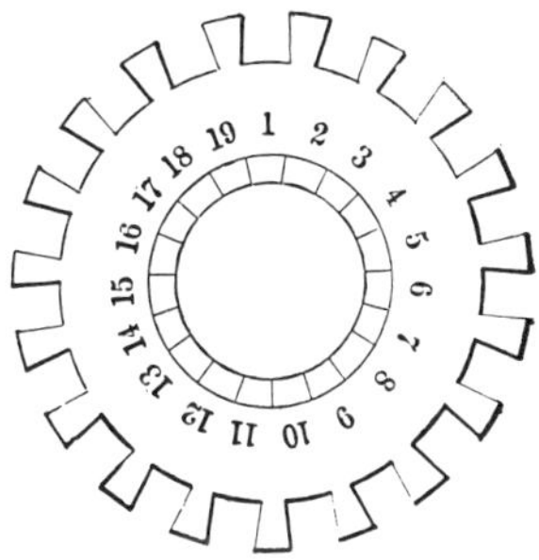

FIG. 36.—ARMATURE DIAGRAM.

of No. 25 wire. The heads of the winding bobbin are 3 ins. apart. The instructions for winding the coils for the bipolar iron-clad machine apply to these also. In connecting the coils on the machine, however, there is a difference. On the bipolar machine the final end of one coil must be connected to the beginning of the other; on the quadripolar the reverse is true. Fig. 37 shows diagrammatically the manner of connecting the field coils of the quadripolar machine. It will be noticed that the exciting current passes around the cores in opposite directions. The connection for the bipolar machine is exactly the reverse of that shown.

The armature core of the four-pole machine has 19 slots 3-10 in. wide and ⅜ in. deep, instead of 18 slots ⅜×⅜. There are 19 coils, each having 28 turns of No. 20 wire, 7 turns wide and 4 turns deep. These may be wound directly on the core, but it will probably be easier for an amateur to wind them in a little frame, tie them at intervals with thread and put them on the core complete. The winding frame will be exactly like the one for the field coils except in size. The "channel" formed between

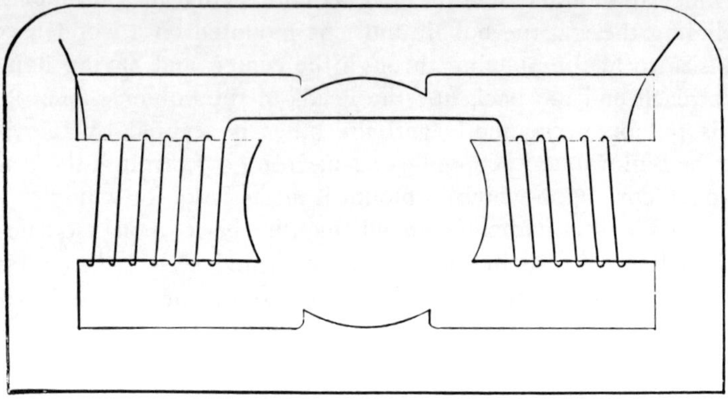

FIG. 37.—DIAGRAM OF FOUR-POLE FIELD-COIL CONNECTIONS.

the heads must be 9-32 in. wide and ¼ in .deep. The body of the frame, which determines the length and width of the coil, is 2¼ ins. one way

TABLE OF WINDING AND CONNECTIONS.

NUMBER OF COIL	IN SLOTS NOS.	BEGINNING END GOES TO SEGMENT NUMBER	FINAL END GOES TO SEGMENT NO.
1	1 and 6	1	11
2	2 " 7	2	12
3	3 " 8	3	13
4	4 " 9	4	14
5	5 " 10	5	15
6	11 " 16	11	2
7	12 " 17	12	3
8	13 " 18	13	4
9	13 " 19	14	5
10	15 " 1	15	6
11	16 " 2	16	7
12	17 " 3	17	8
13	18 " 4	18	9
14	19 " 5	19	10
15	6 " 11	6	16
16	7 " 12	7	17
17	8 " 13	8	18
18	9 " 14	9	19
19	10 " 15	10	1

and 2¾ ins. the other. The coils are put on and connected up as indicated by the accompanying table.

The numbers of the coils indicate the sequence in which they are put on the core, and this order should be observed in order to secure maximum symmetry of the wires across the heads of the core. The numbers of the slots and segments refer to the diagram shown by Fig. 36. Each figure applies to the slot and segment between which it is located.

The brush quadrant for this machine is also different from that of the other two; instead of bearing upon the commutator at diametrically opposite points, the brushes must be 90 deg. apart—corresponding with the relative angular positions of magnet poles of different signs. In the bipolar iron-clad the "north" and "south" poles are, of course, opposite each other; in the four-pole machine the poles directly opposite are of the same sign—if one horizontal pole is "north" the other must also be "north," and the other two, without coils, will be "south."

CHAPTER VI.

ONE HORSE-POWER BIPOLAR MOTOR, WITH DRUM ARMATURE.

The accompanying drawings and description will enable any one with moderate machine-shop facilities to build a 1 horse-power motor to work on a 110-volt or a 220-volt continuous-current circuit. Two types of field magnet are given, the armature and shaft being the same in both cases.

The armature is 4 ins. in diameter, outside, with twenty-four slots, each 7-32 in. wide and ⅝ in. deep. Fig. 38 shows the shaft and a cross-sectional view of the armature core. The discs are compressed by two cast-iron end plates, which are screwed on the shaft; these plates are ½ in. thick at the shaft, and taper to 3-16 in. thick at the outer edge, which

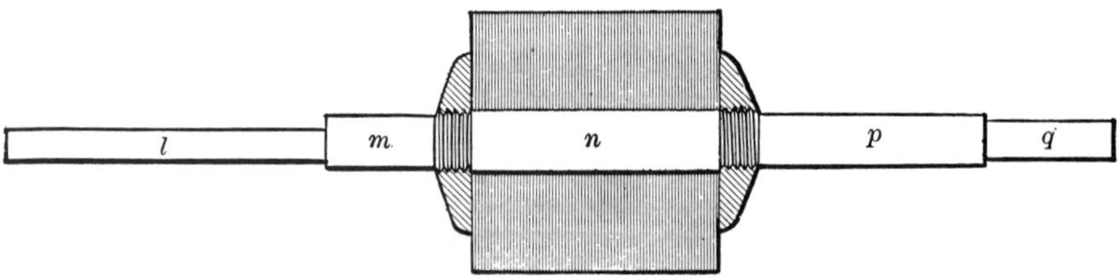

FIG. 38.—ARMATURE SHAFT AND AXIAL SECTION OF ARMATURE CORE.

is rounded as shown, to avoid abrading the insulation between the core and the windings. The full list of armature dimensions is as follows:

	Core		Shaft				
	Body.	Heads.	at l.	m.	n.	p.	q.
Diameter...........	4	2¾	½	¾	1	¾	½
Axial length........	4	½	5	1¾	5	3¾	2

The discs should have a shallow key-seat in the edge of the central hole, and the shaft should be correspondingly key-seated, and a spline, or perfectly straight key, ¼-in. square, should be used to transmit the movement of the discs to the shaft. If this is done, the slots in the periphery of the discs may be milled out; the armature core must be dismantled after the slots are cut, and the burr which is left by the milling center smoothed off. If the key and key-seats are properly fitted the discs will go back on the shaft in precisely the position which the slots were cut, and the sides of the latter will be smooth. If the key is a loose fit, however, it will be advisable to use a straight edge in one of the slots to insure perfect accuracy in re-assembling the discs. It is scarcely necessary to urge a very careful and close fit of the key and its seats. In assembling the core, one of the cast-iron heads should, of course, be screwed to place first; then put on a disc of vulcanized fibre, 1-16 in. thick, 4 ins. in diameter, and next thread on the iron discs. After the last iron disc put on another fibre disc and follow with the end plate or

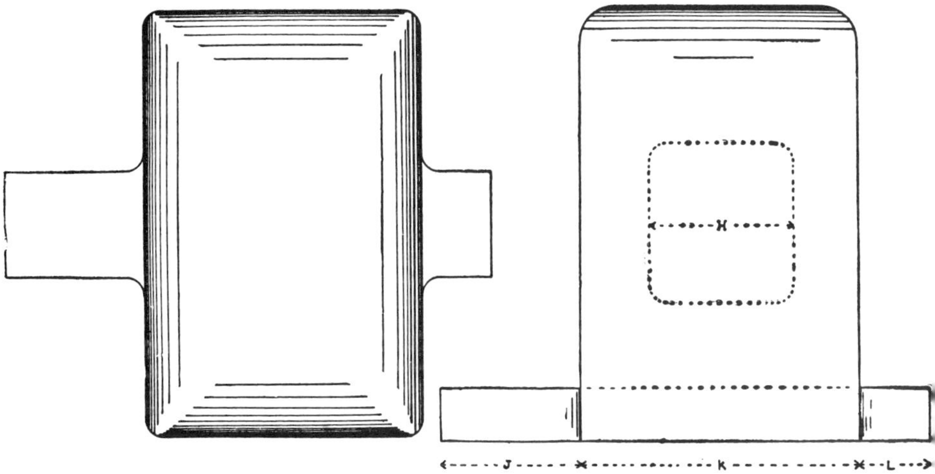

FIGS. 39 AND 40.—PLAN VIEW AND END ELEVATION OF IRON-CLAD MAGNET.

head of cast-iron, which will have to be set up with a pin wrench. If the discs are purchased with the slots already stamped out notches will have to be cut in the fibre end discs to correspond with the armature slots; if the slots are to be milled the fibre discs will, of course, be cut along with the iron ones. The latter must be not over 1-32 in. thick and preferably thinner; care should be taken not to get steel discs, but the very best possible grade of charcoal iron.

Of the two types of field magnets shown, the iron-clad is preferable from a constructional standpoint, as the only operations are boring out the armature chamber and the seats for the journal pedestals, and drilling the bolt holes for the latter. Fig. 39 gives a plan view of the iron-clad magnet, Fig. 40 an end view and Fig. 41 a side elevation. The thickness of the magnet core (the portion on which the coils are placed) parallel with the shaft is 4¼ ins. except right at the pole face, where it is rounded down to 4 ins.; this is necessary in order to reduce the flow of magnetism from the pole to the cast-iron end plates of the armature, which produces waste of energy by heating. The complete measurements of the field magnet are as follows:

	INCHES.
A—Thickness of yoke portion of magnet	1½
B—Inside length of horizontal part of yoke	8
C—Vertical thickness of magnet core	4½
D—Distance from core to yoke	2⅜
E—Total outside width of magnet frame	11
F—Width of journal foot	3
G—Radius to which journal seat is bored	4⅞
H—Horizontal thickness of magnet core (see above)	4¼
J—Length of journal foot, commutator side	4⅜
K—Width of magnet yoke or frame, axially	8¼
L—Length of journal foot, pulley side	2⅜

The bore of the pole pieces is 4 3-16 in. in diameter, and this figure must be rigidly observed for best results, as all the calculations are based upon this length of air-gaps. The above dimensions are intended to apply to a magnet made of the best grade of cast-iron; Scotch pig should be used if it is obtainable, and if not, then the very best grade of soft iron. The casting should be allowed to remain in the mold until it is absolutely cold care being taken not to remove any of the sand from about the magnet proper. The sand can be scraped away from the extreme end of the longer of the two pedestal feet, so as to enable the molder to ascertain when the casting is cold. It is frequently the case that a casting requires as much as two days to thoroughly cool, but it should not be disturbed before it is cold.

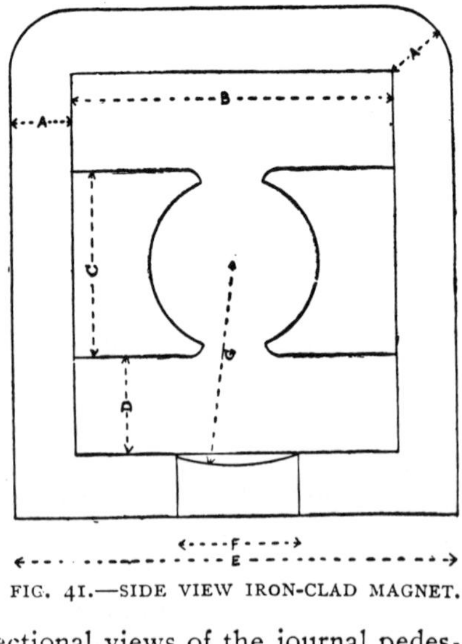

FIG. 41.—SIDE VIEW IRON-CLAD MAGNET.

Fig. 42 gives outside and cross-sectional views of the journal pedestal for this magnet; the two pedestals are alike in every particular, and when in position on the projecting feet of the field magnet frame their outer edges should be exactly flush with the ends of the feet. The pedestals are of iron; the base is curved to conform to the arc of the circle to which the upper surface of the foot is machined, and is ⅜ inch thick. The standard consists of two ribs at right angles with each other, each ⅜ in. thick, with their edges curved as shown. The box is of the ring-oiling type, with a single ring hung midway of the journal; the bushing is easily made from thin brass tubing, ¾ in. outside diameter, and with a very thin wall (not over 1-32 in.), babbitted to fit the shaft and having a slot ⅜ in. wide cut half way through it, midway between its ends. This bushing is shown in Fig. 43, which represents the bearing for the other type of magnet, to be presently described. The bushing is 1¾ ins. long; the oil ring is made of brass, one inch in diameter, inside, 1⅛ ins. diameter outside, and ¼ in. wide along the shaft. Reference to the side views of the journal pedestal will show a slot in the

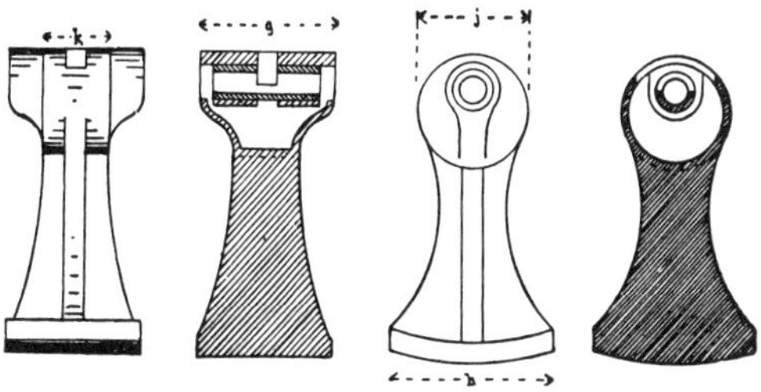

FIG. 42.—DETAILS OF JOURNAL BOX AND PEDESTAL.

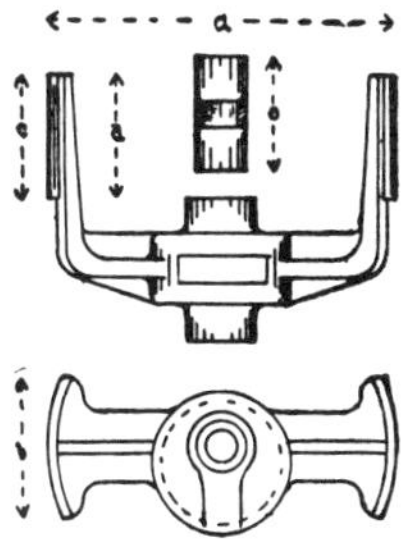

FIG. 43.—JOURNAL YOKE FOR FIG. 44.

upper wall of the box portion, through which the oil ring is inserted before putting in the bushing. A cover should be provided for this slot to keep out dust, etc. The dimensions of the journal pedestals are as follows:

	INCHES.
g—Length of base and journal box	2¼
h—Width of base	3
j—Outer diameter of reservoir	2
k—Axial length of reservoir, outside	1¼
Internal diameter of reservoir	2¾
Internal length of reservoir	1

The bore of the box portion of the pedestal must, of course, be made to fit snugly the outer diameter of the tubing used for a bushing, as the wall of the latter is too thin to admit of turning it down to fit a predetermined bore in the pedestal. After boring the pedestal to fit the bushing it should be mounted on a mandrel and its base turned to fit the circle of the foot on the magnet frame, namely, 9¾ ins. in diameter. Each pedestal should be fastened to the foot with two ¼-in. cap screws.

Fig. 44 gives a side elevation of a much lighter magnet, which may be used in connection with the armature above described, if the builder has sufficient skill and facilities to do the machine work accurately. The magnet core is a round piece of wrought iron, 3¼ ins. in diameter, with its ends turned down to 3⅛ ins. diameter for a distance of 4 ins. from each end; the total length of the core is 12¼ ins., so that the length of the untouched portion will be 4¼ ins. The pole pieces are of cast-iron, only the very best possible grade being suitable. Where the core enters the cast-iron the latter is 4 ins. square, with the corners rounded, and having two ribs or flanges, *f*, *f*, running along one edge; these continue clear up to the top of the pole piece, and are 1 in. thick by 2 ins.

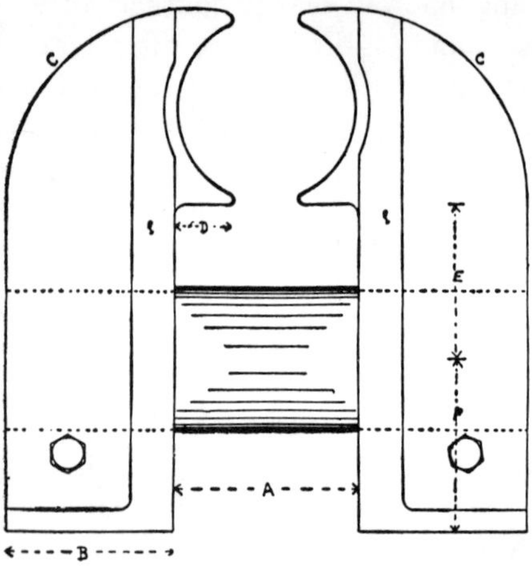

FIG. 44.—SINGLE-COIL MAGNET.

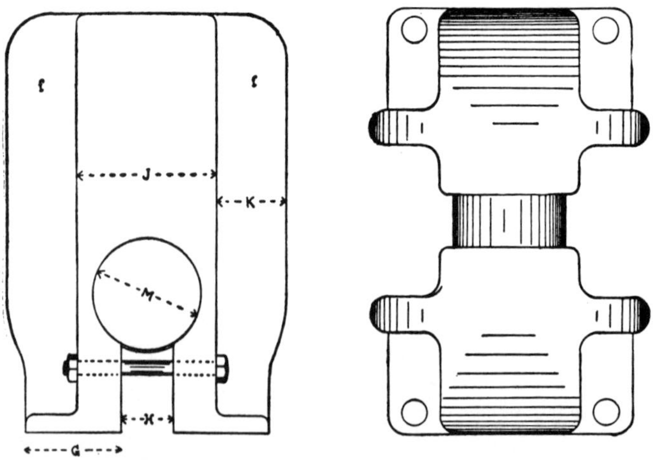

FIGS. 45 AND 46.—END ELEVATION AND PLAN VIEW OF SINGLE-COIL MAGNET.

wide. Fig. 45 shows an end view of the magnet frame, and Fig. 46 a plan view. The hole occupied by the wrought iron core should be cored out to 2⅞ ins. diameter when the casting is made, and afterward bored to a driving fit of the end of the core.

The first operation should be turning off the ends of the core; next, bore the holes in the pole pieces (or, more strictly speaking, the yokes). Then drill a 3/8-in. hole through the yoke just below the lower edge of the big hole and at right angles with it, to accommodate the clamping bolt shown in Fig. 45. Next drive one end of the core into one yoke and set up the nut on the end of the clamping bolt; then put on the other yoke and twist it on the core until the four horns of the pole pieces are exactly opposite each other, tighten up the second clamping bolt, and plane off the bottom surfaces of both yokes. To bring the pole horns into alignment, the simplest method is to cut out two heavy blocks of hard wood, say 3 ins. thick and 3½ ins. square; bore a ½-in. hole through the center of each block, run a ½-in. bolt, 12 ins. long, through the two blocks, and apply them to each side of the pole pieces, the bolt passing through the armature chamber in about the position to be occupied by the shaft. Set up the nut on the bolt until the blocks are hard against all four pole horns, and then tighten up the clamping bolt in the foot of the loose yoke.

After planing off the feet of the frame, bore out the armature chamber 4 3-16 ins. in diameter, and the seats for the journal yokes (in opposite faces of the side flanges, *f*, *f*), 5 ins. in diameter, and then remove one magnet yoke and put on the magnet coil. If the coil is separately wound in a form (which is preferable) only one yoke need come off; if it is wound directly upon the core, both yokes must come off, of course. The base of the machine must be of wood or brass. Wood is better, as, aside from its cheapness, it affords convenient space for the terminal posts and fuse-block of the machine. The base should be 15 ins. × 18 ins., made of two pieces of hard wood each 1½ ins. thick, glued and screwed together with the grains at right angles. The longer dimension of the base is to go parallel with the shaft, and the machine should be so set as to allow the pulley to overhang the edge of the base-board.

The pulley should be 4 ins. in diameter and 2 ins. wide on the face; the latter should be crowned. The pulley should preferably be keyed to the shaft, with a set-screw in the pulley hub on top of the key. If only a set-screw be used to hold the pulley on the shaft, a "flat" must be filed on one side of the shaft under the point of the set-screw.

The journal box and yoke for this magnet is shown by Fig. 43. It must be made of brass or some other non-magnetic composition. The design and dimensions of the oil reservoir, journal box and bushing are exactly the same as those given for the journal box of the iron-clad magnet above. All the dimensions are given in the following list, along with those of the magnet just described.

	INCHES.
A—Distance between yokes	4¼
B—Thickness of yoke	4
C—Radius of outer curve of pole-piece	4¼
D—Length of pole horn	1⅜
E—Distance from pole horn to center of magnet core	3¼
F—Distance from floor line to center of magnet core	3⅞
G—Width of foot	2¾
H—Width of slot under core hole in yoke	1½
J—Width of yoke	4
K—Width of flange	2
a—Diameter of curve of journal yoke ends and seats	5
b—Vertical width of journal yoke arms	2¼
c—Length of machined portion of yoke arms	1¾
d—Distance from end of yoke arm to inner end of journal box, pulley end of shaft	2
—Distance from end of yoke arm to inner end of journal box, commutator end of shaft	4
e—Length of bushing	1¾
g—Length of journal box	2¼
h—Slot to let in the oil ring	⅜x1¼
j—Outer diameter of oil reservoir	2
Outer length of reservoir, axially	1¼

The armature core and field magnet frames may be wound for any voltage desired, but the most efficient windings, as the cores now stand, will be those specified below.

ARMATURE WINDING.

The armature core, after being finally assembled, is to be made ready for windings by applying the insulation. Cut out four discs of heavy canvas, 3 ins. in diameter, with a ¾-in. hole in the center; varnish two of them on one side with shellac varnish, and apply them to the end plates of the armature core, varnished sides in. The edges will turn over to cover the outer edges of the plates, and will have to be slitted at intervals of ⅛ in. all around to prevent bunching up. After putting on these varnish their outer faces, and one face of each of the remaining canvas discs; when the varnish begins to thicken put on the two other discs, one at each end, and apply considerable pressure to them untii they dry. This is best accomplished by boring a hole in a piece of plank, large enough to pass the shaft, and setting the core on the plank, on end, next putting a short piece of board (6 or 8 ins. square) with a hole in its center on the upper end of the armature, and piling any convenient pieces of heavy scrap on the top board.

Next insulate the slots with troughs of oil paper, 1-64 in. thick, such as is used with the ordinary office outfit for copying letters; each trough should consist of two thicknesses of the oil paper, and the floor of the trough should be 4¼ ins. long, so as to project a little beyond the iron of the core and rest upon the edges of the canvas discs, which were previously turned over to cover the edges of the end plates.

The coils may then be wound directly in the slots, each coil consisting of twenty turns of No. 18 wire, four wide and five deep. Each slot will contain, when the windings are complete, half of each of two separate coils. It will facilitate the winding and insure electrical balance (as nearly as a core wound armature can be balanced) if the builder will make a diagram of his armature disc, numbering the slots from 1 to 24 successively around the circumference, as shown by Fig. 47. Then the winding will proceed as follows:

					No.				No.
First	coil	starts	in	slot	1	ends	in	slot	12
2d	"	"	"	"	13	"	"	"	24
3d	"	"	"	"	17	"	"	"	4
4th	"	"	"	"	5	"	"	"	16
5th	"	"	"	"	9	"	"	"	20
6th	"	"	"	"	21	"	"	"	8
7th	"	"	"	"	3	"	"	"	14
8th	"	"	"	"	15	"	"	"	2
9th	"	"	"	"	19	"	"	"	6
10th	"	"	"	"	7	"	"	"	18
11th	"	"	"	"	11	"	"	"	22
12th	"	"	"	"	23	"	"	"	10
13th	"	"	"	"	22	"	"	"	11
14th	"	"	"	"	10	"	"	"	23
15th	"	"	"	"	18	"	"	"	7
16th	"	"	"	"	6	"	"	"	19
17th	"	"	"	"	14	"	"	"	3
18th	"	"	"	"	2	"	"	"	15
19th	"	"	"	"	20	"	"	"	9
20th	"	"	"	"	8	"	"	"	21
21st	"	"	"	"	16	"	"	"	5
22d	"	"	"	"	4	"	"	"	17
23d	"	"	"	"	24	"	"	"	13
24th	"	"	"	"	12	"	"	"	1

After winding the first two coils, thin strips of varnished muslin should be laid over them, across each armature head from slot to slot, so that the next two coils will be insulated from the first pair; each successive pair of coils should receive this treatment, and after the slots are half filled (twelve coils being put on), a strip of oil paper 7-32 in. wide and 4¼ ins. long must be laid in each slot on top of the coil already in place before proceeding to put on the coil which next goes in that slot.

The starting end of each coil should be kept leading out straight from its slot, and the finishing end should be brought across the head and secured to the starting end by a turn around it. When the winding is complete, untwist the finishing end of each coil from its starting end, and twist it and the starting end of the next coil to the right firmly together. This will leave twenty-four terminals to lead out to the commutator lugs.

Before connecting the ends to the commutator, the binding wires should be put on and the winding tested for grounds on the core. The binding wires are put on in two bands, and consist of small tinned iron wire; they should be put on beginning 1 in. from each end of the core, and making each band ½ in. wide. The binding wire should be wound

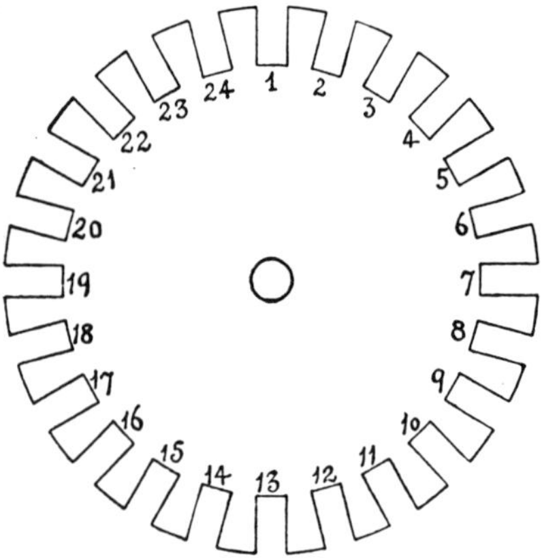

FIG. 47.—WINDING DIAGRAM.

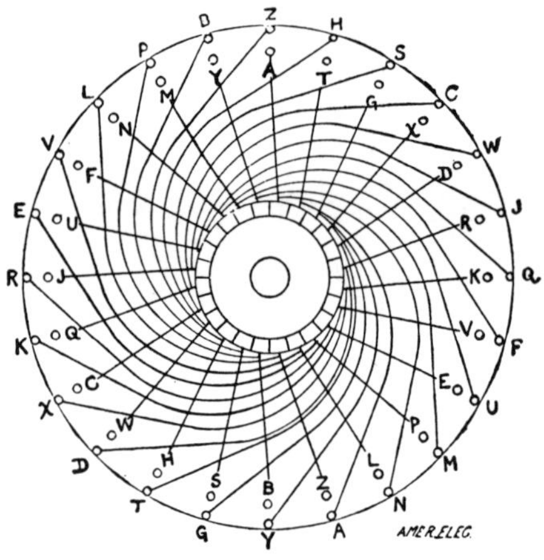

FIG. 48.—CONNECTING DIAGRAM.

on strips of thin varnished muslin laid around the core two layers deep, and the bands should be soldered at four equidistant points around the armature surface, not all the way around. The wire used should be not larger than No. 22 B. W. G. or No. 20 B. S. G.

Unless the machine is likely to be used in very dusty surroundings it is better not to put any covering over the ends of the armature after the winding is complete. If the instructions for insulating each pair of coils from the succeeding pair have been carefully followed out, any ordinary collection of dust will not be liable to cause a breakdown in the heads. The winding just described is intended for a 110-volt machine. If it is desirable to wind the armature for a 220-volt circuit, use No. 21 double cotton-covered wire, making each coil five turns wide and six layers deep.

The commutator had better be purchased from any well-known manufacturer of commutators, as its market price will be less than the cost of material and labor necessary to make one properly. It must have twenty-four segments and be not more than 2 ins. long along the shaft; the diameter does not matter particularly—take one of a stock size from the maker. In connecting up the coils to the commutator carry the ends previously twisted together straight out to the commutator segments. Fig. 48 shows the connections diagrammatically. The slots are omitted and each coil is represented as having only one turn for the sake of simplicity. The coils are lettered, to facilitate identification of opposite ends. The ends leading straight to the commutator are the starting ends; those leading around being the final ends. The diagram is not intended to show the relative radial positions of the coils, and care must be observed to avoid becoming confused. For example, coil *A* may or may not be under coil *Z* at its starting side; they are both in the same slot, but it does not matter which is on top. If coil *A* was the first one put on, it will, of course, be in the bottom of both of its slots, and coil *Z* will come on top of each side of it. The diagram only shows the relative angular positions of the coils and the manner of connecting their ends. The brushes should be of carbon, 3/8 in. thick, and of a width 1/8 in. less than the length of the commutator face, which should be about 1 1/2 ins. The brush holders may be copied from any standard type to which the builder of this motor has access.

The iron-clad magnet requires two magnet coils, one on each pole; for 110-volt circuits these coils consist of No. 22 wire wound in 61 layers, each layer being 50 turns long. If both the coils are wound in the same direction—in other words, if they are precisely alike as to the manner of winding, as they should be—the beginning end of one must be connected to the final end of the other, the two remaining ends being carried to the terminals of the machine. The best arrangement is to connect the two ends that are farthest apart, making this connection on the pulley side of the machine. For 220-volt circuits the wire must be No. 25 gauge, wound to a depth of 75 layers, 65 turns to each layer. These coils should be wound on a block, the cross-section of which is of exactly the same shape as that of the magnet core at its largest part, but which measures 1/8 in. more in each direction. After winding each coil, tie it at each corner with coarse linen thread (cobbler's thread) and cover it with strips of muslin wound at right angles to the direction of the wires, and so put on as to have the edge of each convolution of muslin lay just alongside that of its neighbor—touching it but not lapping it. The muslin must be one-fourth the width of the inner edge of one side of the coil, so that four turns will cover one side evenly. Put the muslin on in two layers, the turns of the second layer covering the joint between the turns of the first layer. Then wind strips over the corners of the coils, two layers deep. After the coil is covered with one layer, varnish the muslin covering heavily with shellac; when this is nearly dry, put on the next layer and the corner strips, and after varnishing the whole, set the coil aside to dry. Do not put any varnish on the wire itself. Next cover the iron cores of the machine with a layer of muslin, this time lapping the edges of successive convolutions; varnish the muslin, and when it and the coils are thoroughly dry, put the latter on. Unless the pattern for the field magnet has been very exactly made, and the casting is an unusually perfect one, it may be necessary to file the corners of the pole pieces slightly to get the coil between them in putting on the magnet core. In filing these corners, be careful to round them, leaving no sharp corners or edges whatever. It is advisable to do this, even if it is not mechanically necessary for the introduction of the coils.

The Jenny type of magnet has only one magnet coil, which consists of No. 25 wire, wound to a depth of 65 layers, with 148 turns to a layer, for 110-volt service. For 220 volts, use No. 28 wire, wound to a depth of 81 layers, 186 turns to each layer. All the wire specified herein for both armature and field winding should be double cotton covered. Circular magnet heads of vulcanized fibre should be used to protect the ends of the coil, as the full voltage of the machine exists between these ends; these heads should be ⅛-in. thick and 6⅜ ins. in diameter, with a hole to fit the magnet core snugly if the coil is wound directly on the core. If not, a bobbin should be made, the center consisting of a tube of 1-32 in. fibre, 4¼ ins. long, and of an internal diameter to go easily over the core; the heads of the bobbin to be of ⅛-in. fibre, as above. If the coil is wound on the core, the latter must be covered with three layers of muslin, each layer varnished with shellac. The whole must dry thoroughly before the wire is wound on.

The data of the machine are as follows:

IRON-CLAD TYPE.

	110 volts,	220 volts.
Resistance of armature winding..........	1 ohm.	3.15 ohms,
Armature capacity, maximum............	8.3 amp.	4.7 amp.
" " normal...............	7 amp.	3.8 amp.
" loss, C^2R, normal..............	49 watts.	45½ watts.
" " hysteresis, normal.........	11.7 "	13 "
" " eddy currents.............	1.3 "	1.5 "
Total internal armature losses............	62 "	60 "
Magnetic flux per sq. in. in armature core.	71,800	72,000
Revolutions per minute, loaded...........	1,800	2,000

Resistance of field winding...............	220 ohms.	682 ohms.
Current in field winding..................	½ amp.	0.322 amp.
Heat loss in field winding................	55 watts.	70.85 watts.
Density per sq. in. in core...............	38,000	38,160
Density per sq. in. in gaps...............	25,500	25,600
Efficiency, approximately.................	70%	75%

SINGLE COIL TYPE.

Armature data same as above.	110 volts.	220 volts.
Resistance of field coil....................	433 ohms.	1,400 ohms.
Current in field coil.....................	¼ amp.	0.16 amp.
Heat loss in field coil...................	27½ watts.	35.2 watts.
Density per sq. in. in core..............	89,750	90,00
Efficiency, approximately...............	78%	79%

An amateur motor builder will be wise not to attempt to make a starting box for this size of machine; one can be purchased for a moderate sum from any of half a dozen reputable manufacturers, and, as either of the motors here described is well worth the outlay necessary to insure its protection in this particular, the writer advises buying the starting rheostat.

If, however, the reader particularly desires to make his own starting rheostat, the arrangement shown by Fig. 49 will be found easier to construct than anything in the shape of a wire rheostat. In the sketch, *L* is the lever, pivoted on a ½-in. metal post, and normally forced downwardly by a coil spring of three or four turns (not shown), which is located under the washer, *W*. A pin through the post secures the washer, spring and lever. *H* is the handle; a wooden handle, such as coffee grinders are given, or a large porcelain knob will answer. *B* is the contact brush of copper, slitted tangentially to the circles of the contact strips, *c, c, c, c, c, c, c*; these circles have their common center, of course, in the center of the post on which *L* is pivoted. An end view of the brush, *B*, is given by *E*, showing the convex shape given the under face of the brush to enable it to pass smoothly over the contact strips. The end of these should be beveled to avoid digging into the brush.

The connections are shown diagrammatically. *R* is a bank of five 32 candle-power incandescent lamps, rated at 110 volts (100 will be better, and they can probably be readily obtained); *C* is the motor commutator; *b, b*, the brushes; *F*, the field winding; *S*, a double-pole combined switch and fuse block, and *M*, the service mains. A glance at the connections will show that the functions of the lever *L* are to first connect in the field, next the armature in series with one lamp; at each successive step a lamp is added in parallel with the first one until all are in, and the last position of the lever cuts out the lamps, leaving the armature in circuit direct.

The lamps should be mounted on the base with the lever and contacts, and it is preferable, though not particularly urgent, that the switch *S* be mounted on that base also. The sketch shows the lever in the "off" position; the switch, *S*, should never be closed, except when the lever *L* is in this position. If the reader desires to make the arrangement automatic he need only add a retractile spring to pull the lever *L* to the "off"

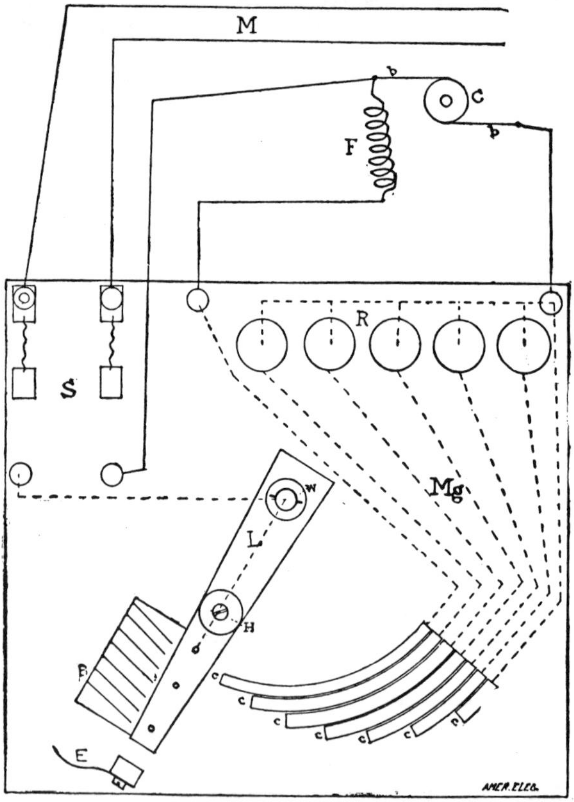

FIG. 49.—MOTOR-STARTING RHEOSTAT AND DIAGRAM OF CONNECTIONS.

position; a bar of iron ⅛ in. by ¼ in. by 2 ins. on the right hand edge of the lever, and a small magnet connected in series with the field, *F*, and located on the base about *Mg*, in such a position that it will hold the lever, by means of the bar of iron, when it is brought to the "on" position.

CHAPTER VII.

ONE HORSE-POWER FOUR-POLAR MOTOR WITH DRUM ARMATURE.

For the four-polar one horse-power motor here described only one type of field magnet is shown, namely, the familiar ring yoke with radial magnet poles. This type combines more good points than any other, hence the limitation to the one type. A choice is given, however, between cast-iron and cast-steel. The armature construction is the same

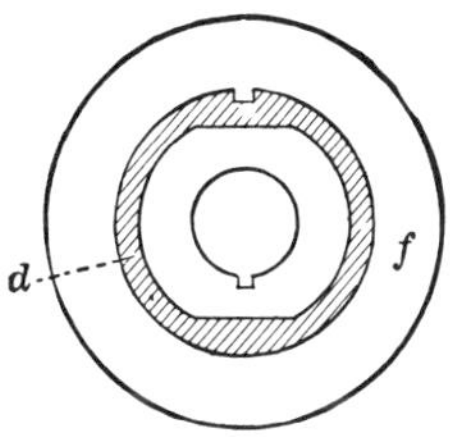

FIG. 51.

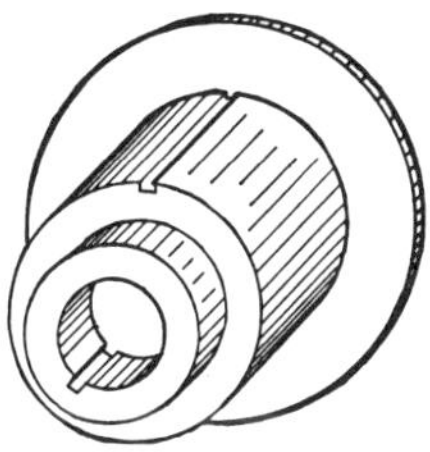

FIG. 52.

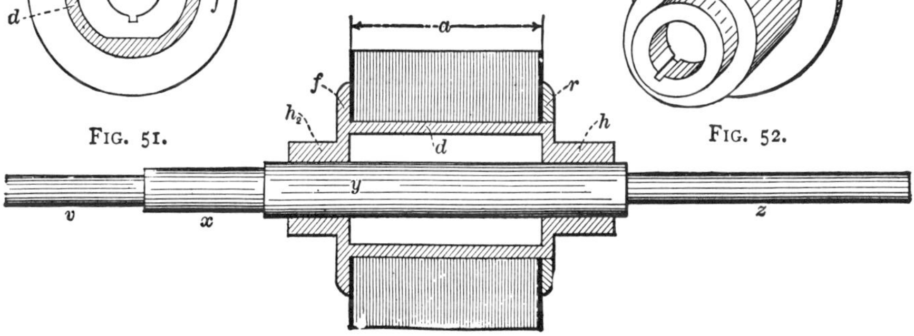

FIG. 50.

DETAILS OF ARMATURE CONSTRUCTION.

for both types of field magnet, the only difference being in the length of the core along the shaft, and, consequently, the length of the shaft.

Fig. 50 shows the shaft and a cross-sectional view of the armature core. The discs are mounted on a cast-iron drum, *d*, which has a flange, *f*, and a hub, h_2, at one end, and a hub, *h*, at the other end. Fig. 51 gives a transverse cross-sectional view of the drum, and Fig. 52 is a perspective view, from the flangeless end. The wall of the drum is thickened at two places, diametrically opposite, as shown in Fig. 51. This is necessary on one side in order to provide sufficient metal under the key-seat; it is necessary on the opposite side to obtain a mechanical balance.

The discs are held endwise by a clamping ring, *r*, which may be either screwed onto the end of the drum, *d*, or held on by four flat-headed screws with large heads. The discs are held from turning by a key. At each end of the magnetic core a disc of fibre, indicated by heavy black lines, should be placed. These discs must be exactly like the iron discs, except that they are 1-16 inch thick.

The iron core discs are 5½ ins. in diameter and 1-40 in. thick, with 32 slots, each ¼ in. wide and 9-16 in. deep. The slots have parallel sides. The discs must be of the best charcoal iron; the hole in the center is 3 ins. in diameter, key-seated. The flange, *f*, and the clamping-ring, *r*, must have their outer edges rounded off to avoid cutting the insulation of the winding. The dimensions of the core drum are as below:

	INCHES.
Length of drum, d	4¼
Inner diameter of d	2½
Outer diameter of d	3
Diameter of flange, f, and ring, r	4⅜
Thickness of flange, f, and ring, r	¼
Thickness of d at thickest point	½
Diameter of hubs, h and h_2	1⅞
Bore of hubs, h and h_2	1⅛
Length of hub, h	1¼
Length of hub, h_2	1
Length, a, of disc portion of core	4

The shaft measurements are as follows:

At	v	x	y	z
Diameter, inches	⅝	⅞	1⅛	⅝
Length, inches	3	2½	7½	6

The shoulders where *v* and *x* meet and where *y* and *z* meet should be slightly rounded off at the corner and filleted in the angle. A key should be used to fasten each hub to the shaft, but the machine will doubtless give satisfaction with only one key, that one being in the hub, *h*, at the pulley end.

Figs. 53 to 56 inclusive show end and side views and cross-sections of a journal pedestal and box. The two bearings are alike in every particular, and are made of cast-iron. The base or foot is tooled to conform to the circle to which the pedestal seat, on the magnet frame, is machined, and is ½ in. thick. The standard or pedestal consists of two ribs at right angles to each other, ½ in. thick and having curved edges, as shown. The box is of the ring-oiling type, with a single ring hung about midway of the journal; the bushing is easily made from thin brass tubing, ⅞ in. outside diameter, and with a very thin wall (not over 1-32d in.), babbitted to fit the shaft and having a slot ⅜ in. wide cut half way through it, nearly midway between its ends; accurately, slot must be ¼ in. nearer one end than the other. The bushing is 2¾ ins. long; the oil ring is made of brass, 1½ ins. in diameter inside, 1 11-16 ins. diameter outside, and ¼ in. wide along the shaft. Reference to the side views of the journal pedestal will show a slot in the upper wall of the box portion, through which the oil ring is inserted before putting in the bushing. A cover should be provided for this slot to keep out dust, etc. The dimensions of the journal pedestals are as follows:

	INCHES.
B—Radius of arc, pedestal seat	5½
g—Length of circular oil reservoir	2
i—Length of journal box	3
Bore of journal box	⅞
j—Diameter of oil reservoir	2⅜
Internal diameter of oil reservoir	2
J—Width of pedestal foot	3
k—Length of pedestal foot	3

The bore of the box portion of the pedestal must, of course, be made to fit snugly the outer diameter of the tubing used for a bushing, as the

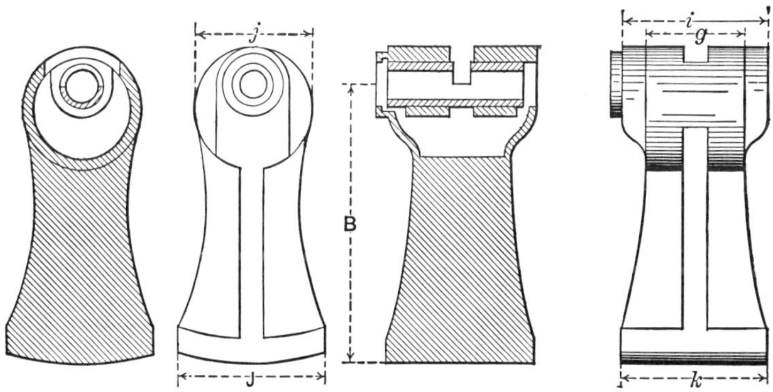

FIGS. 53, 54, 55, 56.—DETAILS OF JOURNAL BOX AND PEDESTAL.

wall of the latter is too thin to admit of turning it down to fit a predetermined bore in the pedestal. After boring the pedestal to fit the bushing it should be mounted on a mandrel and its base turned to the radius *B*, of 5½ ins., which is the same as the radius of the circle of the foot on

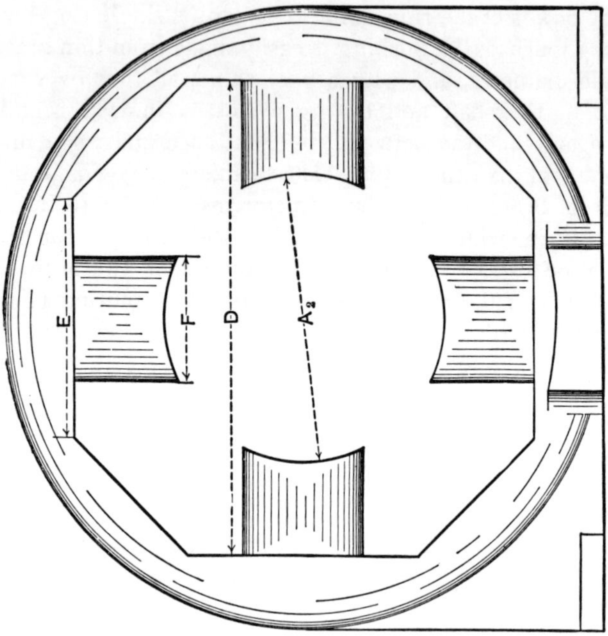

FIG. 58.—END ELEVATION OF CAST-STEEL MAGNET FRAME.

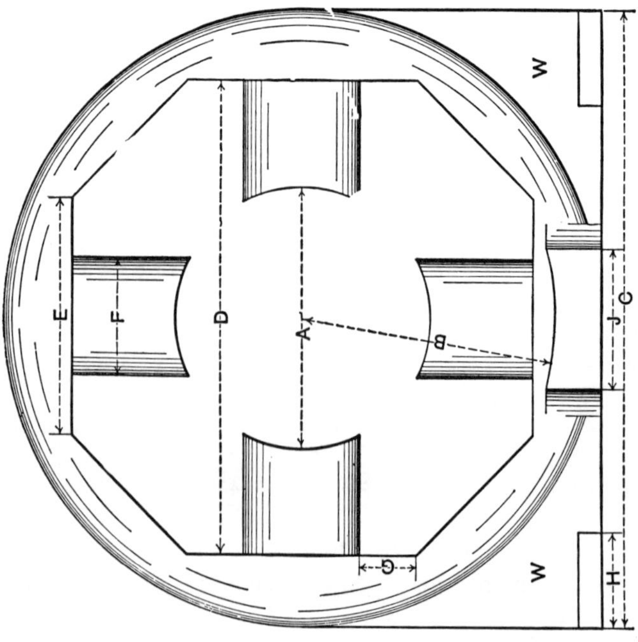

FIG. 57.—END ELEVATION OF CAST-IRON MAGNET FRAME.

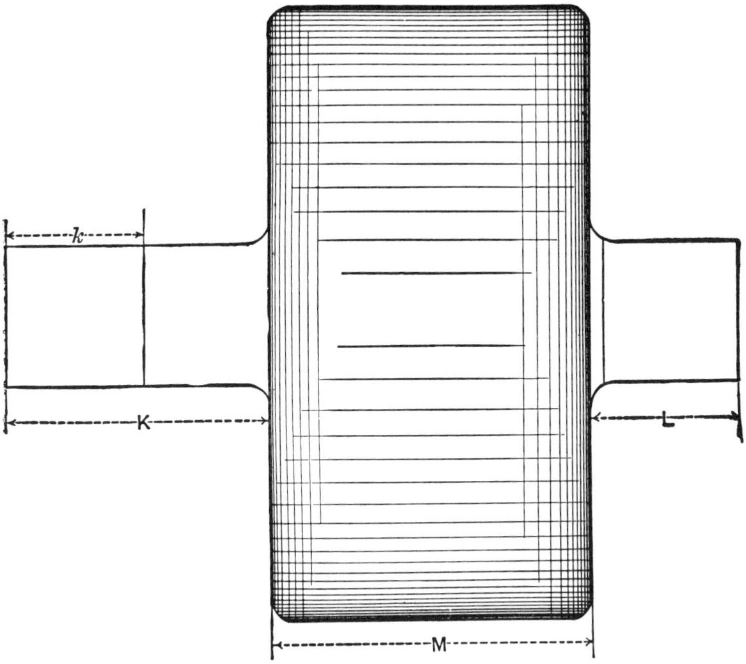

FIG. 59.—PLAN VIEW OF FIELD-MAGNET; EITHER CAST-IRON OR CAST-STEEL.

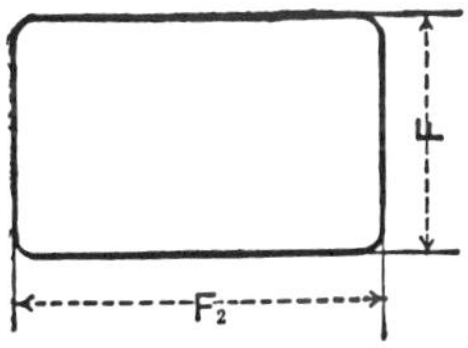

FIG. 61.
SHAPE OF MAGNET CORE.

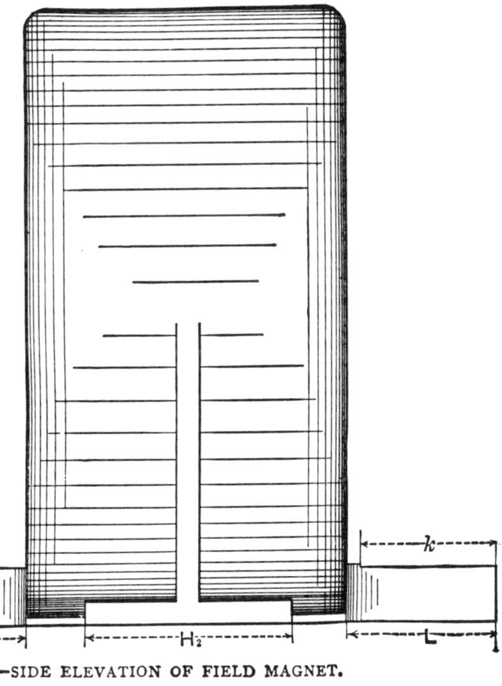

FIG. 60.—SIDE ELEVATION OF FIELD MAGNET.

the magnet frame. Each pedestal should be fastened to the foot with four ¼-in. cap screws.

Of the two field magnets shown, the cast-iron one will be found much easier to make because there is less tooling to be done and iron castings are smoother than steel, requiring little or no finishing elsewhere than the pedestal seats and pole faces. Fig 57 shows the cast-iron magnet frame and Fig. 58 the cast-steel frame. Fig. 59 is a plan view of either frame and Fig. 60 is an edge view.

The measurements for the cast-iron magnet are as follows:

	INCHES.
A—Bore of armature chamber	5⅝
B—Radius to which pedestal seat is bored	5½
C—Outer diameter of yoke ring	13
D—Distance between parallel inner faces of yoke ring	10
E—Width of plane surface behind coil	5
F—Width of magnet coil	2½
F_2—Breadth of magnet core	4
G—Distance from core to angle of yoke	1¼
H—Width of frame foot	2
H_2—Length of double foot	4½
J—Width of pedestal lug and seat	3
K—Length of pedestal lug commutator side	5¾
L—Length of pedestal lug pulley side	3¼
k—Length of pedestal seat	3
M—Axial width of magnet yoke	7

Fig. 61 shows the cross-section of a magnet core, from which it will be seen that the corners of the core are rounded off. The radius of the curve here is ¼ in. The only machining that should be required for this frame is boring the armature chamber and pedestal seats and drilling 12 bolt-holes. The frame should be clamped to a lathe carriage with its center true with the lathe centers, and the boring done at one setting by means of a boring bar and tool. Both pedestal seats should be cut before the frame is moved from its original position.

The magnet must be made of the very best grade of iron obtainable; use Scotch pig if possible. It should be allowed to remain in the sand until it is cold, care being taken not to remove any of the sand around the magnet portion until the casting is ready to come out. The longer of the two lugs might advantageously be placed uppermost in putting the pattern in the sand, and after the casting has been cooling for 24 hours the sand may be scraped away from the end of this lug so that its temperature may be noted .

The steel field magnet is much preferable if the reader has the skill and facilities to make it properly. The difference from the cast-iron magnet consists in making the magnet cores round instead of oblong, and putting on pole-shoes. The length of the machine is thereby reduced one inch, but all the transverse measurements remain unchanged. The magnet ends are machined, exactly as in the case of the cast-iron frame, but the bore, A_2, is greater, namely, 6⅛ ins.

The pole-pieces are made in one piece, called a polar-bushing, like Fig. 62, and this had better be done before the magnet is bored out. This bushing is a simple cylinder of cast-iron with four openings in its wall, equidistant from each other. Fig. 63 shows the exact shape of each of these openings. The measurements of the bushing are these:

	INCHES.
A—Bore of bushing, finished	5⅝
A_2—Diameter of bushing, finished	6⅛
a_2—Length of bushing, finished	3¼
b—Length of openings in wall	3
c—Radius of curve, side of opening	4
e—Maximum width of opening	1¾

The casting for this bushing should be about 3⅝ ins. long, 6⅜ ins. in diameter and 5⅜ ins. bore in the rough. After it has been turned down to the finished diameter, mount the magnet frame and bore out its

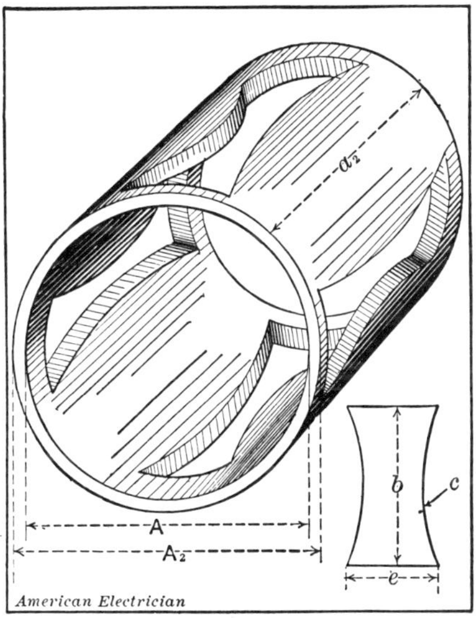

FIGS. 62 AND 63.—MAGNET POLE BUSHING.

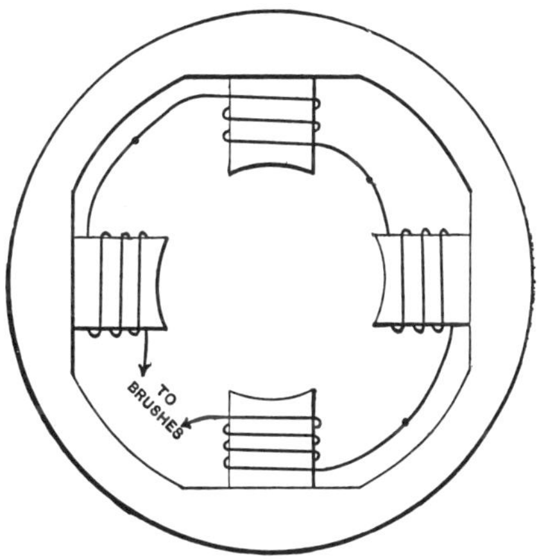

FIG. 64.—FIELD COIL CONNECTIONS.

polar circle to such a size that the bushing is a snug fit—not quite a driving fit, but tight enough to prevent turning by hand. Then insert the bushing so that the openings in its sides come half way between the magnet cores, and scribe the outlines of two opposite cores on its surface.

Remove the bushing and set a steel pin at each extremity of each ellipse scribed on the surface. Then put the bushing back and bore it out for the armature chamber. The pins will take up against the edges of the magnet cores and prevent the bushing from turning. After boring it out, turn on the ends of the bushing so as to leave the connecting webs from pole piece to pole-piece 1/8 in. thick.

The objection to this magnet is the difficulty of fitting the bushing to the magnet with sufficient accuracy to make good magnetic contact and still leave it loose enough to permit removal without breaking the thin connecting webs. This could be obviated by bolting the pole-pieces to the ends of the magnet cores by means of long, slender machine screws, put in from the outside of the yoke through holes in the centers of the magnet cores. Then the connecting webs could be sawed out entirely, leaving each pole-shoe independent of the others. This construction is also magnetically preferable, and if the builder has means for drilling a

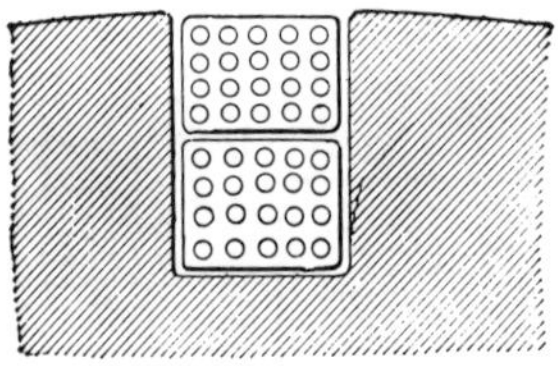

FIG. 65.—CROSS SECTION OF ARMATURE SLOT, SHOWING ARRANGEMENT OF WIRES.

1/4-in. hole from the outside of the ring to the end of the magnet core (a distance of 3 1/2 ins.), the pole-shoes should be held on this way.

With the steel magnet the following measurements must be substituted for those previously given:

	INCHES.
F—Diameter of magnet core	2 5/8
M—Width of magnet yoke	6
a—Length of disc part of core	3
Length of drum, d	3 1/4
Outer diameter of drum, d	2 3/4
Inner diameter of drum, d	2 1/4

The four field coils for the cast-iron magnet frame described in the preceding chapter are of No. 21 single-cotton-covered magnet wire. The depth of the winding must be 1 1/4 ins., as nearly as possible, and the length along the core should be 2 ins. Careful and close winding should give 40 layers of wire, with 58 turns to a layer. Whatever number of turns the reader may obtain, that number must be precisely the same in all four coils. In order to attain uniformity the coils should be wound upon a frame and the turns religiously counted.

It will be found advantageous to tie a knot in the starting end of each coil before taping it so that it may be identified afterward. The coils must be connected up as shown by the diagram, Fig. 64, so that the starting end of one connects to the finishing end of its neighbor. This presupposes that all four are wound in the same direction, as they should be.

The coils for the cast-steel magnet are of No. 24 single-cotton-covered wire, 1⅛ ins. deep and 1⅞ ins. long. Good winding will enable the reader to put on 50 layers of wire and 75 turns to a layer. As in the previous case, however, the depth in inches is the essential point, though it is advantageous to get as many layers in that depth as possible. The coils are, of course, wound, insulated and connected up exactly like the oblong coils of the cast-iron frame.

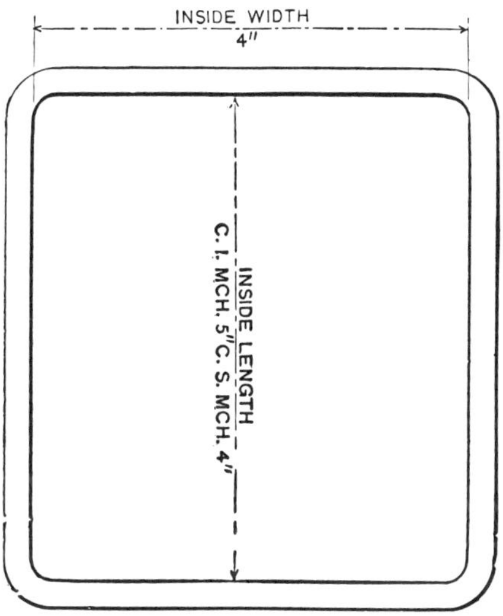

FIG. 66.—DIMENSIONS OF ARMATURE COIL.

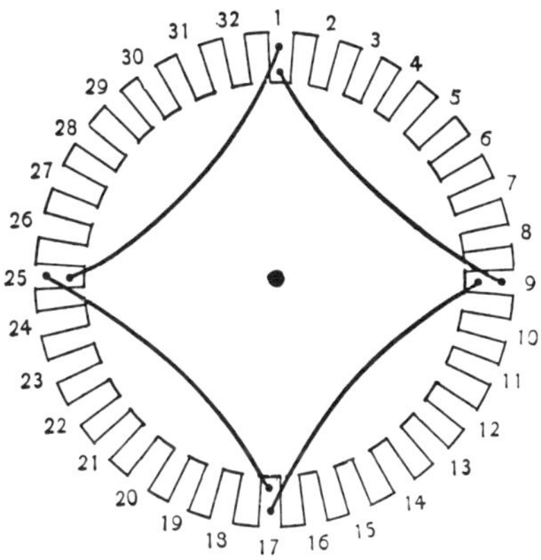

FIG. 67.—WINDING DIAGRAM.

The armature core for either of the magnet frames will contain 32 coils; each coil consists of No. 21 double-cotton-covered wire, wound five turns wide by four layers deep. Each slot contains one side of each of two coils, so that the cross-section of the winding in a slot will be as in Fig. 65, except that the wire will lie closer together than the sketch indicates. All armature coils should be wound on a forming bobbin so that they will all be exactly alike. Fig. 66 shows what the essential dimensions should be. The width of the hollow of the coil is the same for both armature cores. As the armature core to be used with the steel magnet is an inch shorter than the other one, the coils for this core must be an inch shorter; hence the two dimensions for coil lengths.

Fig. 67 is a winding diagram and shows the first four coils in position. The coils are indicated by a single line across the head and dots in the slots for simplicity. The builder should note that the left-hand side of each coil is in the bottom of the slot and the right-hand side is on top; this should be true of every coil. The starting ends should be knotted for identification, and all the knotted ends should occupy the same relative position on the core. For smoothness of finished heads the coils should be put on the core in the following order:

Coils 1, 2, 3, 4 in Slots 1, 9, 17, 25.
Coils 5, 6, 7, 8 in Slots 2, 10, 18, 26.
Coils 9, 10, 11, 12 in Slots 3, 11, 19, 27.
Coils 13, 14, 15, 16 in Slots 4, 12, 20, 28.
Coils 17, 18, 19, 20 in Slots 5, 13, 21, 29.
Coils 21, 22, 23, 24 in Slots 6, 14, 22, 30.
Coils 25, 26, 27, 28 in Slots 7, 15, 23, 31.
Coils 29, 30, 31, 32 in Slots 8, 16, 24, 32.

If put in properly, the coils will give a regular sequence of knotted ends and straight ends, one each projecting from each slot. The con-

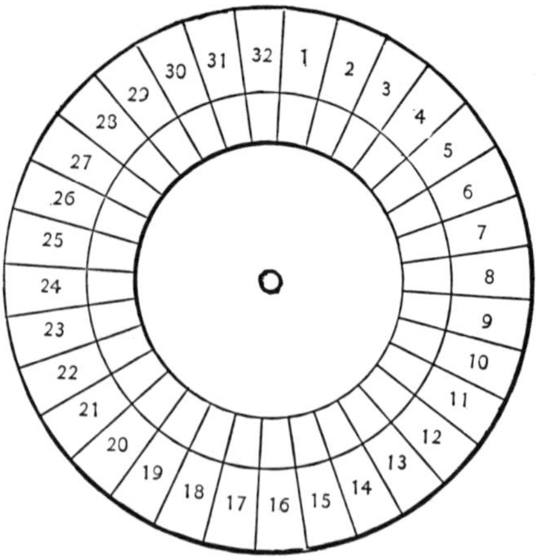

FIG. 68.—COMMUTATOR DIAGRAM.

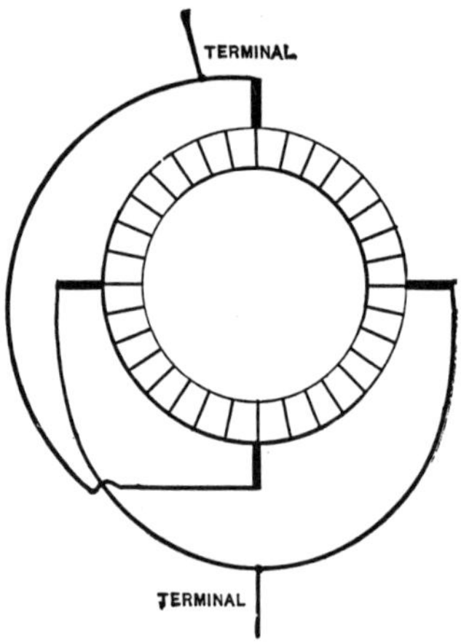

FIG. 69.—CONNECTIONS BETWEEN BRUSHES.

nections to the commutator are then simple. Carry all the knotted ends straight out to the commutator, and the straight ends one segment less than a quarter circle backwards from their corresponding knotted ends. Thus, the knotted end from slot No. 1 goes to commutator segment No. 1 (see Fig 68), and so on, all the way around. Then the straight ends go to the commutator as follows:

From Slot No. 1 to Segment No. 26.
From Slot No. 2 to Segment No. 27.
From Slot No. 3 to Segment No. 28.

From Slot No. 4 to Segment No. 29.
From Slot No. 5 to Segment No. 30.
From Slot No. 6 to Segment No. 31.
From Slot No. 7 to Segment No. 32.
From Slot No. 8 to Segment No. 1.
From Slot No. 9 to Segment No. 2.
From Slot No. 10 to Segment No. 3.
From Slot No. 11 to Segment No. 4.
From Slot No. 12 to Segment No. 5.
From Slot No. 13 to Segment No. 6.
From Slot No. 14 to Segment No. 7.
From Slot No. 15 to Segment No. 8.
From Slot No. 16 to Segment No. 9.
From Slot No. 17 to Segment No. 10.
From Slot No. 18 to Segment No. 11.
From Slot No. 19 to Segment No. 12.
From Slot No. 20 to Segment No. 13.
From Slot No. 21 to Segment No. 14.
From Slot No. 22 to Segment No. 15.
From Slot No. 23 to Segment No. 16.
From Slot No. 24 to Segment No. 17.
From Slot No. 25 to Segment No. 18.
From Slot No. 26 to Segment No. 19.
From Slot No. 27 to Segment No. 20.
From Slot No. 28 to Segment No. 21.
From Slot No. 29 to Segment No. 22.
From Slot No. 30 to Segment No. 23.
From Slot No. 31 to Segment No. 24.
From Slot No. 32 to Segment No. 25.

The commutator must have 32 segments, as indicated by Fig. 68, and should be purchased already built for assured satisfaction. The brush surface of the commutator must be 1¼ ins. long, at least, so that carbon brushes 1 in. wide and ¼ in. thick can be used. The diameter of the barrel of the commutator should be not less than 3, and preferably 4 ins. The brush holders and yoke may be copied advantageously from any of the standard machines now on the market. Four brushes must be used, and the two diametrically opposite are connected together, as shown by Fig. 69.

The windings just described are for machines to work on a 110-115-volt circuit. If windings for 220-230 volts are desired the armature coils should be of No. 25 wire, each coil five layers deep and eight turns wide, making ten layers of wire per slot. The field coils of the cast-iron magnet must be of No. 24 s.c.c. wire, wound to the dimensions specified above, namely, 1¼ ins. deep and 2 ins. long. The coils for the cast-steel magnet will be of No. 27 wire wound to a depth of 1⅛ ins. and a length 1⅞ ins.

The principal magnetic and electrical data of the two machines are as below:

115-VOLT MOTOR.

	Cast-iron.	Cast-steel.
Resistance armature winding	1 ohm	0.9
Normal armature currents	9 amp.	9 amp.
C^6R loss armature	81 watts	73 watts
Hysteresis and eddy currents	4½ watts	3 watts
Approximate speed	1600	1600
Resistance field winding	183 ohms	424 ohms
Normal field current	.628 amp.	.27 amp.
C^2R field loss	72.22 watts	31 watts
Flux per pole		400,000 lines
Density in field cores	48,000	93,000
Density in air gap	27,500	37,500
Efficiency, assuming 10 per cent friction and windage	75 per cent	80 per cent

As in the preceding case, it is by far preferable to buy a starting box from one of the standard rheostat manufacturers. If the reader insists upon having a home-made one, however, the arrangement shown by Fig. 49 and described on pages 55 and 56 will answer.

CHAPTER VIII.

TWO HORSE-POWER FOUR-POLAR MOTOR WITH TWO-PATH DRUM ARMATURE.

For this motor, as in the preceding design, only one type of field magnet is shown, namely, the familiar ring yoke with radial magnet poles; a choice is given between cast-iron and cast-steel field magnets. The armature construction is identical for both types of magnet, there being a difference only in the length of the armature and shaft.

Fig. 70 shows the shaft and a cross-sectional view of the armature core. The discs are mounted on a cast-iron drum, *d*, which has a flange, *f*, and a hub, *h*2, at one end, and a hub, *h*, at the other end. Fig. 71 gives

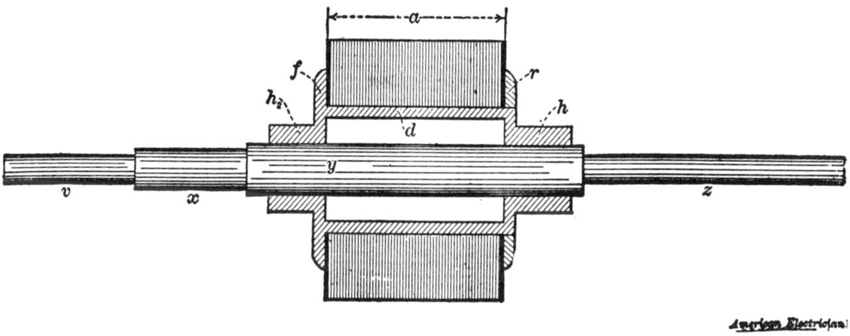

FIG. 70.—ARMATURE SHAFT AND CROSS-SECTION OF ARMATURE CORE.

a transverse cross-sectional view of the drum, and Fig. 72 is a perspective view, from the flangeless end. The wall of the drum is thickened at two places, diametrically opposite, as shown in Fig. 71. This is necessary on one side in order to provide sufficient metal under the key-seat; it is necessary on the opposite side to obtain a mechanical balance.

The discs are held endwise by a clamping ring, *r*, which may be either screwed onto the end of the drum, *d*, or held on by four flat-headed screws with large heads. The discs are held from turning by a key. At each end of the magnetic core a disc of fibre, indicated by heavy black lines, should be placed. These discs must be exactly like the iron discs, except that they are 1-16 in. thick.

The iron core discs are 6⅞ ins. outside diameter and 1-40 in. thick, with 43 slots, each ¼ in. wide and 9-16 in. deep. The slots have parallel sides. The discs must be of the best charcoal iron; the hole in the center is 3⅝ ins. in diameter, key-seated. The flange, *f*, and the clamping-ring, *r*, must have their outer edges rounded off to avoid cutting the insulation of the winding. The dimensions of the core drum are as below:

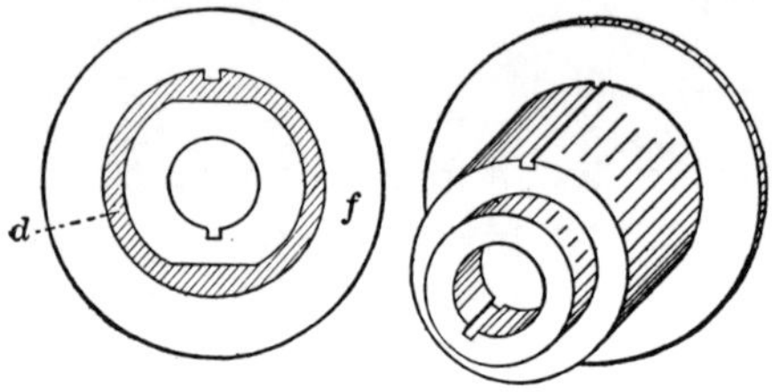

FIGS. 71 AND 72.
CORE DRUM AND HEAD.

Length of drum, d	5 1/16
Inner diameter of d	3
Outer diameter of d	3⅝
Diameter of flange, f, and ring, r.	5¾
Thickness of flange, f, and ring, r.	5/16
Thickness of d at thickest point	½
Diameter of hubs, h and h_2	2
Bore of hubs, h and h_2	1¼
Length of hub, h	1½
Length of hub, h_2	1¼
Length of a, of disc portion of core	4¾

The shaft measurements are as follows:

At	v	x	y	z
Diameter, inches	¾	1	1¼	¾
Length, inches	3⅝	3	9	7¼

The shoulders where *v* and *x* meet and where *y* and *z* meet should be slightly rounded off at the corner and filleted in the angle. A key

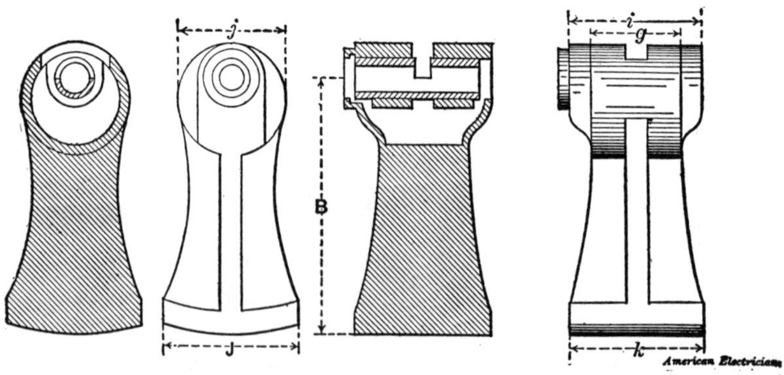

FIGS. 73, 74, 75, 76.—DETAILS OF JOURNAL BOX AND PEDESTAL.

should be used to fasten each hub to the shaft, but the machine will doubtless give satisfaction with only one key, that one being in the hub, *h*, at the pulley end. The hub, *h*, must be exactly ¼ in. from the shoulder on the shaft.

Figs. 73 to 76, inclusive, show end and side views and cross-sections of a journal pedestal and box. The two bearings are alike in every particular, and are made of cast-iron. The base or foot is tooled to conform to the circle to which the pedestal seat, on the magnet frame, is machined, and is ⅝ in. thick. The standard or pedestal consists of two ribs at right angles to each other, ⅝ in. thick and having curved edges, as shown. The box is of the ring-oiling type, with a single ring hung about midway of the journal; the bushing is easily made from thin brass tubing, 1 in. outside diameter, and with a very thin wall (not over 1-32d in.), babbitted to fit the shaft and having a slot 7-16 in. wide cut half way through it, nearly midway between its ends; accurately, the slot must be ¼ in. nearer one end than the other. The bushing is 3⅜ ins. long; the oil ring is made of brass, 2 ins. in diameter inside, 2¼ ins. diameter outside, and ⅜ in. wide along the shaft. Reference to the side views of the journal pedestal will show a slot in the upper wall of the box portion, through which the oil ring is inserted before putting in the bushing. A cover should be provided for this slot to keep out dust, etc. The dimensions of the journal pedestals are as follows:

	INCHES.
B—Radius of arc, pedestal seat	6
g—Length of circular oil reservoir	2
i—Length of journal box	3⅝
Bore of journal box	1
j—Diameter of oil reservoir	2⅞
Internal diameter of oil reservoir	2½
J—Width of pedestal foot	3⅝
k—Length of pedestal foot	3⅝

The bore of the box portion of the pedestal must, of course, be made to fit snugly the outer diameter of the tubing used for a bushing, as the wall of the latter is too thin to admit of turning it down to fit a predetermined bore in the pedestal. After boring the pedestal to fit the bushing it should be mounted on a mandrel and its base turned to the radius *B*, of 6 ins., which is the same as the radius of the circle of the foot on the magnet frame. Each pedestal should be fastened to the foot with four 5-16 in. cap screws.

Of the two field magnets shown, the cast-iron machine will be found easier to make because there is less tooling to be done and iron castings are smoother than steel, requiring little or no finishing elsewhere than the pedestal seats and pole faces. Fig. 77 shows the cast-iron magnet frame and Fig. 78 the cast-steel frame. Fig. 79 is a plan view of either frame and Fig. 80 is an edge view.

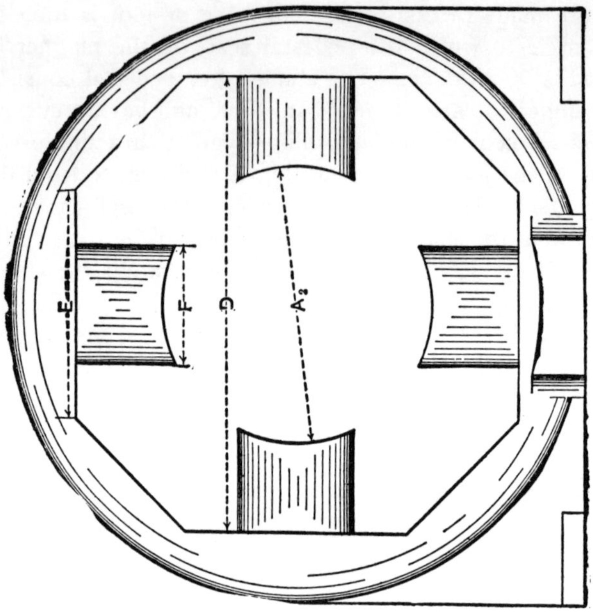

FIG. 78.—END ELEVATION OF CAST-STEEL FIELD MAGNET.

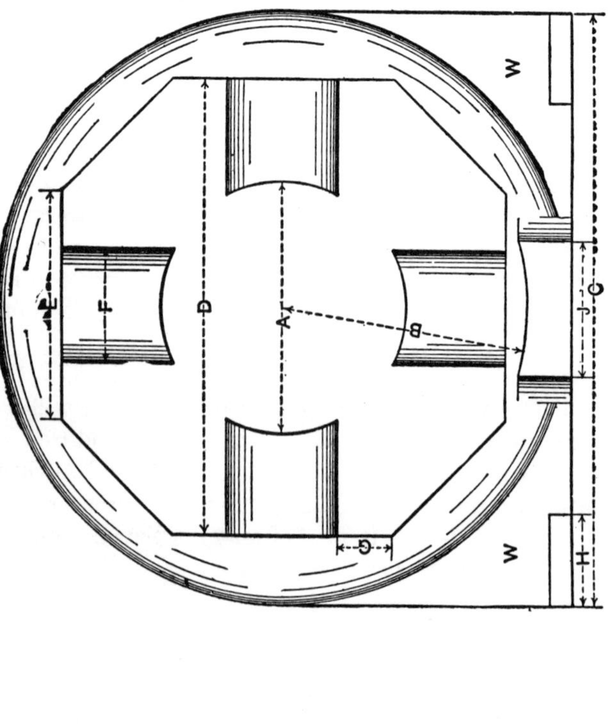

FIG. 77.—END ELEVATION OF CAST-IRON FIELD MAGNET.

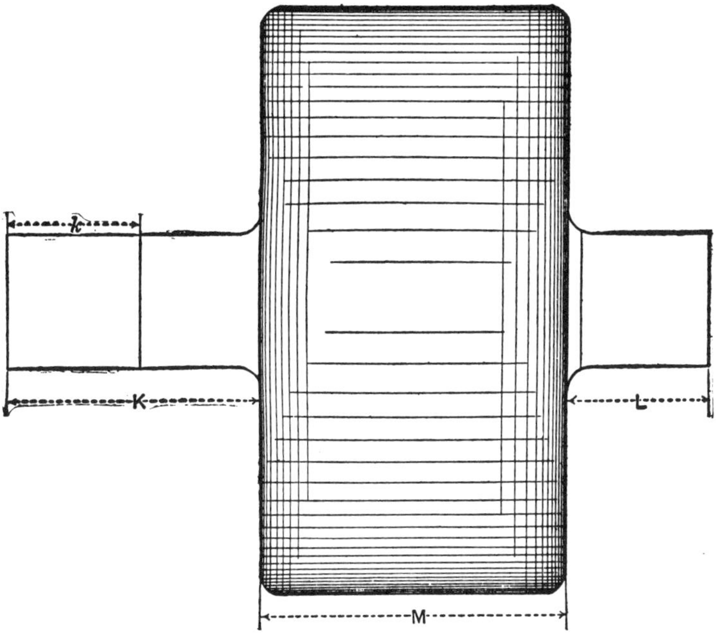

FIG. 79.—PLAN VIEW OF FIELD MAGNET; EITHER CAST-IRON OR CAST-STEEL.

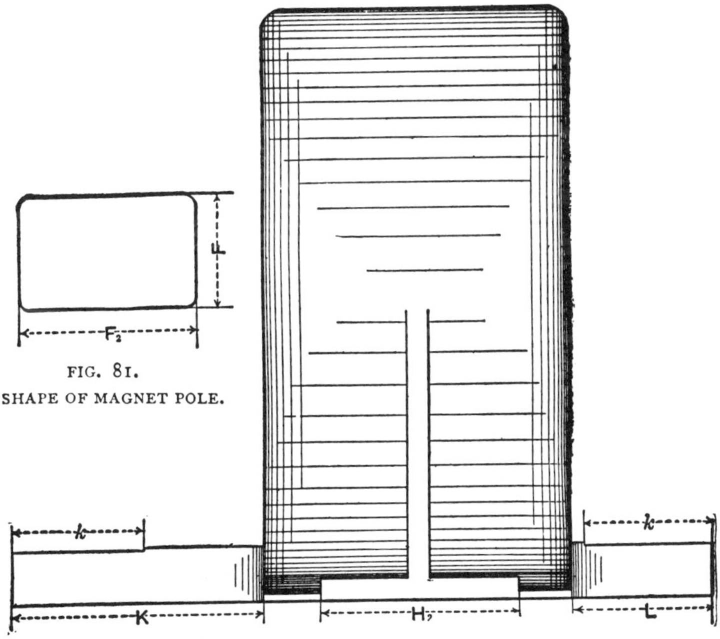

FIG. 81.
SHAPE OF MAGNET POLE.

FIG. 80.—SIDE ELEVATION OF FIELD MAGNET.

The measurements for the cast-iron magnet are as follows:

	INCHES.
A—Bore of armature chamber	7
B—Radius to which pedestal seat is bored	6⅝
C—Outer diameter of yoke ring	15⅝
D—Distance between parallel inner faces of yoke ring	12
E—Width of plane surface behind coil	6
F—Width of magnet core	3
F_2—Breadth of magnet core	4½
G—Distance from core to angle of yoke	1¼
H—Width of frame foot	2½
H_2—Length of double foot	5
J—Width of pedestal lug and seat	3⅝
K—Length of pedestal lug; commutator side	6⅞
L—Length of pedestal lug; pulley side	3⅞
K—Length of pedestal seat	3⅝
M—Axial width of magnet yoke	8 7/16

Fig. 81 shows the cross-section of a magnet core, from which it will be seen that the corners of the core are rounded off. The radius of the curve here is 0.3 in. The only machining that should be required for this frame is boring the armature chamber and pedestal seats and drilling 12 bolt-holes. The frame should be clamped to a lathe carriage with its center true with the lathe centers, and the boring done at one setting by means of a boring bar and tool. Both pedestal seats should be cut before the frame is moved from its original position.

The magnet must be made of the very best grade of iron obtainable; use Scotch pig if possible. It should be allowed to remain in the sand until it is cold, care being taken not to remove any of the sand around the magnet portion until the casting is ready to come out. The longer of the two lugs might advantageously be placed uppermost in putting the pattern in the sand, and after the casting has been cooling for 24 hours the sand may be scraped away from the end of this lug so that its temperature may be noted.

The cast-steel field magnet is much preferable if the reader has the skill and facilities to make it properly. The difference from the cast-iron magnet consists in making the magnet cores round instead of oblong, and putting on pole-shoes. The length of the machine is thereby reduced 1¼ ins., but all the transverse measurements remain unchanged. The magnet ends are machined exactly as in the case of the cast-iron frame, but the bore, A_2, is greater, namely, 7½ ins.

The pole-pieces are made in one piece, called a polar-bushing, like Fig. 82, and this had better be done before the magnet is bored out. This bushing is a simple cylinder of cast-iron with four openings in its wall, equidistant from each other. Fig. 83 shows the exact shape of each of these openings. The measurements of the bushing are these:

	INCHES.
A_2—Diameter of bushing, finished	7½
A—Bore of bushing, finished	7
a_2—Length of bushing, finished	3¾
b—Length of openings in walls	3½
c—Radius of curve, side of opening	4½
e—Maximum width of opening	2½

The casting for this bushing should be about 4 ins. long, 7¾ ins. in diameter and 6¾ ins. bore, in the rough. After it has been turned down to the finished diameter, mount the magnet frame and bore out its polar circle to such a size that the bushing is a snug fit—not quite a driving fit, but tight enough to prevent turning by hand. Then insert the bushing so that the openings in its sides come half way between the magnet cores, and scribe the outlines of two opposite cores on its surface.

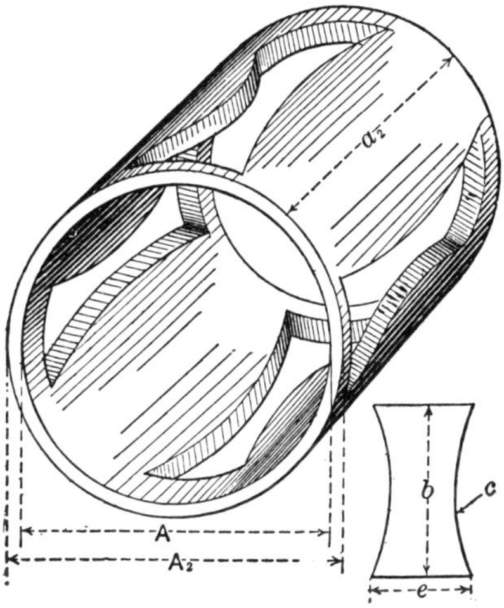

FIGS. 82 AND 83.—MAGNET POLE BUSHING.

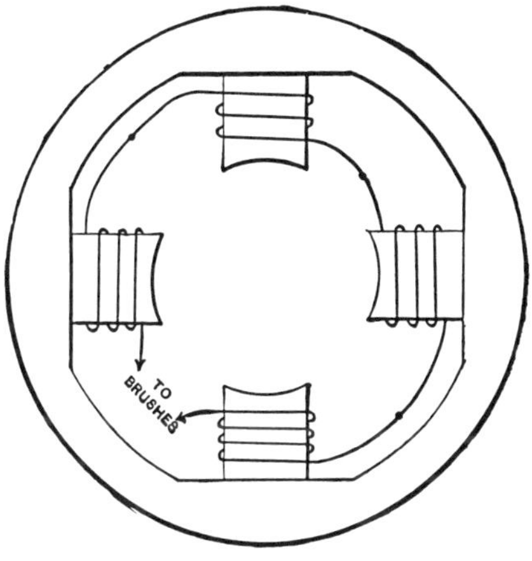

FIG. 84.—FIELD COIL CONNECTIONS.

Remove the bushing and set a steel pin at each extremity of each ellipse scribed on the surface. Then put the bushing back and bore it out for the armature chamber. The pins will take up against the edges of the magnet cores and prevent the bushing from turning. After boring it out, turn off the ends of the bushing so as to leave the connecting webs from pole-piece to pole-piece ⅛ in. thick, measured axially.

The objection to this magnet is the difficulty of fitting the bushing to the magnet with sufficient accuracy to make good magnetic contact and still leave it loose enough to permit removal without breaking the thin connecting webs. This could be obviated by bolting the pole-pieces to the ends of the magnet cores by means of long, slender machine screws, put in from the outside of the yoke through holes in the centers of the magnet cores. Then the connecting webs could be sawed out entirely, leaving each pole-shoe independent of the others. This construction is also magnetically preferable, and if the builder has means for drilling ¼-in. holes through the magnet cores from the outside of the yoke ring to the inside of the bushing (a distance of 4⅜ ins.), the pole-shoes should be held on this way.

With the steel magnet the following measurements must be substituted for those previously given:

	INCHES.
F—Diameter of magnet core	2⅞
M—Width of magnet yoke	7$\frac{3}{16}$
a—Length of disc part of core	3½
Length of drum, d	3$\frac{13}{16}$
Length of y, on shaft	7¾

This is to say, the machine must be exactly 1¼ ins. shorter, axially.

The four field coils for the cast-iron magnet frame are of No. 22 single-cotton-covered magnet wire. The depth of the winding must be 1½ ins., as nearly as possible, and the length along the core should be 2¼ ins. Careful and close winding should give 50 layers of wire, with 70 turns to a layer. Whatever number of turns the reader may obtain, that number must be precisely the same in all four coils. In order to attain uniformity the coils should be wound upon a frame and the turns religiously counted.

It will be found advantageous to tie a knot in the starting end of each coil before taping it so that it may be identified afterward. The coils must be connected up as shown by the diagram, Fig. 84, so that the starting end of one connects to the finishing end of its neighbor. This presupposes that all four are wound in the same direction, as they should be.

The coils for the cast-steel magnet are of No. 23 single-cotton-covered wire, 1½ ins. deep and 2 ins. long. Good winding will enable the reader to put on 56 layers of wire and 70 turns to a layer. As in the previous case, however, the depth in inches is the essential point, though it is advantageous to get as many layers in that depth as possible. The coils are, of course, wound, insulated and connected up exactly like the oblong coils of the cast-iron frame.

The armature core for either of the magnet frames will contain 43 coils; each coil consists of No. 16 double-cotton-covered wire, wound three turns wide by three layers deep. Each slot contains one side of each of two coils, so that the cross-section of the winding in a slot will be as in Fig. 85.

All armature coils should be wound on a forming bobbin so that they

FIG. 85.—CROSS-SECTION OF A SLOT.

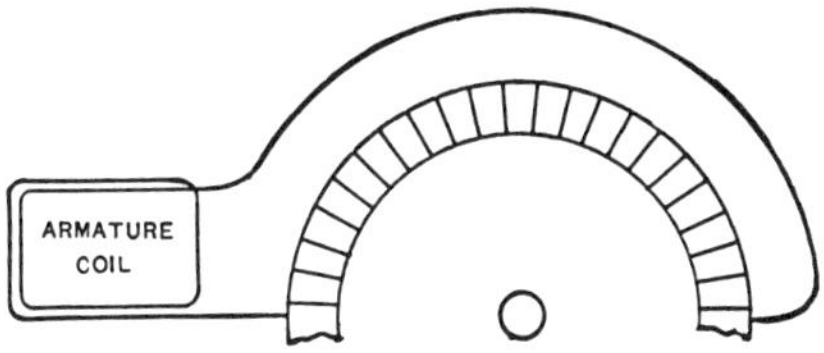

FIG. 88.—CONNECTING DIAGRAM.

will all be exactly alike. Fig. 86 shows what the essential dimensions should be. The width of the hollow of the coil is the same for both armature cores. As the armature core to be used with the steel magnet is 1¼ ins. shorter than the other one, the coils for this core must be correspondingly shorter; hence the two dimensions for coil lengths.

Fig. 87 is a winding diagram and shows four coils in position. The coils are indicated by single lines across the head and dots in the slots for

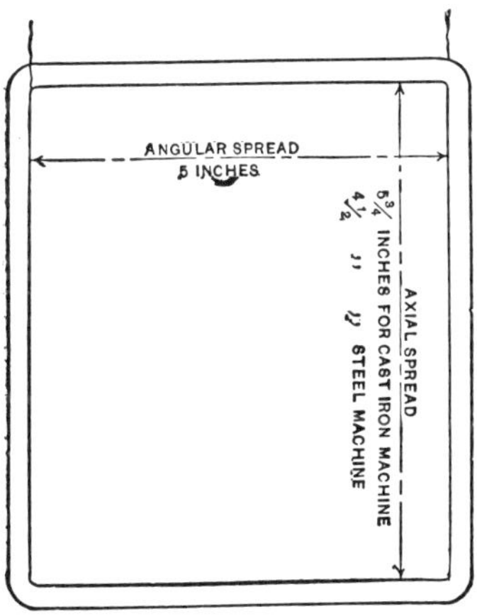

FIG. 86.—DIMENSIONS OF ARMATURE COIL.

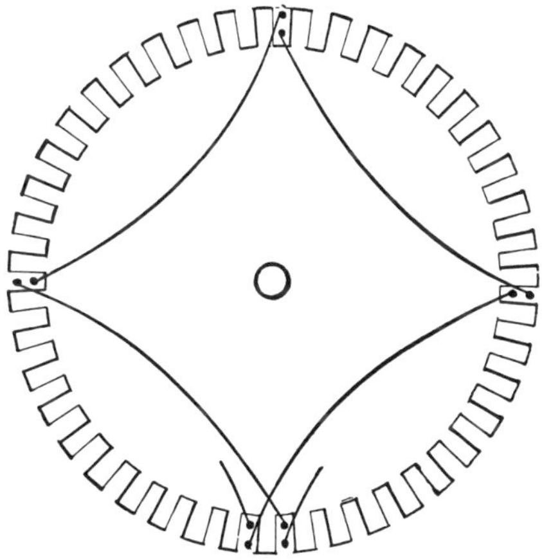

FIG. 87.—WINDING DIAGRAM.

simplicity. The builder should note that the left-hand side of each coil is in the bottom of the slot and the right-hand side is on top; this should be true of every coil, but it is not imperative. The machine will work just as well if half of the coils are put on with both sides bottom and the other half on top of them, but the job will not be so neat on the armature heads. The spacing or pitch of the coils must be exactly as indicated—10 slots in between the two sides of each coil.

The starting ends should be knotted for identification, and all the knotted ends should occupy the same relative position on the core. If put in properly, the coils will give a regular sequence of knotted ends and straight ends, one each projecting from each slot. The connections to the commutator are then simple. Carry all the knotted ends straight out to the commutator, and each straight end to a segment 22 bars from the one to which the knotted end is connected. Fig. 88 represents one coil connected, and shows that there are 21 segments between the two to which the coil ends go, reckoning around that side of the commutator nearest the coil itself. This spacing must be observed throughout.

The commutator must have 43 segments and should be purchased already built for assured satisfaction. The brush surface of the commutator must be 1½ ins. long, at least, so that carbon brushes 1¼ ins. wide and ⅜ ins. thick can be used. The diameter of the barrel of the commutator should not be less than 4, and preferably 5 ins. The brush holders and yoke may be copied advantageously from any of the standard machines now on the market. Only two brushes are to be used, and these set precisely a quarter of a circle apart, reckoning around the barrel of the commutator.

The windings just described are for machines to work on a 110-115-volt circuit. If windings for 220-230 volts are desired the armature coils should be of No. 19 wire, each coil five layers deep and four turns wide, making ten layers of wire per slot. The field coils of the cast-iron magnet must be of No. 26 s.c.c. wire, wound to the dimensions specified above, namely, 1½ ins. deep and 2¼ ins. long. The coils for the cast-steel magnet will be of No. 27 wire wound to a depth of 1½ ins. and a length of 2 ins. For 500-volt service use No. 23 double-cotton-covered wire on the armature, six turns wide and seven layers deep, per coil; 84 wires per slot. On the cast-iron magnet use No. 29 double-covered wire and on the steel magnet No. 30.

The principal magnetic and electrical data of the two machines are below:

115-VOLT MOTOR.

	Cast-Iron.	Cast-Steel.
Resistance armature winding	1 ohm	0.9
Normal armature current	18 amp.	18 amp.
Approximate speed	1325 r. p. m.	1325 r. p. m.
Flux per pole		340,000 lines
Density in field cores	33,700	70,000
Density in air gap	26,000	38,000

CHAPTER IX.

THREE HORSE-POWER MOTOR.

The motor design which forms the subject of this chapter, although somewhat similar to those described in Chapters VII and VIII, differs considerably in the constructional details of the magnet. Here a cast-iron ring and wrought iron cores are employed with a view to simplifying the work as far as possible without sacrificing the efficiency of the machine, and also without making it unduly heavy. The cast-iron ring is preferably made in a single piece and the wrought-iron cores are turned to a very slight taper and drawn into holes in the yoke ring by means of a bolt and heavy washer from the outside. Unless the builder has excellent machine-shop facilities, however, and is an expert machinist, this construction will be found rather difficult, as it is necessary to have a perfect fit between the taper of the magnet core and that of the hole in which it is seated.

As an alternative the magnet frame can be cast in two pieces, the division being along the line, *x*, Fig. 91. If the motor is built in this way, each half must be chucked and the joint faced off fairly smooth, although it is not necessary to have a perfect joint, as no magnetic lines of force cross the break. After truing up the abutting faces of each half of the magnet ring the two halves should be clamped together with 1-32-in. of cardboard in between them and four straight holes bored for the reception of the field-magnet cores.

Fig. 89 is a semi-sectional elevation of the field magnet complete without the journal pedestals; a field-magnet core is shown by Fig. 90. The cast-iron pole-pieces must be accurately fitted to the ends of the cores and pinned permanently in place with iron pins. The magnet ring and pole-pieces should be of the best grade of pig iron obtainable. The

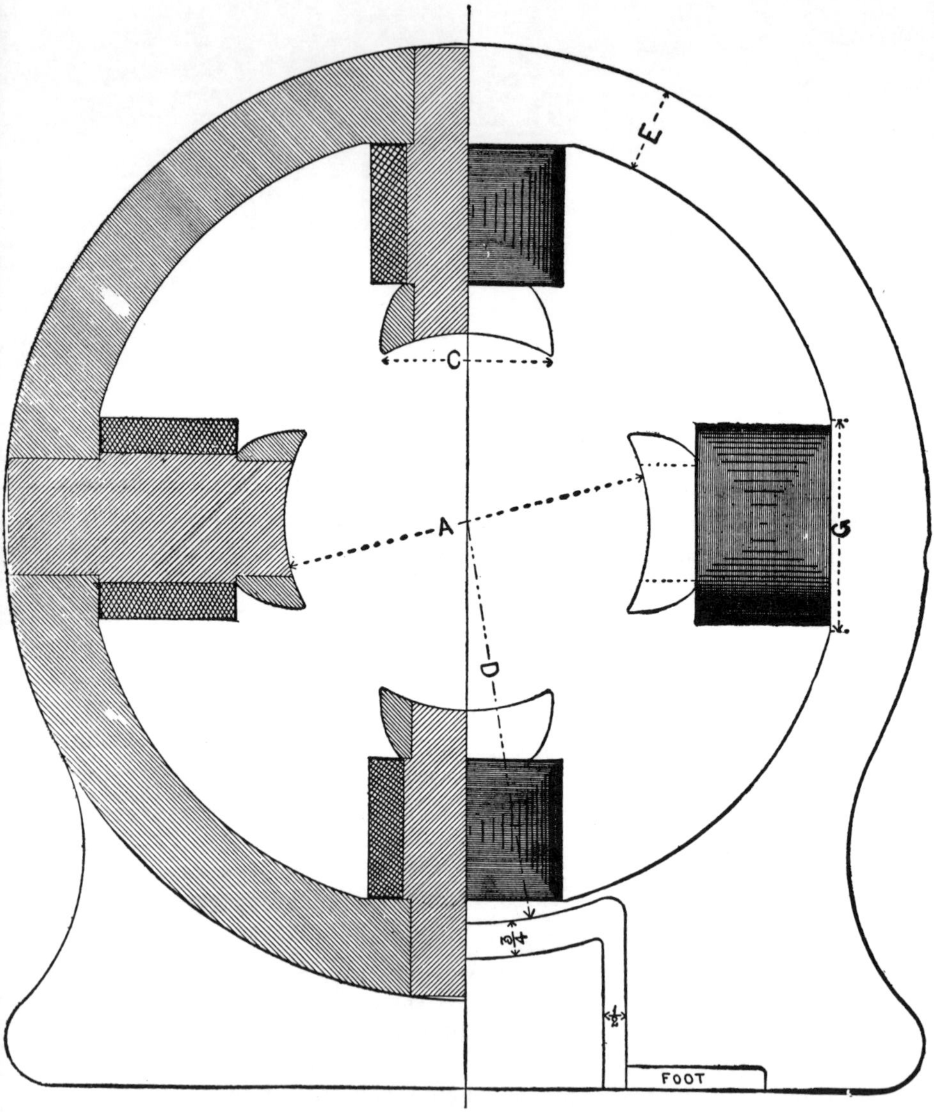

FIG. 89.—SEMI-SECTIONAL ELEVATION OF THE FIELD-MAGNET FRAME.

magnet cores should be made of Norway wrought iron. The corners of the pole-pieces should be heavily rounded so that no sharp edges are left. The length of the pole-piece parallel with the shaft is the same as its width at right angles to this dimension. The outside diameter of the magnet ring

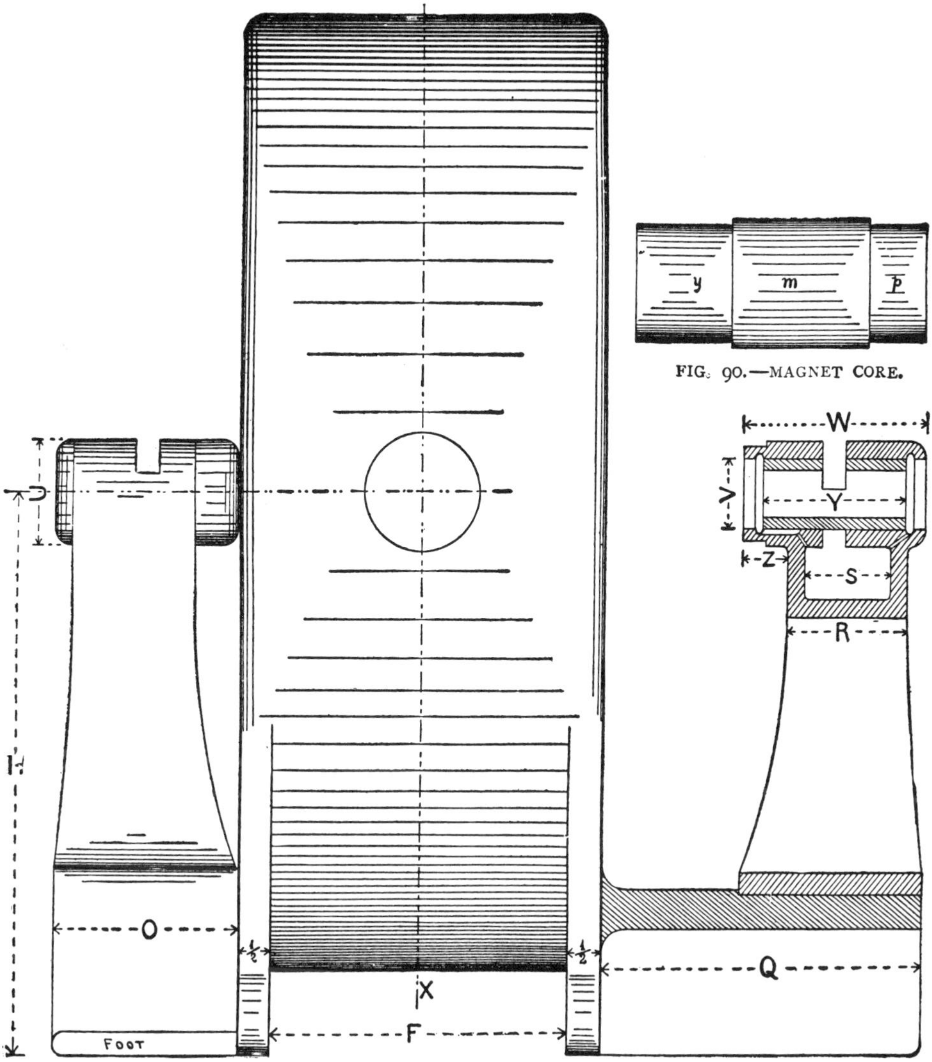

FIG. 90.—MAGNET CORE.

FIG. 91.—SIDE ELEVATION OF MAGNET FRAME AND SECTION THROUGH BEARING.

is 20⅛ ins. The extreme breadth of the ring parallel with the shaft is 8 ins. The other dimensions are as below:

	INCHES.
A—Bore of armature chamber	8
B—Axial length of pole-face	3¾
C—Width of pole-piece, tip to tip, in a straight line	3¾

D—Radius to which pedestal seat is cut	8½
E—Thickness of yoke ring	1½
F—Distance between ribs	7
G—Width of plane surface back of magnet coil	4½
H—Height, base line to armature center	12
O—Axial length of pedestal seat	4
P—Straight-line width of pedestal (width of pedestal seat is the same)	7
Q—Length of pedestal foot, commutator side	7

The dimensions of the magnet core (Fig. 90) are as follows:

	At y	m	p
Diameter, ins	2½	3¾	2½
Length, ins	2	3	1¼

Fig. 91 is an edge view of the field-magnet frame, including one journal pedestal shown in perspective and the other in cross-section. After fitting the magnet cores into place in the ring, the pole-pieces should be bored and the pedestal seats cut, at one setting of the frame. Fig. 92 is a cross-section of the armature core, showing the details of construction. The discs are mounted on a cast-iron drum, which is provided with a flange head and a hub, *J*, at one end, the other end being open. The discs are clamped in place by a cast-iron ring which is provided with a hub, *I*, similar to that at the other end of the drum, and drawn to place by means of six ¼-in. bolts passing through holes in the clamping ring and tapping into the end wall of the cast-iron drum. The center lines of two of these bolts are indicated by *b b*. The cast-iron drum should have a key-seat cut in it so that the discs may be positively driven.

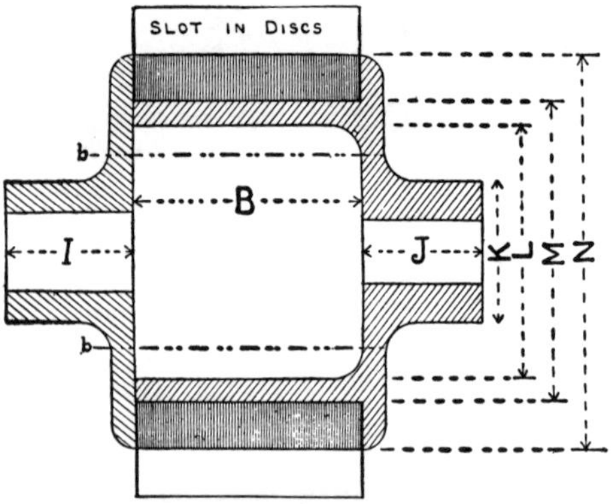

FIG. 92.—SECTION THROUGH ARMATURE CORE.

Both the hub on the end of the drum and that on the clamping ring should be keyed to the shaft so that there will be no opportunity for displacement. At each end of the core structure a disc of fiber 1-16 in. thick should be provided, as indicated by the heavy black lines in the engraving. These discs must be toothed exactly like the core discs so that the ends of the magnetic core will be entirely covered. The iron core discs are 7¾ ins. in diameter, with a central hole 4¾ ins. in diameter, and 47 slots ¼ in. wide and ¾ in. deep; the slots have parallel sides. The discs must be of the best grade of charcoal iron, 1-40 in. thick. The dimensions of the armature core structure are as follows:

	INCHES.
I—Length of hub clear through	2⅛
Bore of this hub	1¼
J—Length of hub clear through	2
Bore of this hub	1
K—Diameter of hubs	2¼
L—Internal diameter of core drum	4
M—Outer diameter of core drum	4¾
N—Diameter of flange and clamping ring	6¼
Thickness of flange and clamping ring	⅜

The two journal pedestals are exactly alike and made of ordinary cast-iron. The base must be turned accurately to conform to the circle to which the pedestal seat on the magnet frame is machined. The standard, or pedestal, is an open frame of ½-in. metal; the box is of the ring-oiling type, with a single ring hung exactly midway of the journal. Fig. 93 is a transverse cross-section of the pedestal and box. The box is bushed; the bushing may consist of a brass casting turned to shape, or it may be made by babbitting a piece of thin brass tubing, the outer diameter of which is a snug fit in the box. The oil slot across the center of the box must be provided with a suitable covering to exclude dust. Each pedestal should be bolted to the magnet frame with two ⅜-in. cap screws. The pedestal and box measurements are below:

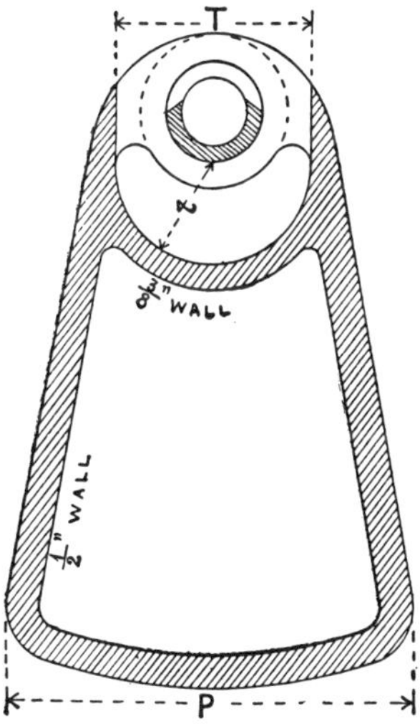

FIG. 93.—CROSS-SECTION OF PEDESTAL AND BOX.

	INCHES.
P—Widest part of standard	7
R—Axial length of oil well	2¾
S—Inner length of oil well	2
T—Inner width of oil well	3
t—Radius line to indicate origin of circle	1½
U—Outside diameter of box	2¼
V—Inside diameter of box	1½
W—Outside length of box	4
Y—Length of bushing	3¼
Bore of bushing	1
Z—Projection of box beyond oil well wall.	$\frac{15}{16}$
Diameter of oil ring	2¼
Bore of oil ring	1⅞
Width of oil ring	⅜

A bearing must be turned on the outside of the inner end of the pedestal on the commutator side of the machine, as indicated in Fig. 91, to accommodate the brush-holder yoke, which may be copied from any of the standard makes. Only two sets of brushes are required, each set comprising two carbon brushes 3/8 in. thick and 1 1/8 ins. wide; the two sets must touch the commutator exactly 90° (11 3/4 segments) apart, center to center. The commutator must have 47 segments, and must measure 3 ins. along the shaft, extreme length. The commutator core must be bored to fit the portion, *c*, of the shaft, and key-seated to correspond. The diameter of the barrel should be not less than 4 ins., and the diameter measured at the connecting lugs must not exceed 6 1/2 ins. The brush surface, measured parallel with the shaft, must be 2 1/2 ins. long. It will be best to buy the commutator complete from one of the several makers of this class of apparatus.

Fig. 94 is the armature shaft. The key-seats are all 3/8 wide and 3-16 deep. The dimensions of the shaft are below:

	At f	At a	At c	At j	Total Length
Diameter, inches	1	1 1/4	1 1/8	1	
Length, inches	6 7/16	6 3/16	5 7/16	3 5/16	21 3/8

The field-magnet coils may be wound directly on the cores or on bobbins made of thin vulcanized fibre. If they are wound directly on the cores, the latter must be wrapped first with three layers of unbleached cottons and painted with shellac varnish, two circular coil heads of hard fibre being first fitted to the large part of each core. The coils consist of No. 22 single cotton-covered wire, wound to a depth of 3/4 in., exactly. The exact number of turns is immaterial, except that all four coils must contain the same number of turns, and as many turns should be put on as can be got in the space available. With careful winding, the builder should get 2,565 turns in each coil. For a 230-volt motor use No. 25 double cotton-covered wire, and for 500 volts use No. 28 double cotton-covered, wound to the depth specified. Should the reader prefer to wind the coils in bobbins, the magnet core need not be wrapped, of course. After each coil is completed, secure the outer end and cover the outside layer with unbleached cottons two layers deep, heavily varnished. Fig. 95 indicates how the field coils should be connected up.

The armature coils of the 115-volt machine are of No. 13 wire, each coil containing eight turns. The winding must be two wires wide and four layers deep, per coil, so that when the coils are in place there will be 16 wires in each slot—two wide and eight deep, as shown in Fig. 96. There are 47 coils, connected up wave-fashion. In winding the coils it will be advisable to bend a hook in each starting end and leave the final ends straight. The armature coils must be wound in a former so that the

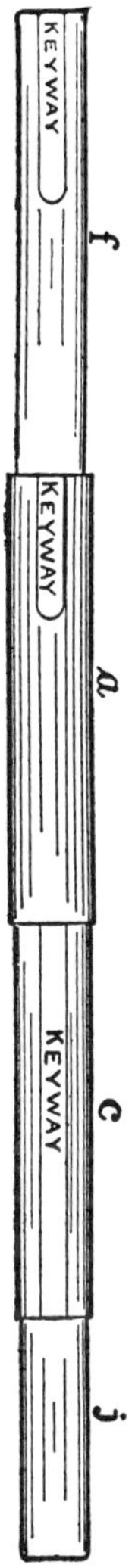

FIG. 94.—THE ARMATURE SHAFT.

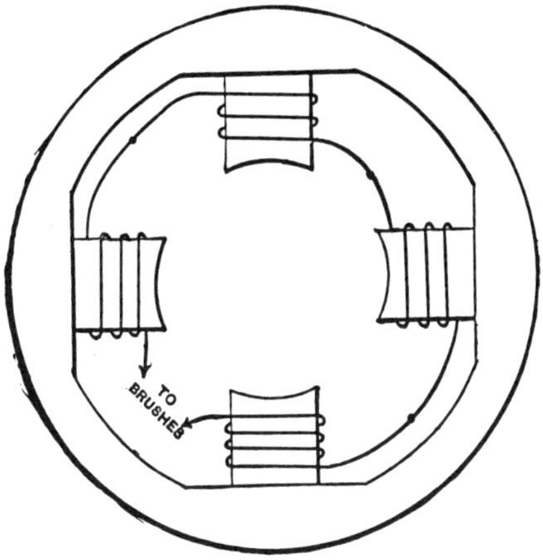

FIG. 95.—FIELD COIL CONNECTIONS.

outline of the opening through each coil is a square measuring 4¾ ins. on each side. The ends should lead out from two corners, and each coil must be wrapped carefully and firmly with two layers of German linen tape, each layer being painted with shellac varnish. The slots in the armature core should be provided with insulating troughs of press board 1-64 in. thick.

Put the coils on the armature all the same way—bent ends to the left and straight ends to the right, facing the commutator. There must be twelve teeth between the two slots in which any given coil is placed, and there must be 22 commutator segments between the two to which the terminals of any given coil are connected, as indicated by Fig. 97. It will be found best to first put on 12 coils in regular right-handed rotation, pressing the ends down closely where they lap, and slipping a bit of thin oiled paper between the crossings. This will put one layer of coils in 24 of the slots. Then put coils in the 23 vacant slots in the same fashion; there will then be 46 half filled slots and one filled.

Continue the second layer of coils right along from the 24th coil, following the same plan as before. At the finish, there will be a bent end and a straight end projecting from each slot.

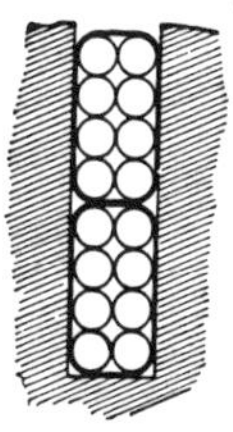

FIG. 96.—SLOT.

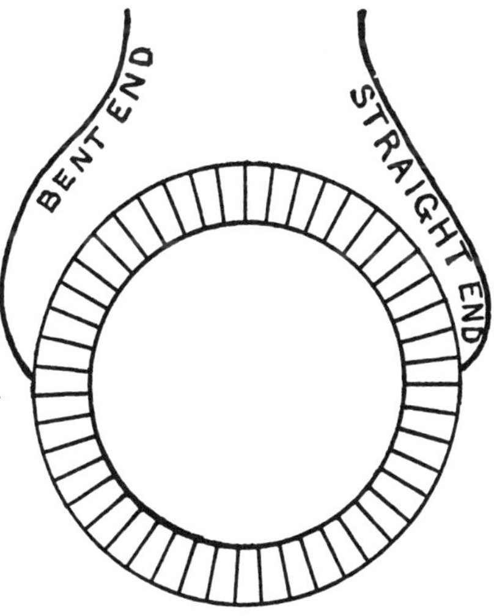

FIG. 97.—CONNECTING DIAGRAM.

Carry the bent ends 11 or 12 segments to the left, around the commutator, and put them all in the segment slots. Then take any one of the straight ends, find the bent end which is the other terminal of its coil, and connect the straight end, as shown in Fig. 97, with 22 segments between it and its mate. The other straight ends may be put in in regular order without tracing, if the coils have been put on the core properly and the bent ends in the commutator lugs in strict sequence.

If it is desired to build the machine for 230 volts, wind the armature with No. 16 double cotton-covered wire, putting 15 turns in each coil—three wide and five deep—so that each slot will contain 30 wires, 3 wide and 10 deep. For 500 volts, use No. 19 wire, putting 28 turns in each coil—4 wide and 7 deep—so that each slot will contain 56 wires, 4 wide and 14 deep. The principal technical data for the 115-volt machine are given below:

Revolutions per minute	1,320
Armature resistance, warm	0.4
Armature current, normal	22
Armature and brush drop, volts about	10
Per cent regulation about	9%
Flux density in armature core and teeth, per square inch	75,000
Flux density in air gap	29,300
Flux density in magnet cores	93,000
Flux density in magnet yoke	48,000
Leakage coefficient	1.25
Resistance of field winding, ohms	169
Exciting current, amperes	0.68
Copper loss in field, watts	78.2
Copper loss in armature, watts	194
Core loss in armature, watts	46
Approximate efficiency, allowing 5 per cent for friction and windage, per cent	82

The starting box should be purchased from any of the standard rheostat builders; a satisfactory home-made one of this size is rarely produced.

CHAPTER X.

ONE KILOWATT COMBINED ALTERNATING AND DIRECT-CURRENT MACHINE.

There are presented in this chapter designs and working drawings for a type of combined alternating and current machine which it is thought will prove generally useful for experimental and laboratory work in alternating and direct currents, and which is applicable on most of the electric-lighting circuits found in practice.

The design contemplates working the machine in a number of different ways:

1. As a direct-current generator or motor.
2. As a single, two or three-phase generator or motor.
3. As a rotary converter, changing single, two or three-phase to direct current.
4. As an inverted rotary converter, changing direct current to single, two or three-phase alternating currents.
5. As a phase transformer, changing alternating current of one phase to that of any other phase.

Some of the foregoing functions may be in operation at the same time; for instance, Nos. 1 and 2 combined would give a "double-current" generator. Also No. 3 or No. 4 may be in operation simultaneously with No. 5.

Three sizes of this type of machine will be described, of 1, 2 and 4 kilowatts capacity, respectively, and in all of these the same scale of voltage has been adopted, namely, 110 volts for the direct current, 80 volts for single or two-phase alternating, and 70 volts for the three-phase alternating. These voltages admit of considerable adjustment, however, by varying the field excitation or speed in case of a generator. The values given represent about the maximum which can be developed continuously.

In operating on single-phase alternating circuits it is necessary to adopt some device which will make the machine self-starting, and this has been provided in the shape of a special switch located in the base of the machine and which, at starting, temporarily changes the connections to those of a series motor which, as is well known, readily starts when alternating current is turned on. The armature is allowed to reach a speed slightly above synchronism, and the switch is then thrown over to the running position, where the machine operates as an ordinary synchronous motor.

In starting on two or three-phase circuits, the same switch is utilized to break up the field winding into a number of short sections on open circuit, thereby avoiding the high induced e.m.fs. which would otherwise be produced on turning the alternating current into the armature winding. It will be understood that where two or three-phase currents are employed the machine is self-starting without any special device, by virtue of the rotary field principle. If the starting current, with this arrangement, is found to be objectionably large, it can be avoided by starting on a reduced pressure supplied from small auto-transformers.

The general features of the design are multipolar field having a circular yoke of cast iron with laminated wrought-iron poles cast in. This type is selected because it admits of high magnetic density and short air gap, and consequently much greater output than does an all cast field, while at the same time it is only slightly more expensive or difficult to construct. An all cast-iron field of the same general design will have only a little more than half the output, and an all cast-steel field, while good magnetically, is scarcely to be considered at present owing to the difficulty in securing steel castings on short notice.

Field coils wound in two or more sections each, and provided with terminals for connection to the starting switch. This is necessary in order to obtain a sufficient reduction in the impedance by connecting the various sections in multiple at the start.

A distributed armature winding, with collector rings tapped in at appropriate intervals on the commutator for alternate-current working. A toothed armature core with deep and narrow slots, and provided with a formed-coil winding, as in direct-current practice.

The minimum number of slots and coils is determined by the number of poles and by the consideration that taps must be made for both two and three-phase working. The quotient obtained by dividing the number of coils or commutator segments by the number of poles must be divisible by two for two-phase working and by three for three-phase working, and hence by two times three for both together. Thus 24 coils and segments are appropriate for a four-pole machine, 36 for a six-pole, and so on.

Six collector rings will be required; ordinarily seven would be necessary, four for two-phase and three for the three-phase. By making one of the two-phase rings the starting point for the three-phase, one ring serves for two, and the total number may be reduced to six.

It would be possible, of course, to use but four rings, obtaining three-phase current by means of two-phase three-phase transformers, but it is preferable to add two rings and obtain all phases directly from the machine.

The hollow base plate, which is cast in one piece with the bearing pedestals, serves as a housing for the starting switch already referred to. This switch is operated by a lever on the outside, at the front or direct-current end of the machine, and has two positions 120 degrees apart, the starting and running positions respectively. In the starting position the various sections of the field winding are in parallel with each other and in series with the direct-current end of the armature.

In the running position the field sections are in series, giving the maximum resistance, and are placed across the direct-current brushes, at the same time alternating current from the single-phase mains is turned into the collector rings.

A pulley having a heavy rim for the purpose of securing a considerable fly-wheel effect will be found advantageous in adding to the smooth running of the machine, particularly when used as a rotary from the alternating-current end.

A pulley of this kind will also be useful where the machine is to be used as a generator direct belted to a gas or gasoline engine. The need for a considerable amount of momentum in the running parts of a rotary is real and genuine, for without it there is a disagreeable oscillation or "pumping," which makes synchronism unstable and sometimes causes the machine to break out of step even before full load is reached.

The bearings are of the ring-oiling type, and of a form which gives good lubrication without the disadvantage of having oil thrown off outside the bearing.

The running qualities of these machines will doubtless prove quite satisfactory. There is not likely to be trouble from sparking, in spite of the fact that the armature is multiple wound, in which, ordinarily, a slight lack of symmetry in field strength would cause heating and sparking. The connections already made to the collector rings for another purpose serve also as equalizers, which permit equalizing currents to flow and thus counteract any slight inequality in the various field poles.

Armature reaction may be guarded against by clipping off the corners of every third lamination in the field poles. This will have the effect of increasing the density in the pole tips to practical saturation, thus avoiding further distortion by armature currents and giving practically a fixed point of commutation for all loads.

Heating in the armature and field windings should not prove serious, for the current densities employed are moderate, considering the size of machine. In the pole pieces, heating would ordinarily be expected, due to the short air gap and high density, but their laminated construction will entirely obviate this difficulty.

While primarily intended for use on 125-cycle circuits, modifications will be indicated enabling these machines to be used on 60-cycle circuits also. This involves either a reduction in speed of one-half, with a correspondingly reduced output and voltage, or a reduction in the number of poles to one-half, keeping the speed and output the same, but necessitating a somewhat more difficult change in connections and winding.

Referring now to the one-kilowatt machine, Fig. 98 shows an end view of the field magnet and base. There are four poles cast into the yoke, which forms a separate casting and is bolted to the base plate by four 7-16-in. by 1½-in. hexagon cap screws. The poles are built up of plain rectangular strips of soft iron about No. 22 gauge, which are clamped between two heavier plates by one or more long flat-head bolts.

The pattern for the field casting should be made just as though it were for an all cast field, the laminated pole pieces being laid in the mould after the pattern has been drawn, and the iron poured in around them. The natural shrinkage of the metal on cooling will cause the poles to be tight and secure. It would give additional security, however, to notch the poles before casting in as indicated by the dotted lines. Still another plan is to leave the end plates short, and to spread the laminations apart where they enter the yoke. This will allow the iron to fill in the interstices and so obtain a good hold on the pole. As the poles have been left with square ends, they must now be bored out 3 5-16 ins. and the corners slightly rounded.

Fig. 99 is a longitudinal half section of the assembled machine, which shows the construction of the armature, bearings, commutator, and collector rings, and also the location of the starting switch in the base.

The armature core is built up of soft-iron discs about No. 27 gauge; two heavier discs of wrought iron, 3-16 in. thick, are provided at the ends as a reinforcement for the teeth, and the whole is clamped between two cast-iron flanges run up on threads cut in the shaft. These flanged pieces serve also as a support for the "straight-out" winding.

Plain round discs may be used in building the core and the slots milled out, being careful, however, to take the discs apart after milling and insulate them with paper or japanning. The keyway in the dics insures their registering when reassembled, in spite of possible slight inaccuracy in milling the slots. It is not necessary to insulate the discs from the shaft if they are fairly well insulated at all other points.

The bearings have a central rib ⅜ in. thick, which supports the brass or bronze sleeve forming the journal proper. The oil pockets at

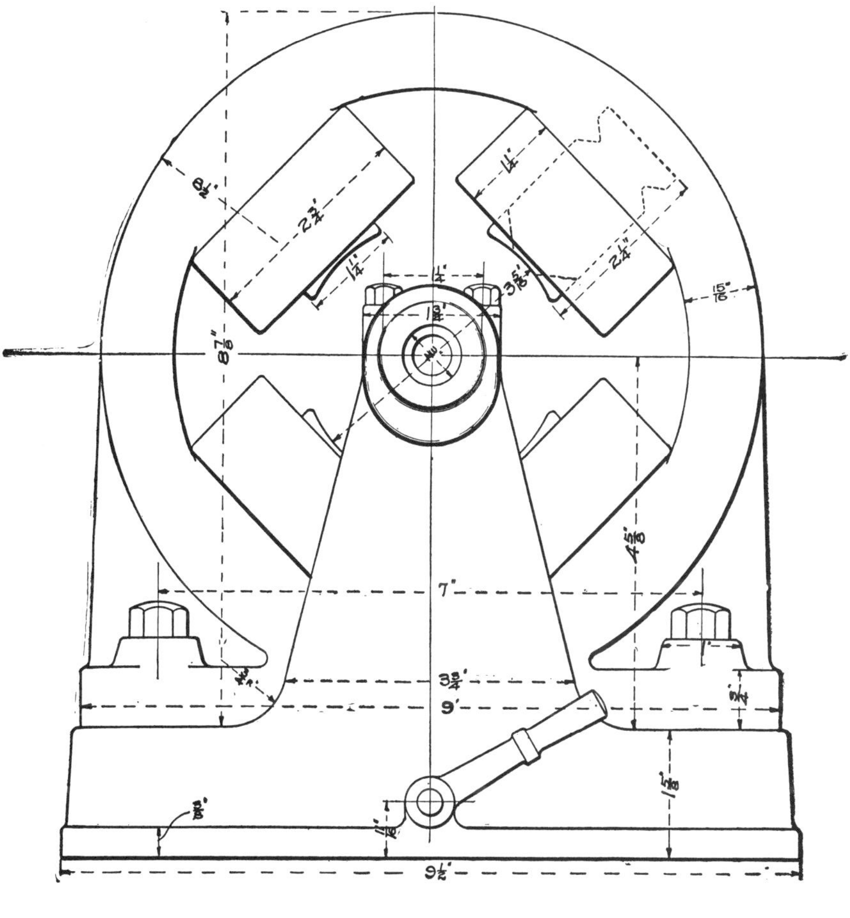

FIG. 98.—END ELEVATION OF THE MACHINE WITHOUT THE ARMATURE.

either side of the web communicate by means of a slot cored out in the web, so that the oil level may remain the same on each side. The oil rings are ⅛ in. wide and ride on the shaft through grooves turned eccentrically in the sleeve.

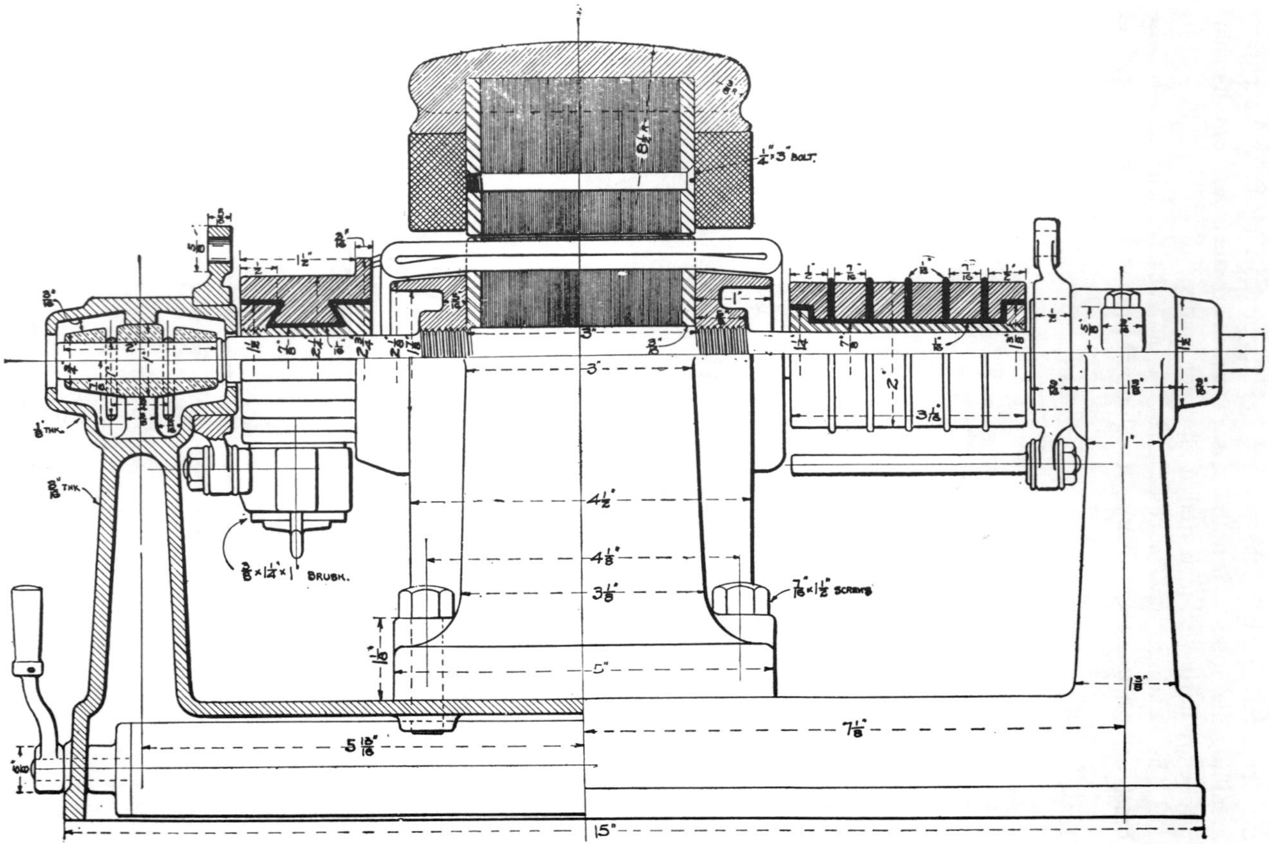

FIG. 99.—SEMI-SECTIONAL SIDE ELEVATION OF THE COMPLETE MACHINE MINUS BRUSH-HOLDERS AND BRUSHES.

The commutator has a steel sleeve fitting the shaft, upon which are two flanges, one solid with the sleeve and the other threaded on it and tightened by means of a spanner wrench applied to holes drilled in its face. Both flanges are undercut at an angle of about 60 degrees, to hold the segments in place.

Probably the best way to construct the commutator is to turn up a copper casting of the required section, and then slit the cylinder into 24 segments by means of a 1-32-in. cutter, in a milling machine. The segments are then built up with 1-32-in. mica between and insulated from the sleeve by 1-16 in. of mica or other good insulation.

The collector rings are similar in construction. The two end rings are counter-bored to let in the flanges of the sleeve, which, in this case, need not be undercut. The other rings are plain round and are simply slipped over the insulating sleeve, and separated from each other by 1-16-in. fiber, or equivalent insulation, which is allowed to project somewhat above the surface of the rings.

Connections to the rings are made by drilling in from the back side and soldering in short wires, No. 12 or No. 14, which should be carefully

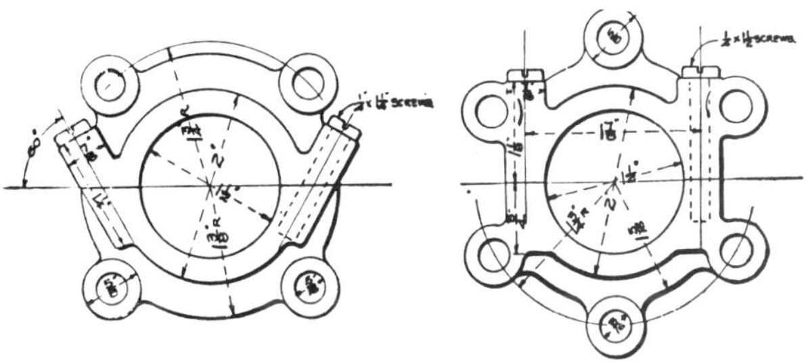

FIGS. 100 AND 101.—BRUSH-HOLDER COLLARS.

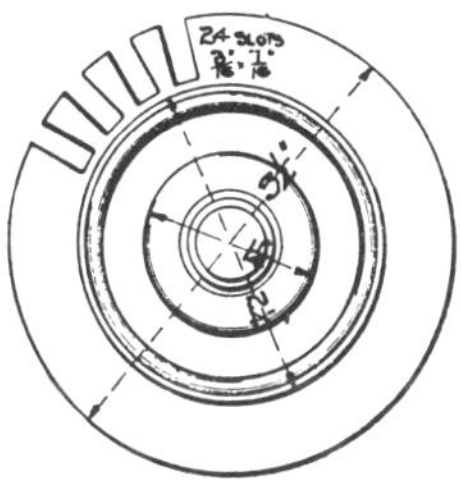

FIG. 102.—ARMATURE HEAD.

insulated where they pass through other rings by small fiber or rubber tubes. These wire leads are made only just long enough to project a short distance from the back ring and are there soldered to some thin copper strips taped and laid in the bottom of the armature slots, six of which have been cut 1-16 in. deeper than the rest to accommodate these connections. It will be the more convenient to make all these connections permanently and test them before laying on the armature coils.

Fig. 100 shows the brush ring for the alternating current end, and Fig. 101 the one for the direct-current end of the machine. They are made in halves, held together by screws, which will facilitate in assembling the machine. The direct-current ring has four lugs for supporting the brush holder and the alternating-current ring has six, one for each of the six collector rings.

Fig. 102 shows an end view of the armature core and Fig. 103 a development of the armature winding. The core has 24 slots 3-16 in. wide and 7-16 in. deep. Every fourth slot is made ½ in. deep to allow space for connections to the rings. The teeth are plain straight and the armature must be banded after the coils are in place.

The armature winding is of the type known as "straight out" and is composed of form-wound coils of No. 20 double cotton-covered wire, each coil consisting of 16 turns arranged four wide and four deep. The coil is wound as a simple straight loop, and after receiving a wrapping of tape it is bent until it will span one-quarter of the armature circumference. One side of a coil occupies the top of a slot and the other side of the same coil occupies the bottom half of a slot 90 degrees, or six slots, in advance of the first. Thus arranged, the coils interleave in a very compact manner and the space required for cross connection is reduced to a minimum.

The terminals are brought out at the apex of the coil and are connected directly to the commutator segments, the beginning of one coil

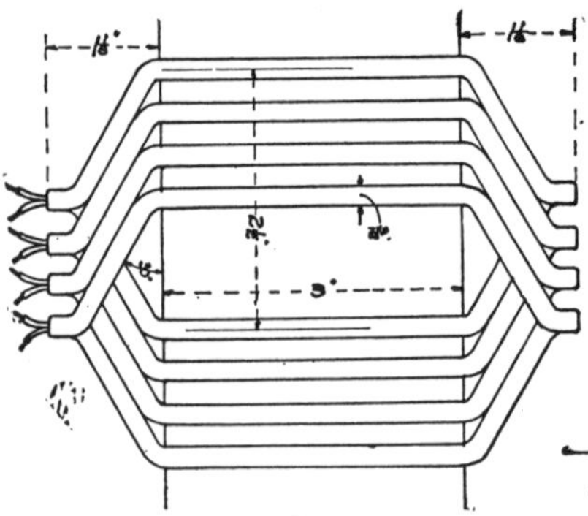

FIG. 103.—BARREL WINDING.

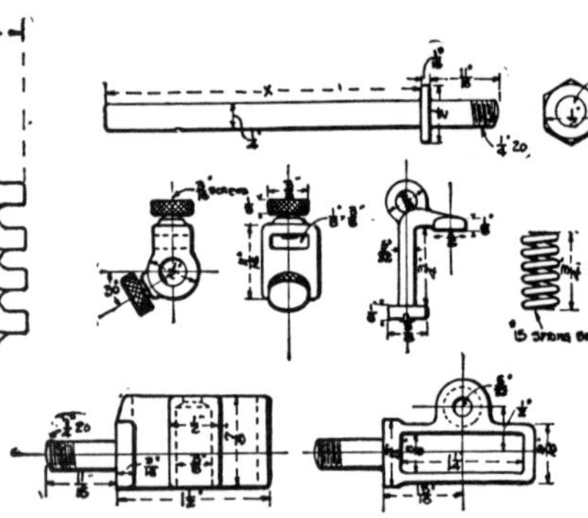

FIG. 104.—BRUSH-HOLDER DETAILS.

and the ending of the adjacent coil connecting to the same segment. The advantage in bringing the terminals straight out to the commutator in this way is that, in addition to being more convenient, it permits the brushes to be placed opposite the poles, where they are more accessible than when placed between the poles.

Fig. 104 shows details of the brush holders. The direct-current holders are of simple construction, but neat in appearance, and are intended for radial graphite or carbon brushes ⅜ in. thick, 1¼ ins. wide and 1 in. long. The necessary tension on the brush is supplied by an open-coil spring concealed in a hollow lug cast on the side of the holder, and acting on a small pressure foot shown separately in the drawings.

By lifting the pressure foot by means of the eye at its top and turning it half around, a brush may be readily removed from or inserted into the holder.

The alternating-current brush holders are carried upon studs supported from the brush ring, and have slots ⅛ in. by ⅜ in. for copper-leaf brushes. There need not be any spring tension provided, as the natural spring of the brush will be sufficient to insure good contact. Two thumb screws are provided, one to hold the brush and the other to clamp the holder upon its stud in the desired position. The studs are of different lengths, the dimension marked X having the values 3⅛ ins., 2⅝ ins., 2⅛ ins., 1⅝ ins., 1⅛ ins. and ⅝ in. for the six studs. Quarter-inch brass rod may be used to make these from, the collars being soldered or threaded on and the ends threaded for a hexagon nut. All brush holders and parts should be made in brass or bronze.

Figs. 105 and 106 are diagrams to be followed in making taps to the collector rings. The 4-pole arrangement, Fig. 105, is intended for operating on 125-cycle circuits and the two-pole, Fig. 106, for 60 cycles. These connections should be made at the back of the commutator before it is placed in position on the shaft. In the four-pole arrangement, for

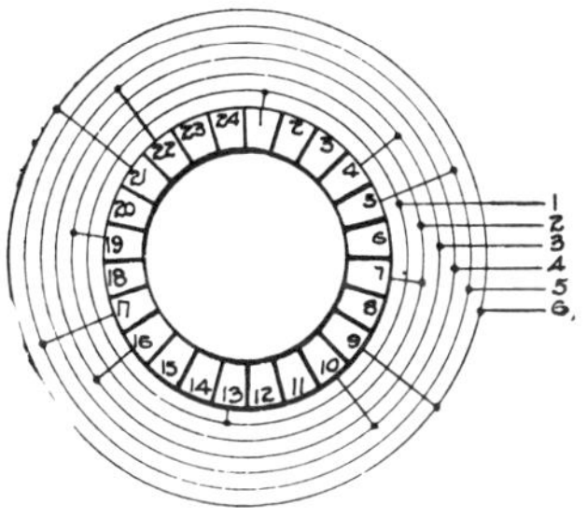

FIG. 105.—FOUR-POLE CONNECTIONS.

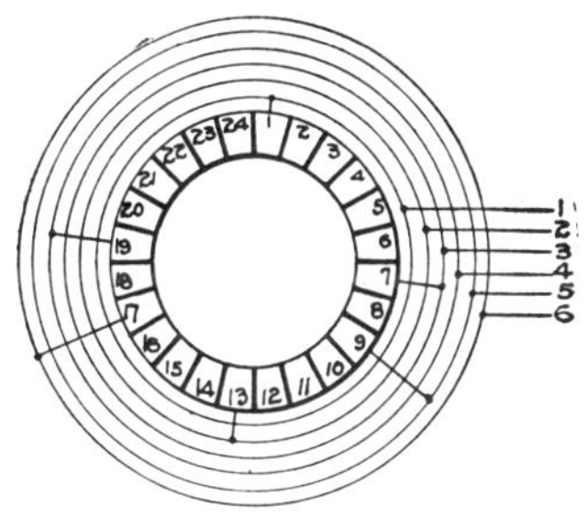

FIG. 106.—TWO-POLE CONNECTIONS.

instance, segments No. 1 and No. 13 are connected together and to a lead marked No. 1, which goes to collector ring No. 1, and similarly for the others. Thus connected, single-phase current may be obtained from rings 1-2 or 3-4. Two-phase current from 1-2 and 3-4 and three-phase current from 1-5-6. The output and voltage with these various connections are as follows: Direct current, 10 amperes at 110 volts; single-phase alternating, 10 amperes at 80 volts; two-phase alternating, 7 amperes per phase at 80 volts; three-phase alternating, 6 amperes per phase at 70 volts.

Fig. 107 shows a form of fly-wheel pulley which is recommended as conducing to smooth running, for reasons already referred to. This pulley is of cast-iron and should be turned perfectly true all over and carefully balanced, as should also the armature. These rotating parts will be required to run at 3750 r.p.m. on 125 cycles, and unless precautions are taken the vibration will be excessive.

Fig. 108 is a detail of the armature shaft. This is designed to be turned from a piece of ¾-in. cold-rolled steel, and for this reason the customary collar at one end has been omitted, and instead threads are cut on both ends for receiving the end plates of the core. This does away with expensive forgings and provides a shaft requiring only a

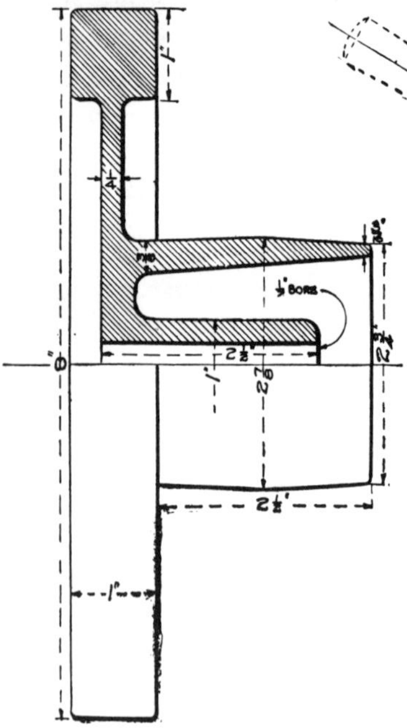

FIG. 108.—FLY-WHEEL PULLEY.

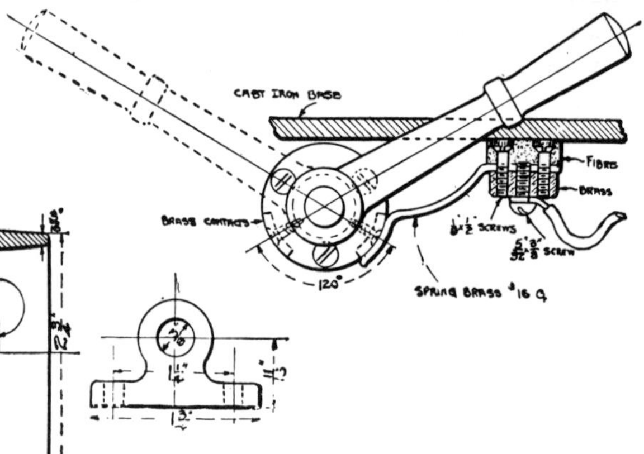

FIG. 109 A.—DETAILS OF THE STARTER SWITCH.

minimum amount of turning. Small grooves are provided at the journals which prevent oil from creeping along the shaft and being thrown off outside the bearing. There are two keys on the shaft, one for the core punchings and the other to hold on the pulley.

Fig. 109 shows the arrangement of the switch cylinder and contacts for the single-phase starting device. There are 30 contact fingers, each 5-16 in. wide, fastened to a strip of fiber ¼ in. thick, which in turn is screwed to the under side of the cast-iron base of the machine. Upon a cylinder of hard wood or fiber 1¼ ins. in diameter are arranged two rows of brass pieces, sunk in grooves cut on the cylinder and upon which the stationary contact fingers press.

The cylinder may be rotated through an angle of 120 degrees by means of a handle on the outside. The contacts on the cylinder are 120 degrees apart, which allows sufficient space for the first set to leave

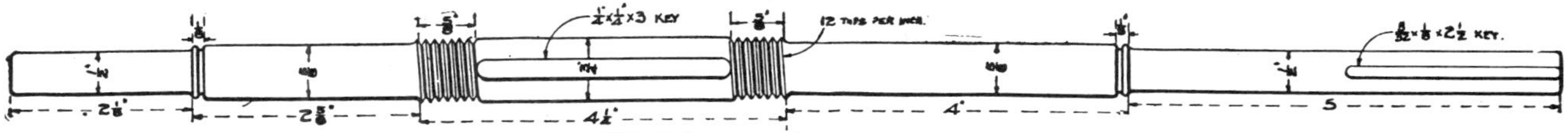

FIG. 107.—THE ARMATURE SHAFT.

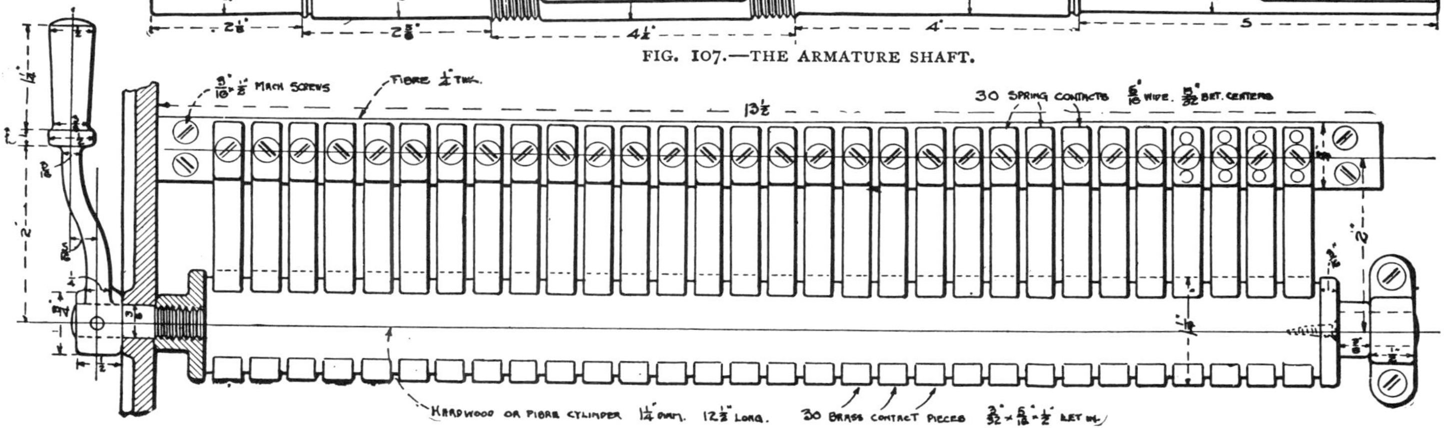

FIG. 109.—CYLINDRICAL STARTING SWITCH AND CONTACT FINGERS.

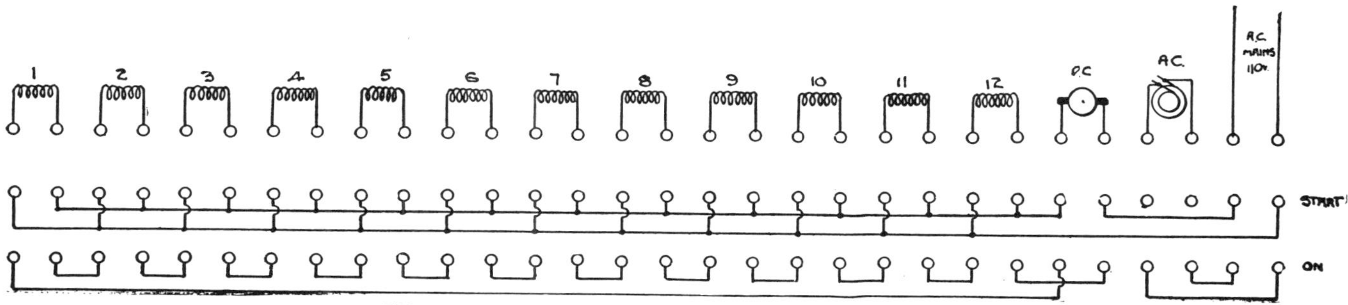

FIG. 110.—DIAGRAM OF CONNECTIONS FOR THE STARTING SWITCH.

contact before the second comes into contact, this being essential to avoid short circuit.

Fig. 110 shows a diagram of connections for the starting switch, by means of which its action may be readily traced out. Numbers 1-12 represent the sectional field winding, there being four coils, each of which is wound in three sections of approximately equal resistance. There are then twelve pairs of ends which lead down into the base of the machine and are connected to the stationary contact pieces, which are represented by the upper row of small circles. The remaining three pairs of contacts connect to the d.c. brush leads, the single-phase rings and the single-phase mains respectively.

The lower rows of circles represent the contact pieces mounted on the cylinder, and these are connected, as here indicated, by means of wires laid in grooves upon the cylinder and occupying that portion of the cylinder over which the contact fingers do not pass.

To operate the machine at 110 volts direct current or 125-cycle alternating, no changes are necessary. For 60-cycle alternating, however, the number of poles is reduced one-half by reversing the terminals of any two successive field coils, and the armature winding must be changed to a bipolar one.

Another plan is to reduce the speed one-half, thus halving the voltage and output and connecting the field coils in series-multiple so that they will still take the same current as at the higher voltage. In operating the machine as a converter, if it is desired that the direct-current output be at 110 volts, the single or two-phase input must be at 80 volts. This relation of voltage is fixed and can be expressed by d. c. volts $\times$ .707 $=$ a.c. volts, and for three-phase by d. c. volts $\times$ .612 $=$ a.c. volts. So that if the alternating circuit is of 52 or 104 volts the machine should be supplied at the proper voltage through a transformer. An old 15-light transformer will serve for this purpose, and it should be arranged so that its secondary voltage can be varied to some extent by changing the number of secondary turns in circuit, thus giving a means of adjusting the direct-current voltage.

The following is a brief summary of the data for winding and general dimensions, and shows the method of calculating same:

Four-pole machine, 3750 r.p.m.; armature, 3¼ ins. diameter, 3 ins. long; 24 slots, 3-16 in. wide, 7-16 in. deep; total number of conductors, 768; 24 coils, No. 20 wire, 4 wide, 4 deep; No. 20 has 1021 circ. mils, diameter d.c.c., .042 in.; direct-current output at 400 c.m. per ampere, 10 amperes; useful lines per pole $\frac{115 \times 10^8}{768 \times 62.5} = 240{,}000$; total lines, 330,000.

Part	Material	Total lines	Cross sect.	B.	H.	L.	Amp.turns
Armature	Wrought iron	120,000	2.25 sq. ins.	53,300	14	1.4 in.	20
2 air gaps	Air	240,000	4.5 "	53,300	16,800	.06 "	1,000
4 teeth	Wrought iron	240,000	2. "	120,000	180	.45 "	80
2 cores	Wrought iron	330,000	3.75 "	88,000	20	1.5 "	30
1 yoke	Cast iron	165,000	4. "	41,300	74	4.25 "	315
Total........							1445

The table above gives a total of 1445 ampere-turns or 725 ampere-turns per coil; mean length, 1 turn, 11 inches.

Circ. mils shunt wire $\frac{11 \times 11 \times 725}{25 \times 12} = 290.$

Use No. 25 wire, 320 c.m., .028 inch d.c.c. 1155 turns (approximate) per coil; 25 layers, 45 turns wide.

Wind in three sections. Bring out terminals from each section.

Resistance of shunt field $= \frac{4 \times 1155 \times 11 \times .97}{1000 \times 12} = 136$ ohms.

Normal shunt current, .63 ampere. Use a rheostat of about 50 ohms total resistance in shunt-field circuit.

Weight of wire in shunt coils $= \frac{4 \times 1115 \times 11 \times .97}{1000 \times 12} = 4.1$ pounds.

Length of wire, each armature coil $= \frac{16 \times 13}{12} = 17.4$ feet.

Total length of wire, armature, $= 24 \times 17.4 = 417$ feet.

Total weight of armature wire $= \frac{417 \times 3.09}{1000} = 1.3$ pounds.

Resistance of armature $= \frac{417 \times 10.1}{1000 \times 16} = .26$ ohm.

Drop in armature at full load $= 10.63 \times .26 = 2.76$ volts.

CHAPTER XI.

TWO KILOWATT COMBINED ALTERNATING AND DIRECT-CURRENT MACHINE.

The 2-kw. machine shown in the accompanying drawings is similar in design, construction and operation to the four-pole machine described in the preceding chapters. The present machine is somewhat larger, runs at a slower speed, and has about double the output capacity of the four-pole machine. Fig. 111 gives an end view of the field-magnet frame. There are six poles of laminated iron cast into a circular yoke of cast-iron, which, in turn, is bolted to the base plate by four hexagon cap screws. After the poles are cast in and it is seen that all of them are tight and firm in the yoke, they may be bored out to the proper diameter, 4.04 ins. The armature is to be finished 4 ins. in diameter, so that the air-gap will be .02 in. across at each pole; this will be ample for clearance if care is taken in lining up the machine.

Fig. 112 is a section of the assembled machine which shows the construction and relation of the various parts. The armature is of the usual laminated construction, the core discs being held between two cast-iron flanges screwed upon the shaft. If the armature slots are milled out, the discs must be taken apart, cleaned up, and insulated before being finally assembled on the shaft. If this is not done the eddy current loss will be excessive, causing heating and seriously reducing the available output. Fig. 113 shows a detail of the armature shaft. This is intended to be made from 1-in. cold-rolled steel. Threads are cut at both ends of the core portion to receive the cast-iron flanges which clamp the core punchings. Two keys are provided, as shown in the drawing.

The bearings are made with a brass sleeve fitting the shaft, supported at its center by a projecting web cast in the bracket. Although it is preferable to bore the bracket for this sleeve, the machine work may be avoided by coring the bracket somewhat larger and then babbiting the sleeve into its support when the parts have been lined up in their proper position. The oil rings are of brass ⅛ in. wide and 1⅛ ins. inside diameter; grooves are cut eccentrically in the sleeves to receive the rings, the grooves being made about 5-32 in. wide in order to allow the ring a small amount of play.

The commutator is built up on a machine steel sleeve, with the flanges undercut at 60 degrees. The segments may be cast separately or cast as a solid cylinder and afterward cut into segments on a milling

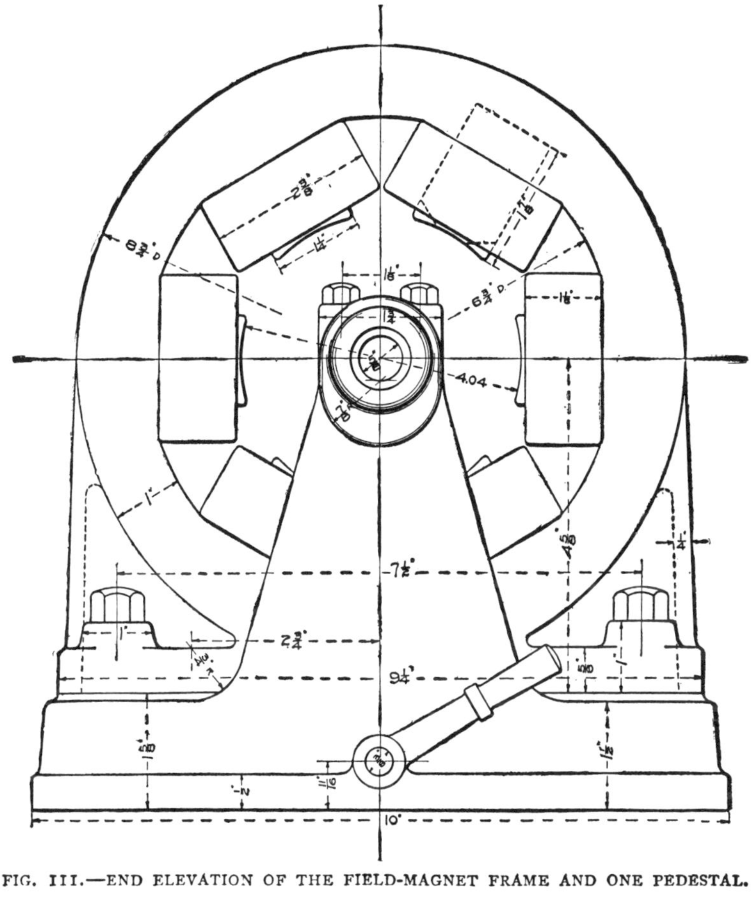

FIG. 111.—END ELEVATION OF THE FIELD-MAGNET FRAME AND ONE PEDESTAL.

machine. The segments should be of copper, with 1-32-in. mica between them, and 1-16-in. micanite, or equivalent insulation, between the sleeve and segments. The number of segments is 36. The collector rings are six in number and mounted upon a sleeve with 3-32 in. fiber discs between the rings. Connections are made by drilling in from the back and soldering in short leads of No. 8 or No. 10 wire, one to each ring. Each of the

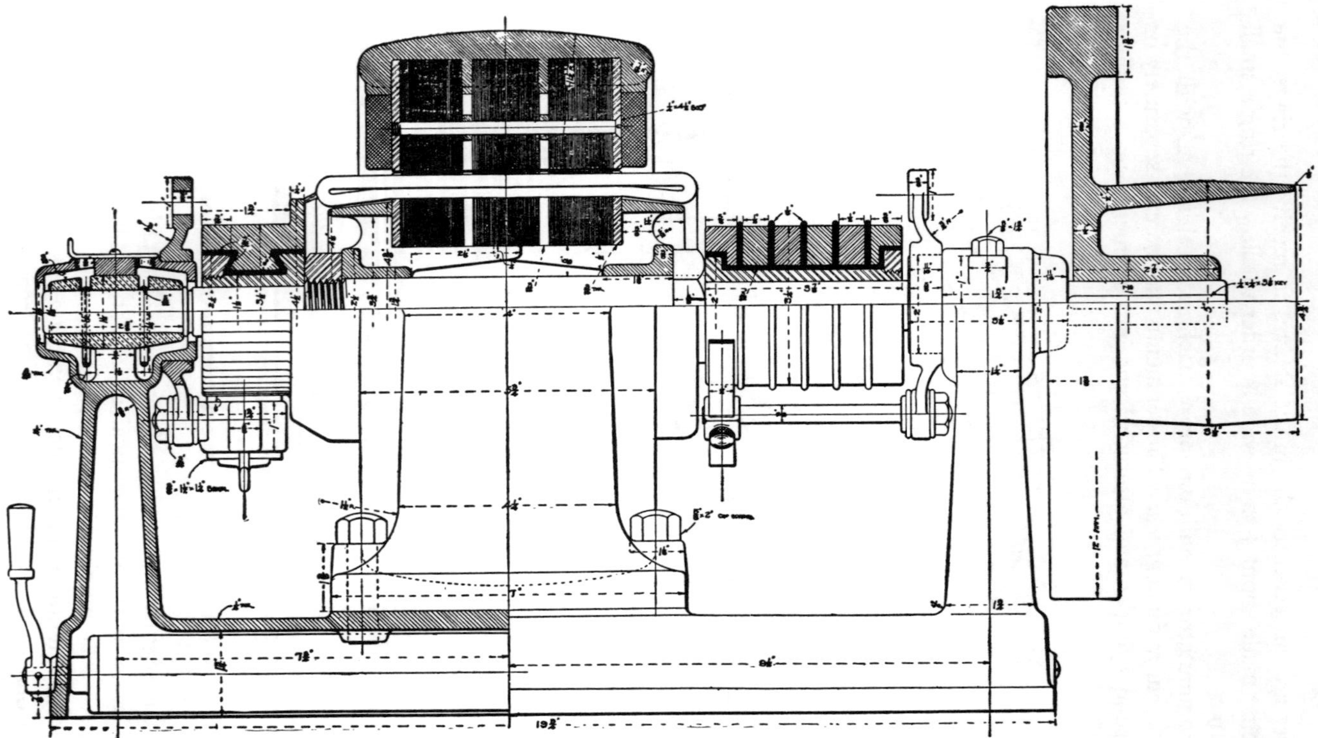

FIG. 112.—LONGITUDINAL ELEVATION OF THE COMPLETE MACHINE, ONE-HALF IN CROSS SECTION.

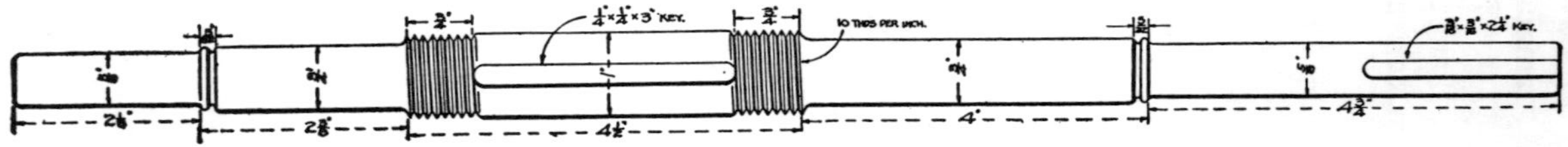

FIG. 113.—THE ARMATURE SHAFT.

leads must be carefully insulated from all rings, except the particular one to which it is electrically connected. Some thin copper strips are to be provided with a wrapping of tape and laid in the bottom of the armature slots, six of which must be made 1-16 in. deeper than the rest to accommodate the strips. These strips carry the current across the armature and are connected to the commutator at the proper intervals.

At the alternating-current end of the machine the fly-wheel pulley is shown in position on the shaft. This style of pulley will be found advantageous in operating the machine as a rotary converter or in driving it by means of a gas engine. If the machine be used as a motor an ordinary pulley will answer. The pulley is for a 2½-in. belt, and is 3½ ins. in diameter.

Fig. 114 shows the brush-holder collar. This answers for both the alternating-current and the continuous-current ends of the machine, as

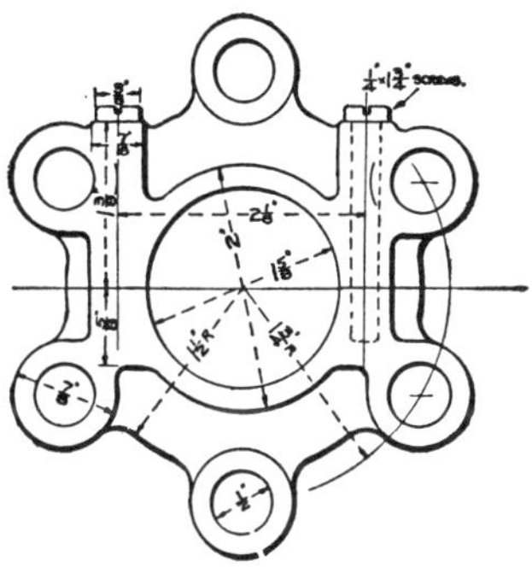

FIG. 114.—BRUSH-HOLDER COLLAR.

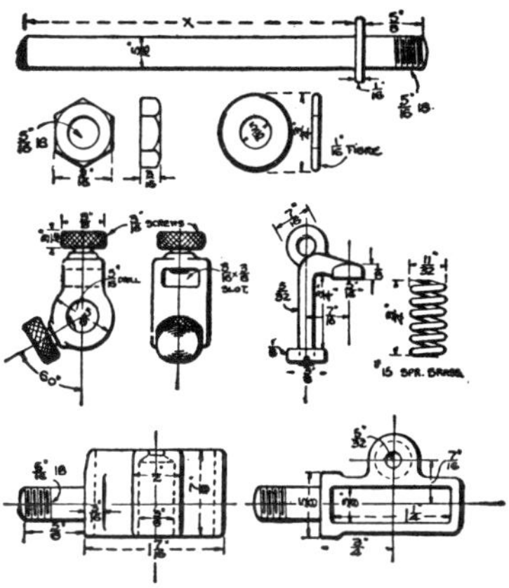

FIG. 115.—DETAILS OF BRUSH-HOLDERS.

there are six collector rings and also six brush holders. Care should be taken in drilling the holes for brush holders to have them equidistant, for upon this the accuracy in spacing the brushes around the commutator depends. At the alternating-current end this does not matter particularly. The brush-holder collars are necessarily made in halves, as it would be difficult to assemble the machine with a one-piece collar.

Fig. 115 shows details of the brush holders. These are of the same type as those already described in connection with the four-pole machine. The alternating-current brush holders have no spring tension and are designed for leaf-copper brushes ⅛ in. thick and ⅜ in. wide. The studs are of different lengths to suit the position of the rings; the dimension, *X*, is 3⅜ ins., 2⅞ ins., 2¼ ins., 1¾ ins., 1 3-16 ins. and 11-16 inches for the six studs. They are made of 5-16-in. brass rod. The continuous-current brush holders are designed for radial carbon brushes 3/8 in. thick, 1 1/4 ins. wide and 1 1/8 ins. long.

Fig. 116 is an end view of the armature core. There are 36 slots, each 3-16 in. wide and 7-16 in. deep; every sixth slot is made 1/2 in. deep to allow space for the connection strips referred to above. After the coils are in place the armature must be banded at three points, one band to go around the center of the core, and one around each end of the winding where it projects beyond the core. A groove must be turned in the periphery of the core to accommodate the central band, so that the thickness of the band will not be added to the length of the air-gap. This groove may be turned on the core before the slots are milled out, or it may be done afterward by filling in the slots temporarily with hard-wood strips. It should be about 1-16 in. deep and 3/8 or 7-16 in. wide.

Fig. 117 shows a development of the armature winding. This is of the "straight-out" type, and is composed of 36 form-wound coils of No. 20 wire, 16 turns per coil. One side of a coil occupies the top half of slot No. 1, and the other side of the same coil occupies the bottom of slot No. 7; that is to say, each coil spans one-sixth of the circumference of the core. The terminals are brought out at the apex of the coil, and each is

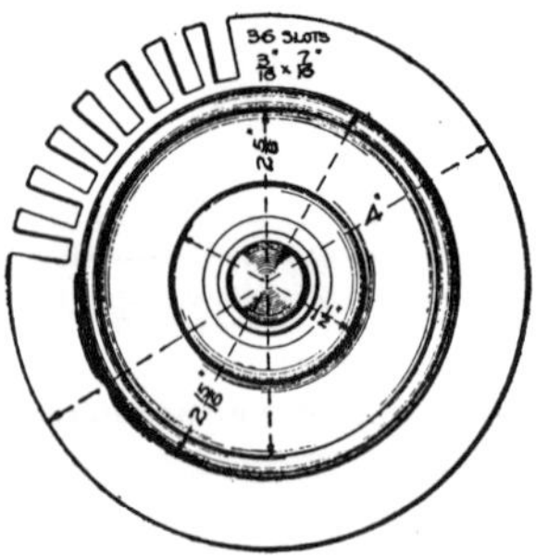

FIG. 116.—END OF ARMATURE CORE.

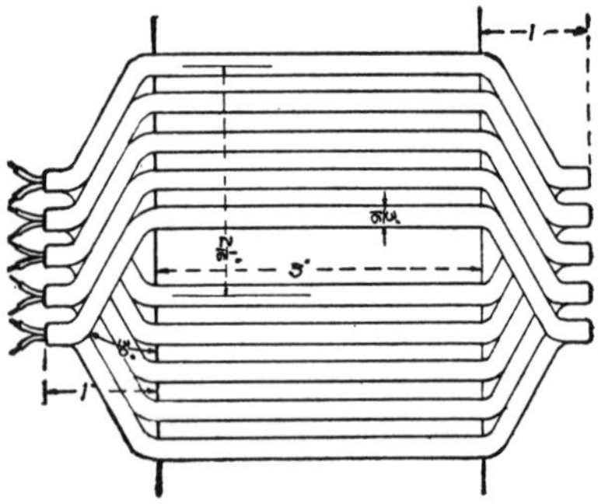

FIG. 117.—DEVELOPMENT OF WINDING.

connected to the nearest commutator segment; the inside terminal of one coil and the outside terminal of the adjacent coil connect to the same segment. The point of commutation will be found at or near the center line of the pole pieces.

Fig. 118 is a diagram of the connections for the collector rings. This arrangement is for a six-pole field. The leads numbered 1 to 6 pass across the armature and are connected to the six collector rings at the alternating-current end of the machine. Connected in this way, single-phase current may be obtained from rings 1 and 2 or 3 and 4, two-phase currents from rings 1 and 2 and 3 and 4, and three-phase currents from rings 1, 5 and 6.

The output and voltage with each of these various methods of working are as follows: Direct current, 15 amperes at 115 volts; single-phase alternating, 15 amperes at 80 volts; two-phase alternating, 11 amperes per phase at 80 volts; three-phase alternating, 9 amperes per phase at 70 volts.

Fig. 119 shows the outline of one of the field coils. These are wound on a form, and each coil is divided into two sections of approximately

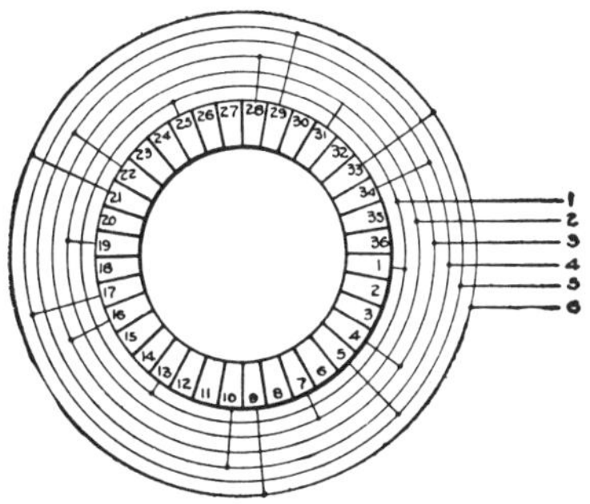

FIG. 118.—DIAGRAM OF TAP CONNECTIONS.

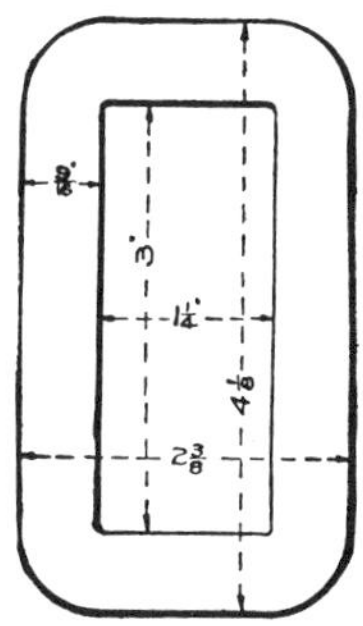

FIG. 119.—FIELD COIL.

equal resistance, with separate terminals brought out from each section. The size of wire is No. 23, B. & S. gauge.

The arrangement employed for starting the machine as a motor on single-phase circuits is as shown in the description of the four-pole machine (see Fig. 109 and 110, and the description on pages 96 and 97, with the single exception that in the present machine there are six coils of two sections each instead of four coils of three sections each.

To operate the machine at 110 volts, continuous current, or 125 cycles alternating, the speed should be 2500 r.p.m. For 60 cycles the only method available is to reduce the speed to 1200 r.p.m and to connect the field winding in series multiple. This is most conveniently done at the starting switch by changing the wiring of the last row of contacts on the switch cylinder, so that when the switch is in the running position the two sections of each field coil will be in multiple and the six multiplied pairs in series and connected across the continuous-current brushes. This will reduce the voltage to about one-half of its value at the higher speed, and the output will then be as follows: Continuous current, 15 amperes at 55 volts; single-phase, 15 amperes at 40 volts; two-phase, 11 amperes at 40 volts; three-phase, 9 amperes at 35 volts. It is probable that by adjusting the field excitation, the voltage could be brought up to 45 or 47 volts, and thus admit of working directly on single-phase circuits of 50 or 52 volts as a motor or rotary without the use of an individual transformer. For other voltages a transformer will be necessary.

The following is a summary of the data for winding and general dimensions: Speed, 2500 r.p.m. on 125 cycles, or 1200 r.p.m. on 60 cycles. Cast-iron yoke, laminated-iron poles cast in. Armature, 4 ins. in diameter, 3 ins. long; 36 slots 3-16 in. wide, 7-16 in. deep; every sixth slot ½ in. deep, 36 coils of No. 20 wire; 16 turns per coil, four wires wide and four deep. Total, 1152 conductors. At 15 amperes continuous-current output the cross-section of armature conductors is 400 circ. mils per ampere.

Useful lines per pole, at 115 volts and 2500 r.p.m.:

$$\frac{1152 \times 41.6}{115 \times 10^8} = 240{,}000.$$

TOTAL LINES PER POLE, 320,000

Part	Total lines	Cross sect. sq. in.	B.	H.	Length	Ampere turns
Armature	120,000	3.	40,000	10	1.5″	15
2 air gaps	240,000	4.	60,000	18,800	.04	750
5 teeth	240,000	2.	120,000	180	.44	79
1 yoke	160,000	3.5	46,000	102	3.	106
2 cores	320,000	3.75	85,000	18	2.5	45

Total ampere turns in field winding, 995. Circ. mils field wire (No. 23) $= \frac{11 \times 12 \times 500}{16 \times 12} = 345$. Mean length per turn, 12 inches. Turns (approx.) per coil, 500; 16 layers of 32 turns each. Resis. of field winding (coils in series), 63 ohms. Normal shunt current, 1 ampere (nearly). Use rheostat of about 40 ohms total in field circuit.

Mean length of wire per armature coil, 16 feet. Total length of armature wire, $36 \times 16 = 610$ feet. Total weight of armature wire, 2 pounds. Resistance of armature, 0.17 ohm. Drop in armature winding at full load, 2½ volts.

CHAPTER XII.

FOUR KILOWATT COMBINED ALTERNATING AND DIRECT-CURRENT MACHINE.

The machine here illustrated is the largest of the machines of the same general type of which this is the third to be described in this book. The present machine has 8 poles; its speed is from 1800 to 1875 r.p.m., and it has an output capacity of four kilowatts.

Fig. 120 shows an end view of the field, base, and bearing pedestals. The field has a circular yoke of cast-iron with pole-pieces of laminated wrought iron cast in. About No. 20 gauge iron may be used in the poles and they are bored out to 5 9-16 ins. diameter after being cast in. Fig. 121 is a section of the assembled machine, which shows the construction and relation of the various parts. The armature core is built up of soft iron discs about No. 27 gauge, having an external diameter of 5½ ins., with a 2½-in. hole in the center. The discs are mounted upon three-arm spiders, one at either end of the core, and the arms of which intermesh about ½ in. at the center of the core, so that all the discs are supported at least three points, and at the same time the air has free access to the interior of the core. Distance pieces are provided at two points in the core which divide the laminations into three groups with 3-16-in. ventilating ducts between them. Two hexagonal nuts upon the shaft provide means for clamping the core discs and spiders.

The commutator, which is shown partly in section, is 3½ ins. in diameter and 1¾ ins. wide on the face. There are 48 segments of copper with 3½-in. mica between them, and 3-32-in. insulation separates the segments from their supporting sleeve. The sleeve is of machine steel and has flanges undercut at an angle of 60 deg. The collector rings are six in number and are made of copper. The rings at the ends are ⅝-in. wide; the rest are ½ in.; ⅛-in. insulation separates the rings from each other. The bearings have a brass sleeve fitting the shaft, and this is slotted to allow oil rings 5-32 in. wide to revolve freely with the shaft.

Fig. 122 is a detail of the armature shaft. This is designed to be made from a piece of 1¼-in. cold-rolled steel. Threads are cut at both ends of the core portion to receive the hexagonal nuts which compress the core discs.

At the alternating-current end of the machine the fly-wheel pulley is shown in position upon the shaft. A pulley of this kind will be found very useful in operating the machine as a rotary or in connection with a gas

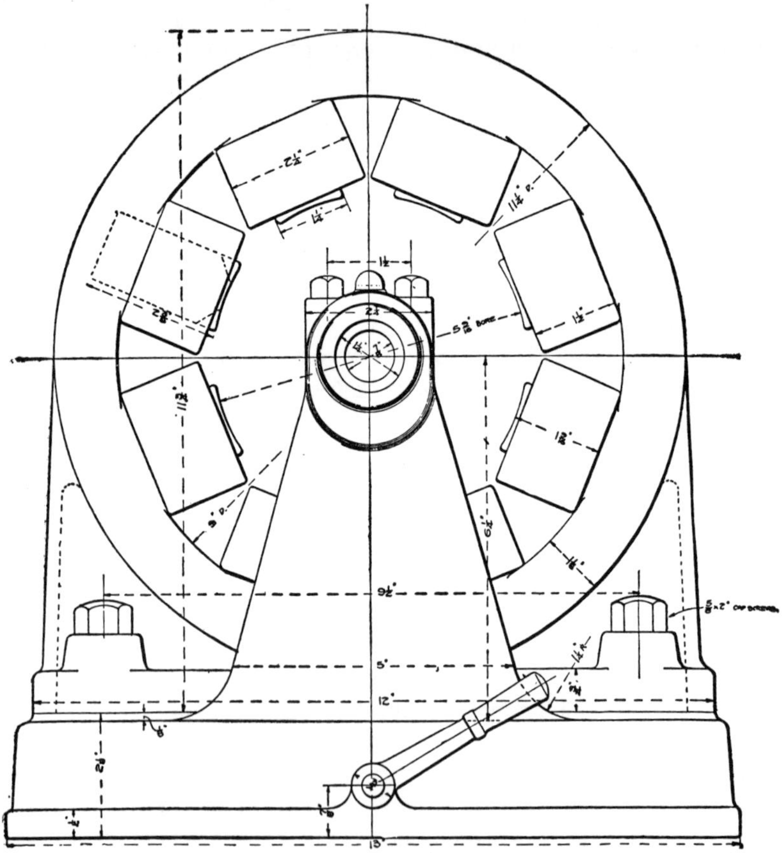

FIG. 120.—END ELEVATION OF FIELD-MAGNET FRAME AND ONE PEDESTAL.

engine for driving, as it assists the armature in maintaining a uniform rate of rotation.

Figs. 123 and 124 show the brush-holder collars or so-called "quadrants" for the alternating-current and direct-current ends of the machine. The one at the direct-current end has eight lugs for supporting the eight brush holders and the alternating-current end has six, one for each of the

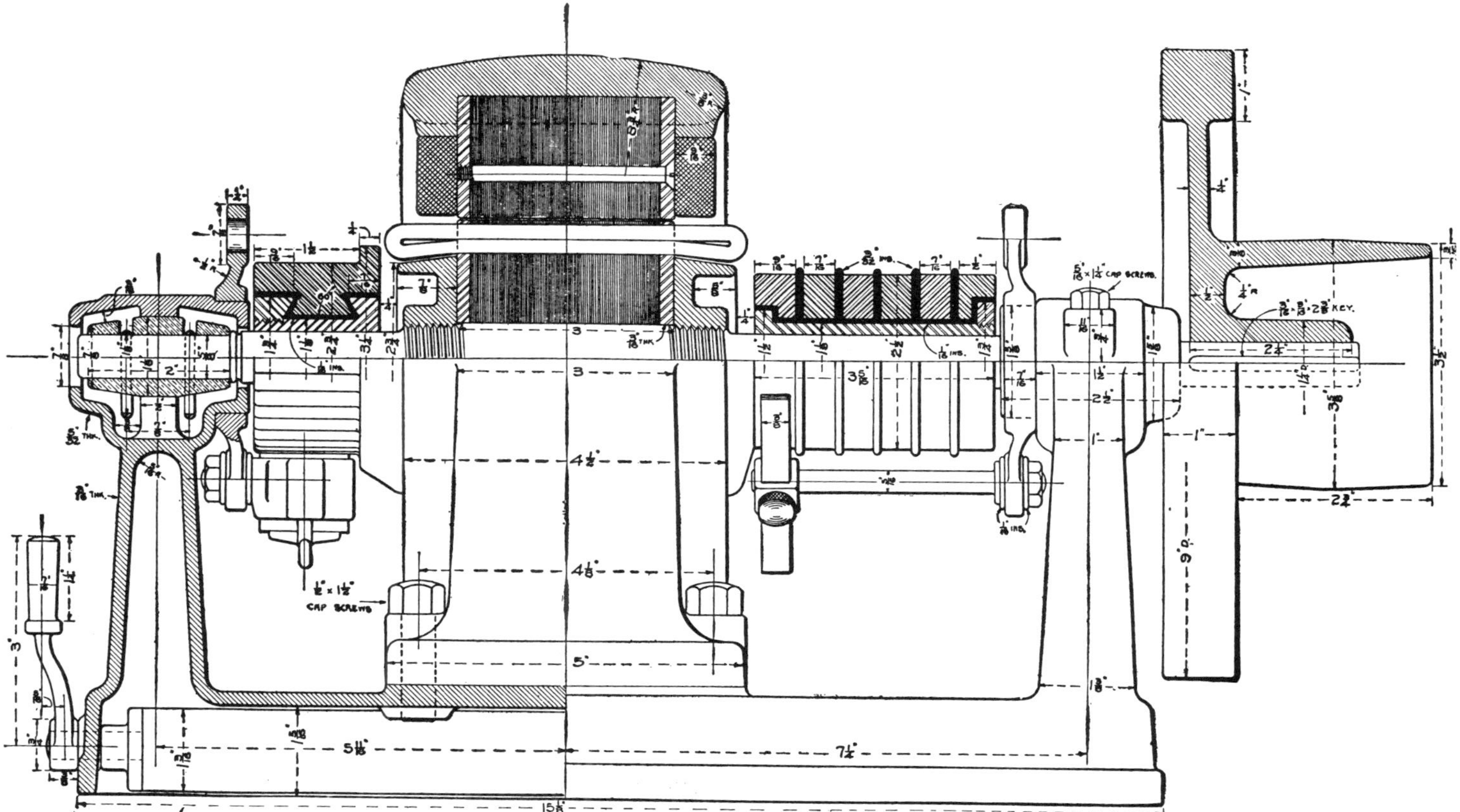

FIG. 121.—LONGITUDINAL ELEVATION OF THE COMPLETE MACHINE, ONE-HALF IN CROSS SECTION.

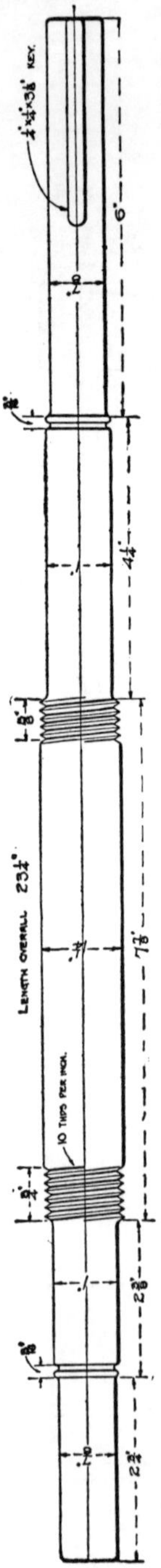

FIG. 122.—THE ARMATURE SHAFT.

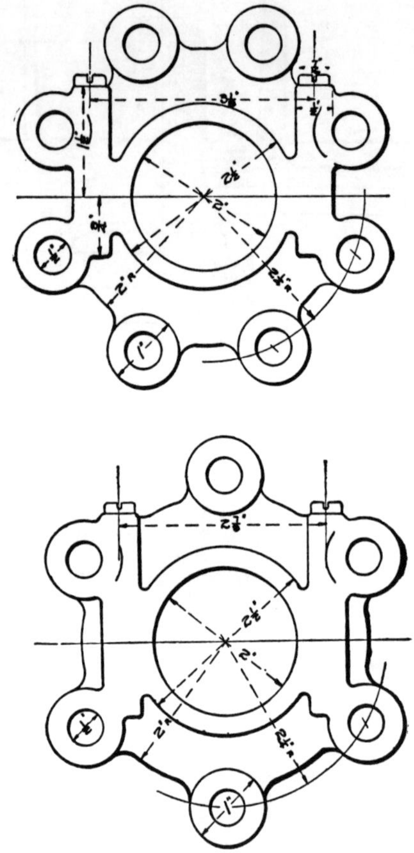

FIGS. 123 AND 124.—BRUSH HOLDER COLLARS.

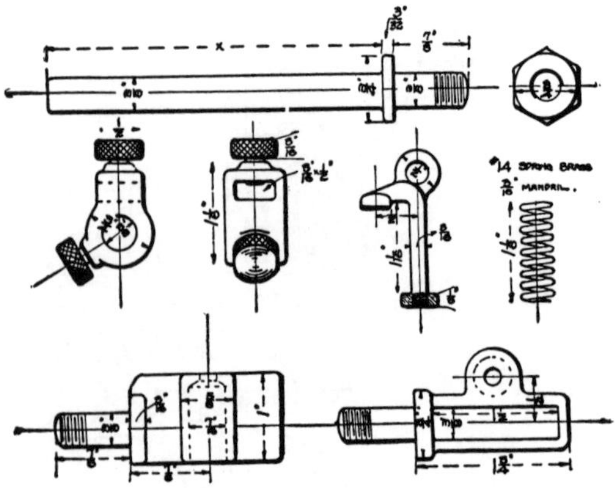

FIG. 125.—DETAILS OF BRUSH HOLDERS.

six rings. Both are made in halves and screwed together after being placed in position on the bearings.

Fig. 125 shows details of the direct-current and alternating-current brush holders. The direct-current holders are for radial brushes, ⅜ in. thick, 1½ ins. wide and 1¼ ins. long, and are provided with a spring tension arrangement, the details of which are shown in the engraving. The alternating-current brush holders are for copper-leaf brushes ⅛ in. thick and ½ in. wide; the spring of the brush itself will be found sufficient to give proper contact with the collector rings. The studs which support these holders are of different lengths to suit the position of the various collector rings. The dimension marked x on the drawings has the values 3⅞ ins., 3¼ ins., 2⅝ ins., 2 ins., 1⅜ ins. and ¾ in. for the six studs respectfully. They are best made of ⅜-in. brass rod.

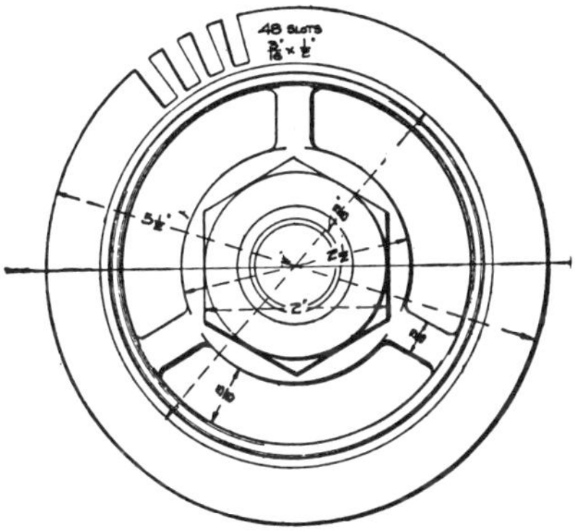

FIG. 126.—END OF ARMATURE CORE.

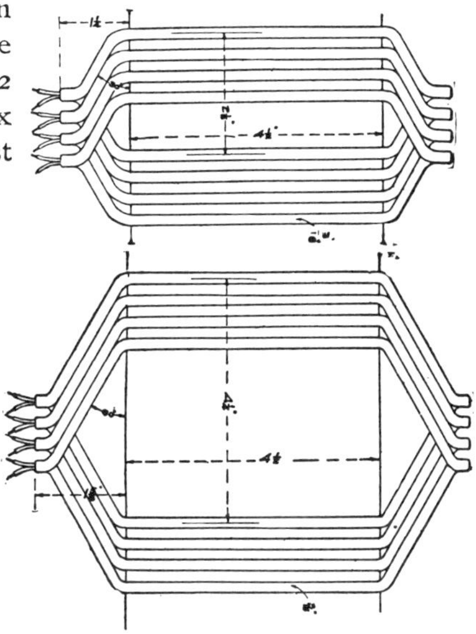

FIGS. 127 AND 128.—DEVELOPMENT OF WINDINGS.

Fig. 126 shows an end view of the armature core. There are 48 slots, 3-16 in. wide and ½ in. deep. If these slots are milled out, it will be necessary to take the discs apart after this operation in order to anneal and insulate them before the final assembling. Annealing will improve the discs, which will have become somewhat hardened from the machine work which has been done upon them.

Figs. 127 and 128 show developments of the armature winding. This is of the "straight-out" type and is of two forms, known as the "short coil" (Fig. 127), and "long coil" (Fig. 128). The long coil is for use in a four-pole field and for 60-cycle work. The short coil is for eight poles and 125 cycles. These coils are form wound of No. 17 double cotton-

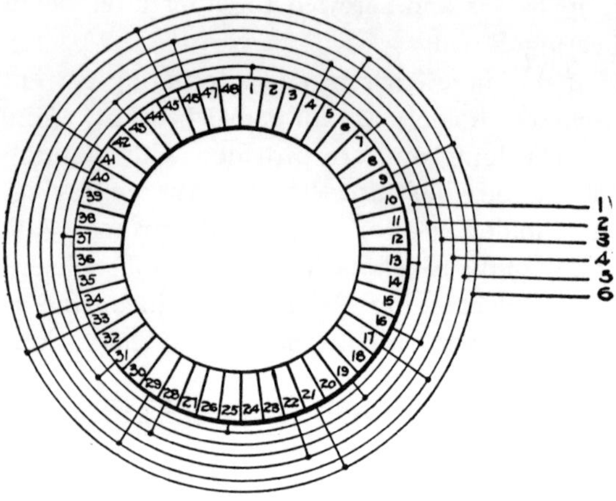

FIG. 129.—EIGHT-POLE TAP CONNECTIONS.

covered wire, 12 turns per coil. The terminals of each coil are brought out at its apex, and are connected to the two nearest commutator segments, inside terminal of one coil and the outside terminal of the adjacent coil connecting to the same segment. This will bring the neutral point or line of commutation at or near the center of the pole-pieces.

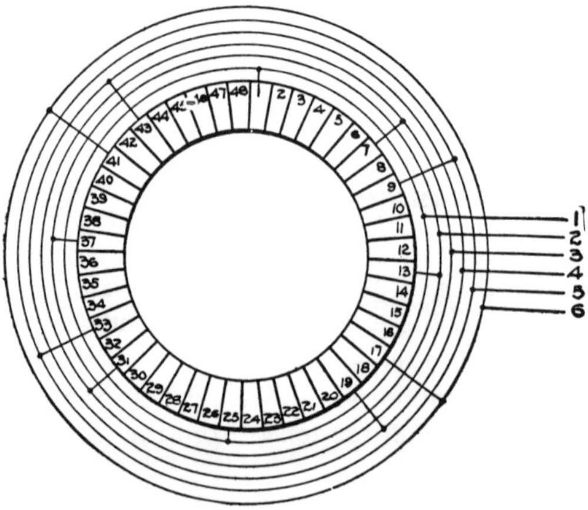

FIG. 130.—FOUR-POLE TAP CONNECTIONS.

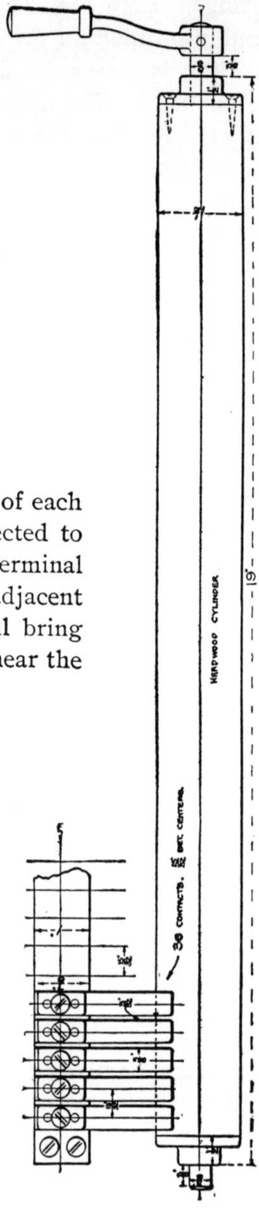

FIG. 131.—STARTING-SWITCH CYLINDER AND FIVE OF THE CONTACT FINGERS.

Figs. 129 and 130 are diagrams to guide in making taps to the collector rings. The eight-pole arrangement is intended for operating on 125-cycle circuits, and the four-pole for 60-cycle circuit. The necessary interconnections between segments may be made at the back of the commutator before it is placed in position on the shaft. For instance, in the eight-pole arrangement, segments 1, 13, 25 and 37 are connected together and to a lead marked No. 1. The leads numbered 1 to 6 inclusive pass through the air space in the center of the core and connect to the corresponding rings. Thus connected, single-phase current may be obtained from rings 1—2 or 3—4; two-phase currents from 1—2 and 3—4, and three-phase currents from 1—5—6. The outputs and voltages are as follows:

	Amperes.		Voltage.
Direct current...........	35		115
Single-phase alternating..	35		80
Two-phase alternating...	25	per phase	80
Three-phase alternating..	21	" "	70

Figs. 131 and 132 show the details of the switch cylinder and contacts which are located

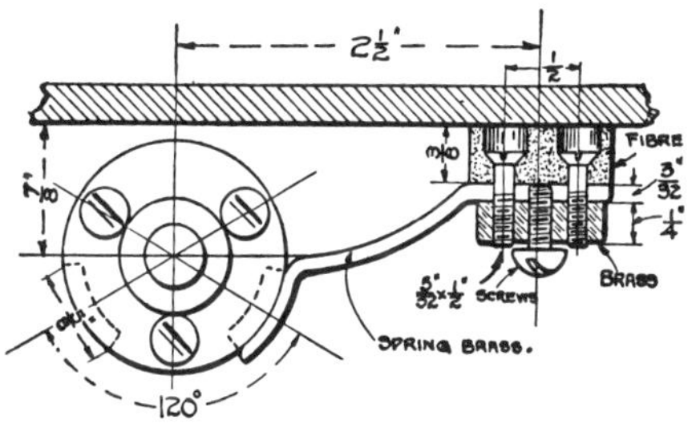

FIG. 132.—DETAIL OF STARTING SWITCH.

in the base of the machine and serve as a starting device for operating on single-phase circuits.

A strip of fibre 19 ins. long, 1 in. wide and ⅜ in. thick is fastened to the under side of the cast iron base and upon this are mounted 38 contact springs, each ⅜ in. wide. The cylinder is of hard wood 1½ ins. in diameter, and is of the proper length to just go inside the base and have its ends journaled therein. The free ends of the contact springs press upon the cylinder and make connections with the

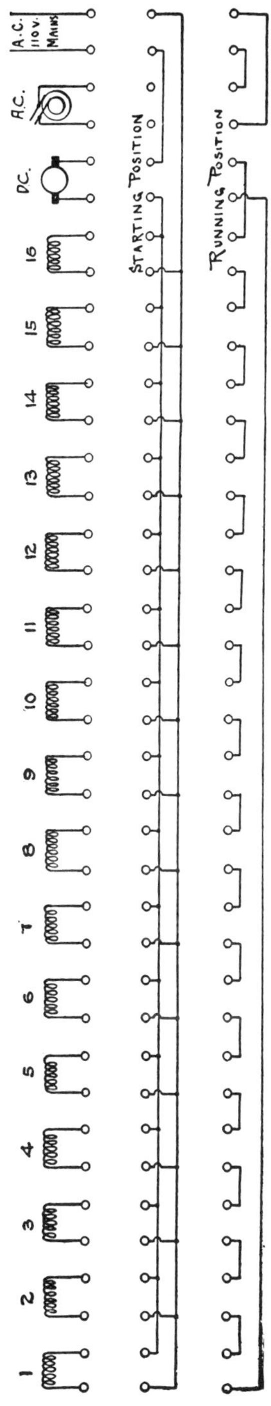

FIG. 133.—DIAGRAM OF CONNECTIONS FOR THE STARTING SWITCH.

contacts fastened upon the cylinder, but they can only be in connection with one row at a time. Fig. 132 is an end view showing details.

The diagram, Fig. 133, shows how the connections are made to the switch contacts. The uppermost row of small circles indicates the contact springs to which the sectional field winding is connected. There are eight coils of two sections each, making 16 pairs of terminals to be connected to the switch. The remaining three pairs of contacts connect with the direct-current brush leads, alternating-current rings, and single-phase mains in the manner indicated. The lower two rows of circles represent the contact pieces upon the revolving switch cylinder, these being simply connected in groups by means of short wires or metal strips fastened upon the cylinder itself and occupying that position of the cylinder over which the contact springs are not required to pass.

The action of this device may readily be followed by assuming that the "starting" row of contacts has been moved up to engage the contact springs when it will be seen that all the field sections are in multiple, the armature in series with them, and the whole placed across the single-phase alternating-current mains, so that the machine starts as a series motor. When the second row of contacts engages the contact springs the field sections will be in series and placed in shunt across the direct-current brushes, while at the same time the single-phase supply current is connected to the collector rings, and the machine is now operating as a synchronous motor, exciting its field from the direct-current end, which is the normal running condition.

To operate the machine at 110 volts direct current or 125-cycle alternating-current no changes are necessary and the speed will be 1875 r.p.m. For 60-cycle alternating-current work the number of poles is halved by reversing the terminals of any two successive field coils, skipping the two coils and reversing the next two. This being most conveniently done by changing the connections at the starting switch. This change, together with the "long coil" winding and the diagram of connections for four poles fits the machine for operating at 60 cycles. The speed will now be 1800 r.p.m. and the voltage and output practically the same as before.

The following is a brief summary of the general dimensions and data for winding:

Eight-pole machine: 1800 r.p.m. at 60 cycles; 1875 r.p.m. at 125 cycles. Armature 5½ ins. diameter, 4 ins. long, 48 slots 3-16 in. × ½ in.; 3-16 in. = 188 in. width of slot, allowing three No. 17 wires and insulation of 18 mils; 5 in. depth slot, taking eight No. 17 wires and insulation and bands of 45 mils., giving 24 conductors per slot. Total, 1,152 conductors. Direct-current output at 470 circ. mils per ampere = 36 amps.

There will be 455 ampere-turns in each field coil. Mean length of 1 turn = 13.5 inches circ. mils shunt wire $= \frac{11 \times 13.5 \times 455}{12 \times 12} = 465.$

Use No. 23, having 509 circ. mils, .034 in. diam., d.c.c.; 620 turns (approx.) per coil, in 14 layers, 44 turns wide. Wind in two sections and bring out individual terminals from each section.

Resistance of shunt field $= \frac{8 \times 620 \times 13.5 \times 20.3}{1{,}000 \times 12} = 114$ ohms.

Normal shunt current $= \frac{450}{620} = .75$ amp. Use rheostat of about 40 ohms in shunt field. Weight of wire in shunt coils:

$$\frac{8 \times 620 \times 13.5 \times 1.54}{1{,}000 \times 12} = 8.5 \text{ lbs.}$$

Total length of wire on armature:

$$\frac{48 \times 18 \times 12}{12} = 864 \text{ feet.}$$

Total weight of wire on armature:

$$\frac{864 \times 6.2}{1{,}000} = 5.37 \text{ lbs.}$$

Resistance of armature:

$$\frac{1{,}000 \times 64}{864 \times 5.04} = .068 \text{ ohms.}$$

Drop in armature at full load:

$$36 \times .068 = 2.45 \text{ volts.}$$

Part	Material	Total lines	Cross-section	B.	H.	L.	Ampere turns
1 armature..	Wrought iron	170,000	4. in.	42,500	4	1.5	6
2 air gaps ..	Air	340,000	6.	56,600	17,880	.0313	566
5 teeth......	Wrought iron	340,000	3.	113,000	108	.5	54
2 cores......	Wrought iron	420,000	5.	85.000	18	1.7	31
1 yoke......	Cast iron	210,000	4.7]	42,000	68	3.7	252
Total ampere-turns......							909

CHAPTER XIII.

SINGLE-PHASE RECTIFIER.

The accompanying drawings and description constitute a design for a machine to "rectify" single-phase alternating current; that is, to change it into a pulsating direct-current without changing its e.m.f.

Fig. 134 shows an end view of the field magnet with coils in place,

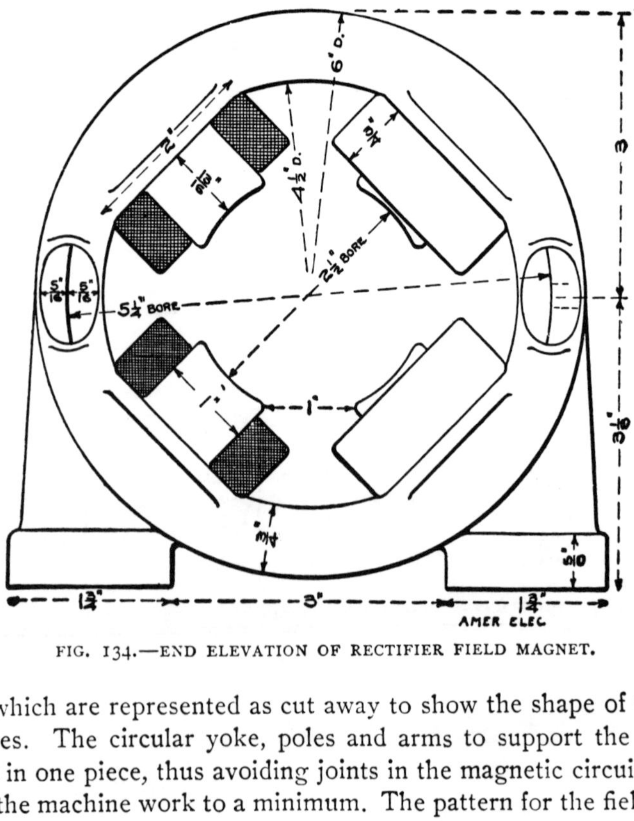

FIG. 134.—END ELEVATION OF RECTIFIER FIELD MAGNET.

two of which are represented as cut away to show the shape of the poles and cores. The circular yoke, poles and arms to support the bearings are cast in one piece, thus avoiding joints in the magnetic circuit, and reducing the machine work to a minimum. The pattern for the field casting should be made in two pieces, the parting being made along the line *AB* in Fig. 135.

The field casting is mounted on the face plate of the lathe—one of 9 ins. swing will do—and the poles are bored out to the required diameter, 2½ ins., at the same time the arms for the bearings are finished to the arc of a circle 5¼ ins. in diameter. These may, however, be filed and fitted by hand, if this is more convenient, but the method of boring is the more accurate.

Fig. 137 shows a section through the armature, commutator and bearings. The armature core is built up of laminated iron in the usual way, and held together by heavy washers of wrought iron threaded on the shaft at either end. A cast-iron core may be used, though it will, of course, heat up more than the laminated one.

The commutator and collector rings are preferably made of copper, though brass may be used. A piece of ingot copper can be obtained and

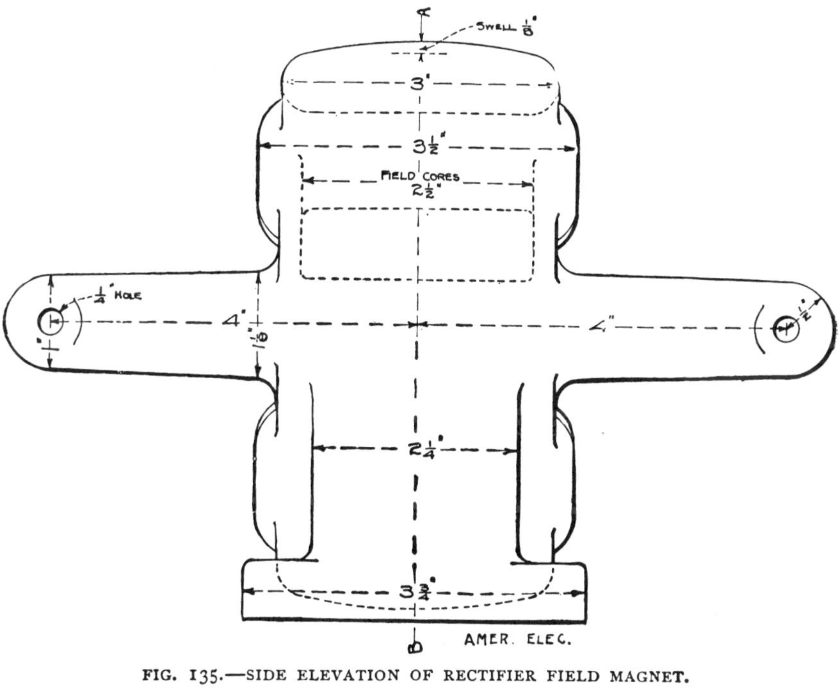

FIG. 135.—SIDE ELEVATION OF RECTIFIER FIELD MAGNET.

forged into a circular blank suitable for turning up into rings and a commutator. The commutator is turned as a solid cylinder of the required section, and it is then cut into four equal segments with a milling machine, or with a sharp hacksaw and hand power, if a milling machine is not available. The segments are then built up with mica insulation and clamped in a brass sleeve fitting the shaft.

It is well to follow the design of arc machine commutators to some extent and to allow about 1-16 in. air insulation between the segments at the top. Thus the mica insulation will not be injured if sparking occurs. Oil and copper dust should not be allowed to accumulate in the air spaces thus formed, as this would cause a severe short circuit.

Connections are made between the collector rings and commutator bars with some strips of copper about 1-32 in. thick and ⅜ in. wide, laid in the armature slots, which have been made about 1-16 in. wider than

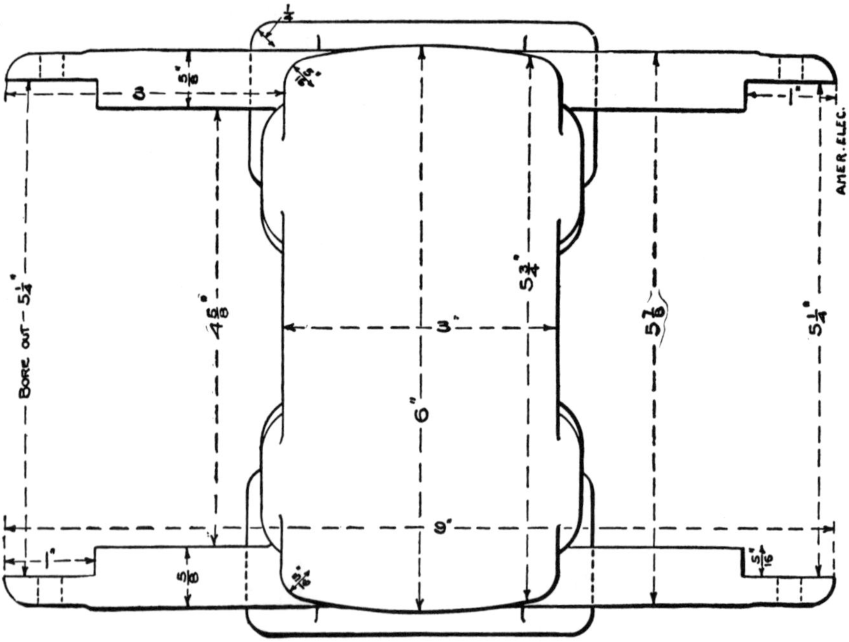

FIG. 136.—PLAN VIEW OF RECTIFIER FIELD MAGNET.

would otherwise be necessary in order to accommodate both the coils and the connection strips. The back collector ring is drilled with ⅛ in. holes at four equidistant points, two of these holes, the opposite ones, go through the ring and part way into the front ring, the other two holes are drilled only part way in. Some short pieces of copper wire, about No. 10, are soldered one into each hole, the two wires from the front ring being insulated from the other ring where they pass through it. The four wires are soldered one to each strip, and these carry the current across the armature and connect to the commutator segments.

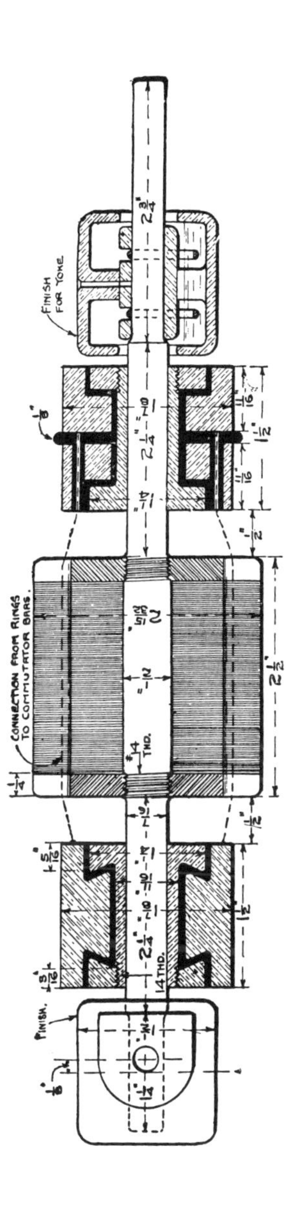

FIG. 137.—ARMATURE SHAFT AND AXIAL SECTION THROUGH ARMATURE CORE, COMMUTATORS, COLLECTOR RINGS AND ONE JOURNAL BOX.

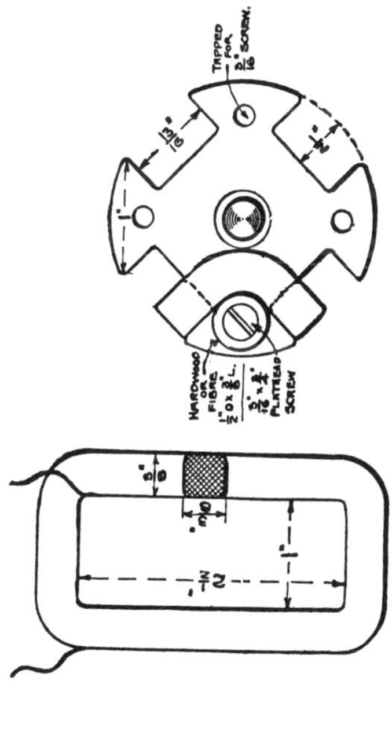

FIG. 138.—AN ARMATURE COIL AND METHOD OF MOUNTING SAME ON THE CORE.

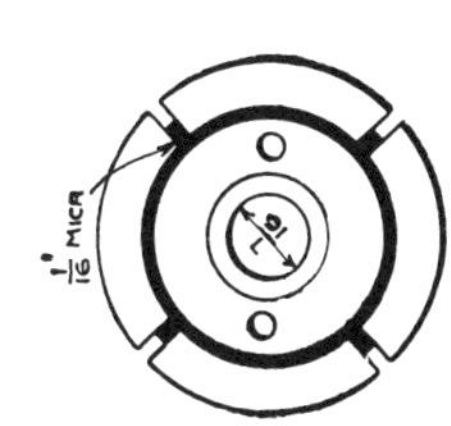

FIG. 139.—RECTIFYING COMMUTATOR.

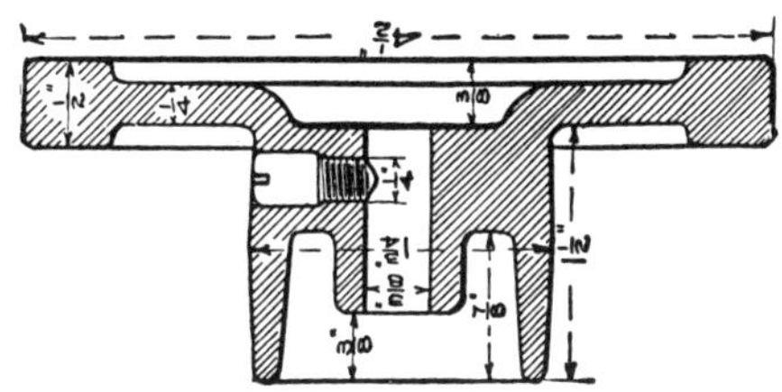

FIG. 140.—FLY-WHEEL PULLEY.

Diametrically opposite segments of the commutator are thus connected to the same collector ring, and neighboring segments have between them the whole potential difference of the alternating circuit. It is much better not to connect the copper strips permanently to the commutator until the builder has decided where he wishes to place the brushes; the commutator may then be twisted around on the shaft to the correct position and the connections made permanently. The brushes may be placed wherever they will be most convenient, the only restriction being that they must be 90° apart and must pass from one segment to the next at the same instant that an armature tooth is exactly under a pole.

Fig 138 shows an armature coil and the method of placing the coils upon the core. The coils are wound on a form, and, after being taped,

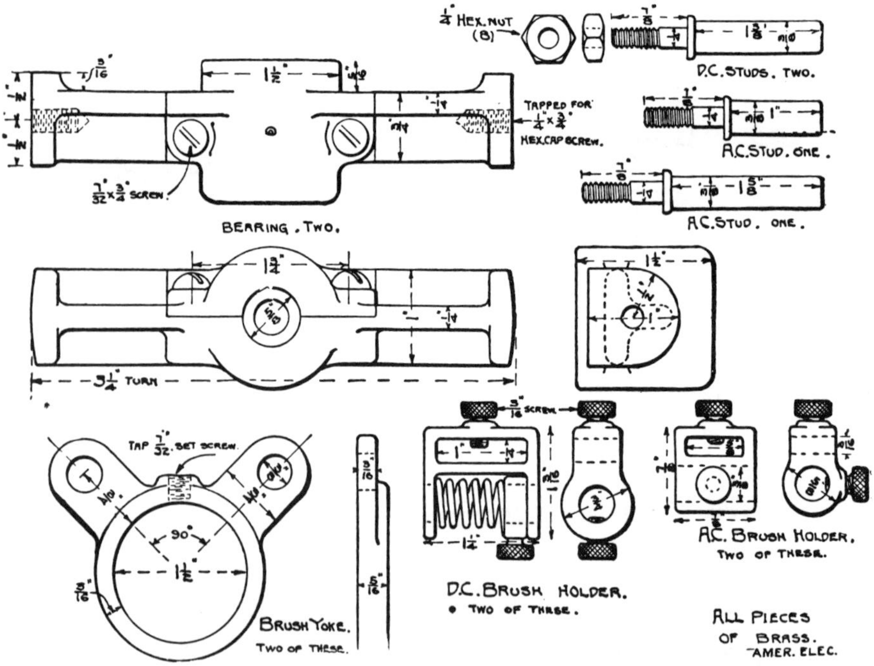

FIG. 141.—DETAILS OF JOURNAL YOKES, BRUSH COLLAR AND BRUSH-HOLDERS.

are slipped over the top of a tooth. The slack is then taken up by bending the ends down in a semi-circular shape and fastening them in this position by screws which carry small fibre or hardwood bushings.

In addition to this the armature should be banded at one or two points with No. 28 brass or German silver wire, small notches having been turned in the core to receive the bands and allow them to come flush with the surface of the core.

With an iron-clad armature like this, the clearance need not be more than from 1-64 in. to 1-32 in.; just how much it will be depends somewhat on the builder's skill; 1-64 in. clearance has been indicated on the drawings, and with care taken in adjustment it should not be difficult to obtain this figure. Fig. 139 is an end view of the rectifying commutator. Fig. 140 is a flywheel pulley which it will be found advisable to use in order to obtain smooth running.

Fig. 141 shows the bearing, brush yoke and brush holders. The bearings, while not so simple in construction as some other designs, have proven very satisfactory. The rings carry up a plentiful supply of oil, and what runs out at the end returns to the well and does not fly off outside the bearing and spatter the surroundings with grease spots.

In finishing the bearings the cap is first fitted and fastened by two machine screws. The bearing is then placed in the chuck and bored out to a diameter of 5/8 in. clear through. At the same time the ends are turned off 5¼ ins. in diameter to fit the arms on the field casting, and the outside of the boss is turned off 1½ in. diameter where it is to receive the brush yoke. The sleeve which forms the bearing proper is turned a tight fit for the central part of the box, then when the cap is screwed down it will be held firmly in place. The grooves for oil rings can be cut in the sleeve conveniently by mounting it eccentrically in the chuck and using a thin cut-off tool.

The brush yoke is shown with two arms 90 deg. apart. Another pair of arms and brushes might be added if it is desired to have more current carrying capacity.

The direct current brushes had better be larger than the alternating current brushes, and the holders should have spring tension, unless a very springy brush is used. Copper brushes are better for this purpose than carbon, as they make better contact and cause less sparking.

Figs. 142 and 143 show the field and armature coils respectively, with the forms upon which they are wound. The former is best made of hard wood, and consists of a block and two flanges, all held together by two wood screws and having a ½ in. hole through the center for placing on a mandrel in the lathe. The block should be made a trifle larger than the pole over which the finished coil is intended to go, and it should be given a slight taper of about 1-16 in., so that it can be readily slipped out of the finished coil.

Before beginning the winding a short piece of tape is laid in the long sides of the former, with the ends left sticking out. When the form is wound full these pieces of tape are tied tightly over the coil, and will hold it in shape while the former is taken apart and the coil is receiving its wrapping of tape. After being shellacked and dried, the coil is placed on the poles and hard wood wedges driven in between coil and pole, thus holding it securely in place.

The same form may be made to serve for both field and armature coils, if the field coils are wound first, and then the block reduced in thickness from ¾ in. to ⅜ in., the armature coils having the same inside dimensions as the field coils, but being only half as thick.

The fields are wound with No. 28 double cotton covered wire and the armature with No. 23. If the coils are wound to the specified dimensions they will have nearly enough the required number of turns.

Fig. 144 shows a diagram of connections and Fig. 145 some e. m. f. curves. For 100 volts the field and armature coils are connected four in series, as shown. For 50 volts the coils may be connected two in series and the twos in multiple. The armature terminals are tapped onto

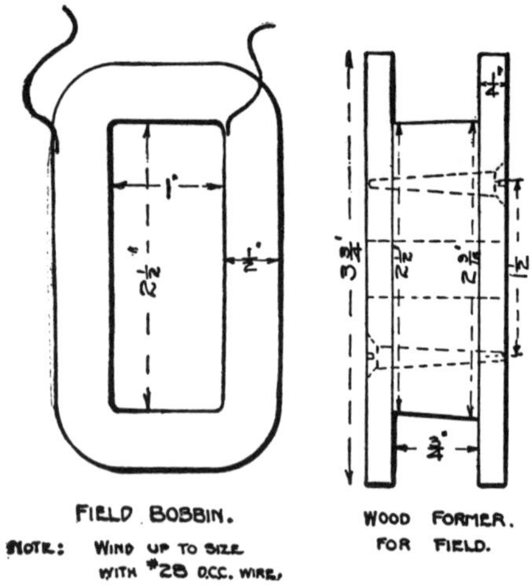

FIG. 142.—A FIELD COIL AND THE WINDING FRAME.

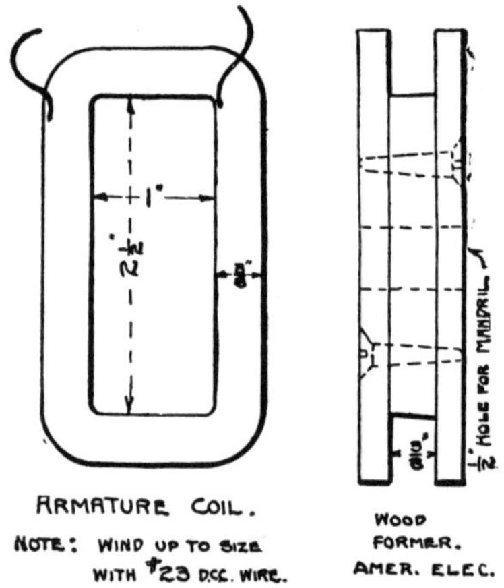

FIG. 143.—AN ARMATURE COIL AND THE WINDING FRAME.

the collector rings, or what amounts to the same thing, placed across any two successive commutator segments.

The connections on both armature and field should be such as to produce alternate north and south polarity all the way round. If all the coils have been wound in the same direction and placed on the poles the same way, connect beginning to beginning and ending to ending, and the polarity will be right. The machine will run at 1,800 r. p. m. on a 60-cycle circuit, and on a 125-cycle circuit it will have to make 3,750 r. p. m. This it can easily do if the armature is well balanced as it should be.

Since the strength of the field has a considerable effect on the behavior of a synchronous motor, it is best to have an adjustable resistance in the field circuit of this machine, so that the field can be adjusted until the minimum armature current is obtained.

Referring now to the curves in Fig. 145, it is clear that if the rectifier is running in synchronism and the angular position of the brushes is correct, the brushes will pass from segment to segment at the instant when the e. m. f. curve reaches its zero value at the points, *a*, *a*, *a*, etc. As the brushes in passing from segment to segment overlap two segments for a brief interval, they form a dead short circuit on the alternating current mains during the interval. This will not, however, result in any damage if the e. m. f. becomes zero at the same instance. If, however, the brushes had been incorrectly placed and commutation occurred at the points *b*, *b*, *b*, etc., an e. m. f. of value equal to the ordinate at *b* would be short circuited four times in a revolution, and serious sparking would result.

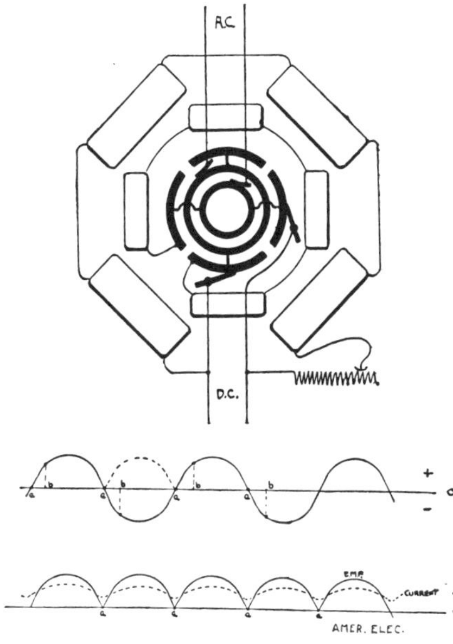

FIG. 144.—RECTIFIER CONNECTIONS.
FIG. 145.—SOME E. M. F. CURVES.

This state of affairs is easily remedied by shifting the brushes, which corresponds to changing the angular position of the point of commutation until a position such as *a a* is reached, when all sparking will disappear. If the armature falls out of step, or if it is thrown into circuit before complete synchronism is reached, a short circuit travels over every portion of the e. m. f. wave, at a slow rate equal to the difference between synchronous speed and the actual speed at that instant, the result being a magnificent display of fireworks and probably a fuse blown.

To obviate this latter difficulty a resistance or choking coil should be placed in series with the alternating current end at the moment of starting, and cut out when it is seen that the machine has settled down to steady running. Such a resistance will not have any appreciable effect on the small current drawn by the armature and field windings, but should be of such a value as to limit the current to about 10 amperes, should a short circuit occur.

The ordinary method of a synchronizing lamp is not easily applicable here on account of the small size of the machine; and, moreover, a little practice will enable the operator to judge by ear the proper instant for closing the circuit.

Thus by making slight changes in the connections, as already pointed out, this machine may be used as a rectifier on single-phase circuits of 50 or 100 volts and 60 or 125 cycles. The amount of rectified current which may be drawn is not limited in any way by the horse-power capacity, but will generally be limited only by the capacity of the transformer which is supplying the current. Thus from 50 to 100 amperes may be drawn, depending somewhat on the nature of the load into which the rectifier is feeding current.

The machine may also be used as a self-exciting synchronous motor,

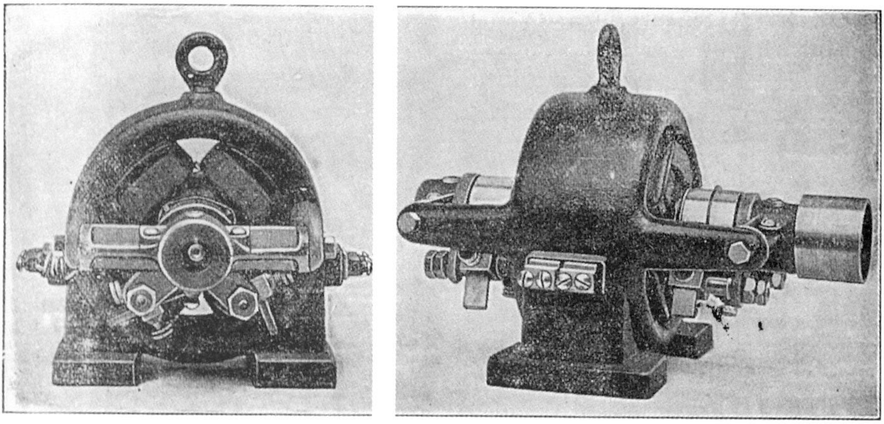

FIG. 146.—END AND SIDE VIEWS OF THE COMPLETE MACHINE.

developing from 1-10 to ⅛ horse-power, according to the strength of field; and finally it may be driven by belt as a self-exciting alternator, supplying either an alternating current or a rectified direct current, or both, up to about 100 watts output.

Fig. 146 was reproduced from photographs illustrating a rectifier built from these designs.

CHAPTER XIV.

UNIVERSAL ALTERNATOR FOR LABORATORY PURPOSES.

The design of the machine illustrated in the accompanying engravings was adopted for the following reasons:

1. Simplicity of construction by students in the engineering shops, without special tools or dies; 2. Its similarity to a bi-polar dynamo so as to illustrate one, two or three-phase-current generation, but without a

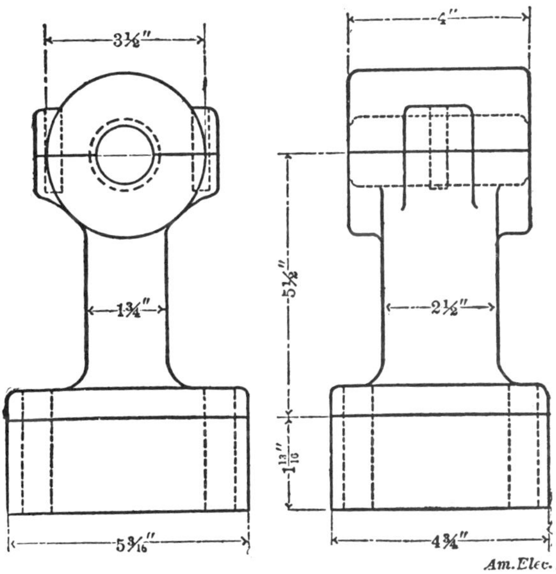

FIG. 147.—JOURNAL PEDESTALS AND BOXES.

low limit to the frequency; 3. To illustrate practically the effect of combining e. m. fs. differing in phase, in a variety of ways.

To accomplish these objects, both the field and the armature were made with poles, the latter having two more than the former. The field revolves and is of the C. E. L. Brown type. The armature is made of sheet-iron rings held together by bolts between cast-iron plates. The spaces between the poles of the armature were milled out after the rings had been bolted together. The field is made of two identically similar steel castings, each with five poles symmetrically spaced and pointing in the same direction parallel to the axis. The field has thus ten poles, alternating in sign, and the armature twelve. The following are some of the dimensions:

Diameter of armature pole-faces, 10¼ ins.; length of faces parallel to shaft, 4 ins.; width of pole-faces, 1½ ins.; pitch of poles on armature, 2.68 ins.; depth of poles, ½ in.; net cross-section of pole in square inches, 5.4. The armature poles have forty turns of No. 16 wire each.

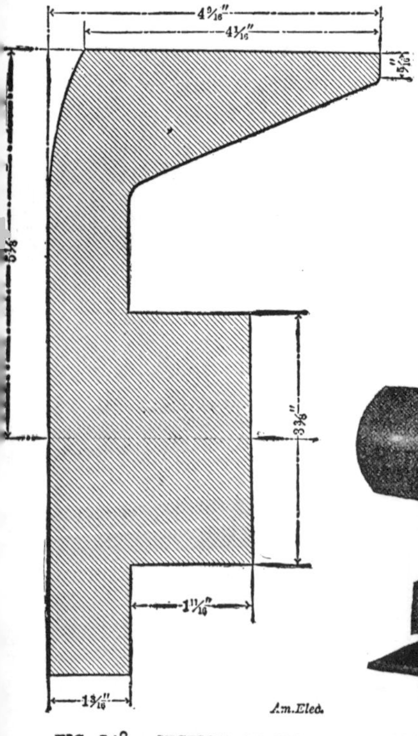

FIG. 148.—SECTION OF ONE MAGNET POLE.

FIG. 151.—COMPLETE MACHINE WITHOUT BASE.

The double air gap is ¼ in. The field coil contains 1012 turns of No. 16 double-cotton-covered wire.

The armature coils were wound in reverse order from pole to pole in the usual way. They are connected in pairs, and the terminals of each pair are brought up to binding posts on a board bolted to the top of the machine.

The machine will give 1,500 watts when connected in three-phase zig-zag mesh fashion and driven at 1,650 r. p. m.

The completed machine is shown in Fig. 151, although the engraver has left off the base-plate on which the armature and pedestals are mounted.

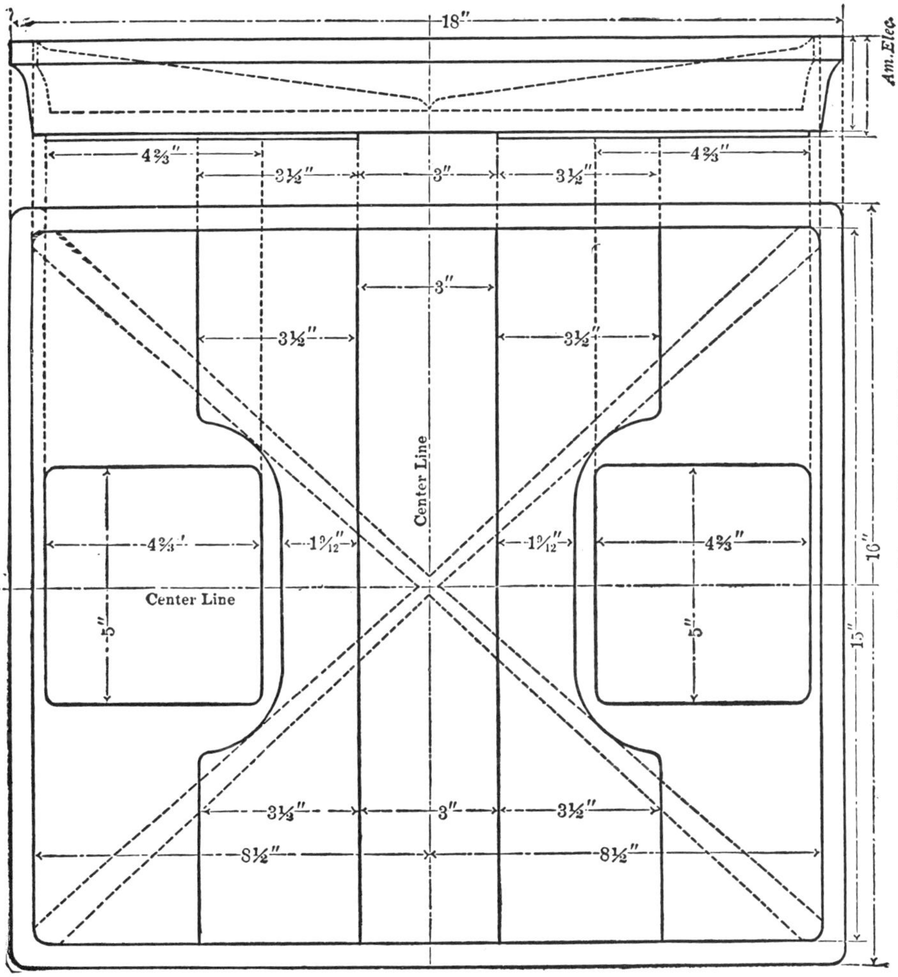

FIG. 149.—THE BASE PLATE.

An inspection of the diagram (Fig. 152) will show that diametrically opposite poles of the field are of opposite sign, while the corresponding coils of the armature are similarly wound. Hence with a closed-coil armature the e. m. fs. balance exactly as with a bi-polar dynamo. If, therefore, connection be made with the armature at two opposite points, the current in the external connecting circuit will be alternating. Further, two such circuits connected at right angles will convey currents in quadrature. By connecting at points 120 degs. apart, three-phase currents will be obtained.

Again, the coils may be joined either in mesh or star fashion by means of the binding posts at the top, and we may zig-zag across either with two or three-phase connections, so as to connect opposite coils by

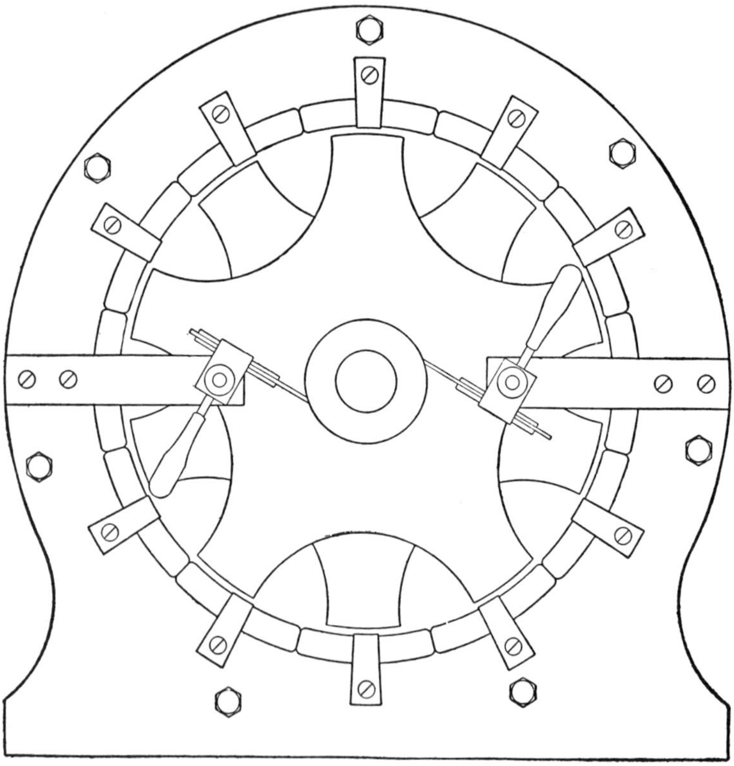

FIG. 152.—SIDE ELEVATION OF THE MACHINE.

twos or threes with no-phase difference between the two opposite groups. This connection, of course, gives the highest e. m. f.

It is evident that the phase difference from coil to coil is 30 degs. or one-twelfth of a period. Hence the voltage for a given magnetic flux cut per second, calculated in the usual way, must be first divided by $\sqrt{2}$ to reduce from maximum to virtual volts, and then the equal e. m. fs. generated by the several coils must be added geometrically with a phase difference of 30 degs. from coil to coil. Since the e. m. fs. of the several coils are equal and differ in phase by one-twelfth of a period, the series may be represented by a regular polygon of twelve sides (Fig. 150). Hence, if E be the e. m. f. of one coil, the following will be the e. m. fs. of the several groups of coils:

AC, E. M. F. of two coils, $2E \cos 15° = 1.93\,E$.
AD, " " three " $E + 2E \cos 30° = 2.73\,E$.
AE, " " four " $2E(\cos 15° + \cos 45°) = 3.346\,E$.
AF, " " five " $E + 2E(\cos 63° + \cos 30°) = 3.73\,E$.
AG, " " six " $4E \cos 15° = 3.86\,E$.

The phase difference between the coils reduces the e. m. f. of the six in series on either side to $3.86 \div 6$ or 0.643 of what it would be were there no such phase difference.

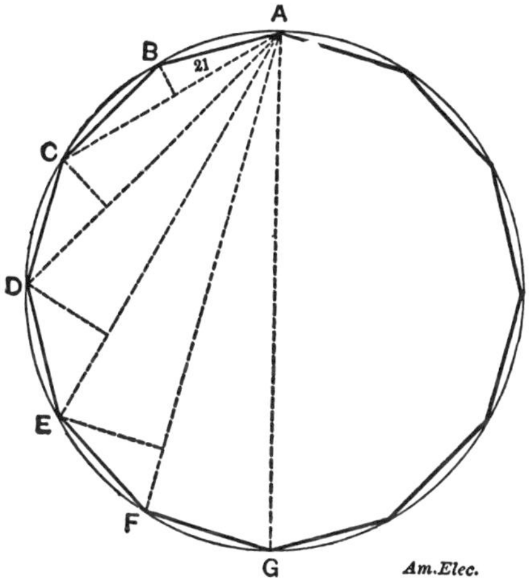

FIG. 150.—DIAGRAM OF E. M. FS.

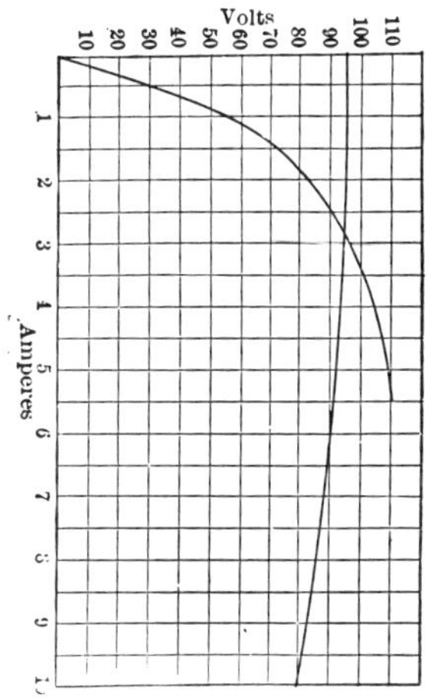

FIG. 153.—MAGNETIC CURVE AND EXTERNAL CHARACTERISTICS.

It will be seen from the subjoined table that the observed e. m. fs. agree very closely with those computed from the foregoing equations.

	Observed.	Computed.
One coil	21	20.5
Two coils	40.3	39.6
Three coils	55.5	56
Four coils	69	68.6
Five coils	75.5	76.5
Six coils	77.5	79.1

With about 2000 ampere-turns on the field coil, the following observed voltages were obtained for the several connections described in the first column. The computed values are readily obtained from the preceding expressions.

Connection.	Observed.	Computed.
Two-phase mesh	79	79
" " star	112	112
Three-phase mesh	69	68.6
" " star	121	118.8
" " zig-zag	136	136.8

The alternator was driven by a motor on a power circuit and the voltage varied a good deal. Some of the irregularities of voltage in the generator are accounted for by the variation in speed of the motor.

Fig. 153 shows the curve of magnetization and the characteristic with 3 amperes in the field. The armature was connected as a closed coil, and only a single alternating current was drawn from it. The total drop for full load is 11 volts; of these, about 4.5 volts are due to drop in the armature, and the rest must be set down to self-induction.

CHAPTER XV.

ONE-QUARTER HORSE-POWER SINGLE-PHASE INDUCTION MOTOR.

The induction motor described in this article was designed to be built by amateurs, and the aim has been to make it simple and easy to construct. It is designed for a single-phase alternating circuit of 104 volts and a frequency of 60 cycles per second. It has four poles, and, therefore, its synchronous speed would be 1,800 revolutions per minute. The actual speed of the motor at load will be about 10 per cent. less than the synchronous speed.

The primary or stator has a plain ring winding, and the secondary or rotor a so-called squirrel-cage winding, consisting of bare copper conductors, embedded without insulation in an iron core, all conductors being connected at the ends. The bearing supports are provided with an oil chamber, and either a ring or felt self-oiler may be used.

In making the calculations for the motor we will follow the method much used in transformer calculation, which consists in assuming the various losses, and from these losses determine the dimensions of the parts in which they occur. We will set down the efficiency of our machine at 60 per cent. and the power factor at 75, figures which obtain in machines of this size on the market. The output being ¼ horse-power or 186 watts, the intake will be $\frac{186}{.6} = 310$ watts, and the total losses are, therefore, 124 watts. We will make a preliminary division of this loss as follows:

Primary C^2R loss, 25 watts.
Secondary C^2R loss, 15 watts.
Hysteresis, 40 watts.
Friction and eddy currents, 34 watts.

The primary current at full load will be $\frac{310}{104 \times .75} = 4$ amperes. From the C^2R loss and the current strength we may now find the resistance of the primary circuit, $R = \frac{25}{4^2} = 1.56$ ohms.

Allowing a current density in the primary winding of 2,000 amperes per sq. in., we find that the wire to be used is No. 16 B & S. The length of this wire, which will have a resistance, when warm, of 1.56 ohms, is 354 ft.

The ohmic component of the e.m.f. in the primary circuit is $1.56 \times 4 = 6.24$ volts, and the induction or counter e. m. f., therefore $\sqrt{104^2 + 6.24^2 - 2 \times 104 \times 6.24 \times .75} = 99.42$ volts.

In a single-phase induction motor the strength of the rotating field is not constant, but fluctuates. We will base our calculations, however, on an equivalent rotating field of constant strength.

Fig. 154 shows a diagram of the magnetic field in the motor. Let Φ be the flux of magnetism that passes through the teeth of the stator, between A and B. Then, as the magnetic field rotates with a velocity of

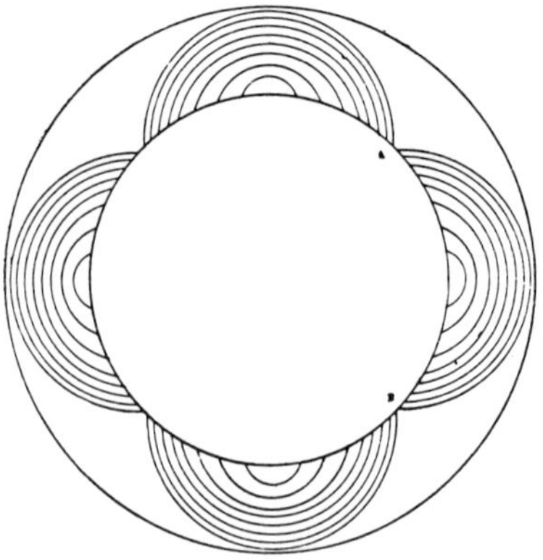

FIG. 154.—MAGNETIC CIRCUIT.

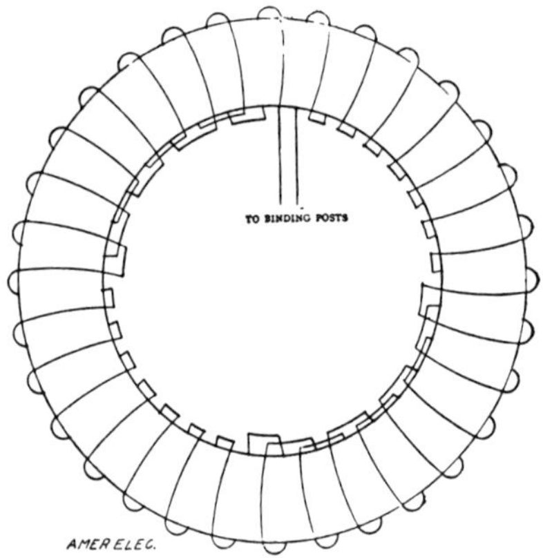

FIG. 155.—DIAGRAM OF STATOR WINDING.

1,800 r.p.m., and the conductors are stationary, each conductor cuts this number of lines 120 times per second. The mean value of the e.m.f. induced in one conductor will be $\frac{\Phi \times 120}{10^8}$, and the square root of the mean square, $\frac{\Phi \times 120 \times 1.1}{10^8}$. But this value, multiplied by the number of conductors, is the maximum e.m.f. induced in the primary winding. Representing by n the number of cycles per second (60), we may write:

$$E\sqrt{2} = \frac{2.2\, n\, \Phi\, C}{10^8} \quad . \quad . \quad . \quad . \quad (1)$$

We will allow a maximum magnetic density in the stator core of 25,000 lines per sq. in., and make the axial width of the core double its radial depth. The maximum magnetic flux at any cross section of the core is $\frac{\Phi}{2}$. The area of the cross section will, therefore, be $\frac{\Phi}{50000}$. The periphery of this area is $6\sqrt{\frac{\Phi}{100000}}$ and taking the length of turn 10 per cent. greater than the periphery of the core, we have for it the value, $L = 6.6\sqrt{\frac{\Phi}{100000}}$.

The number of conductors is equal to the total length of wire divided by the length of one turn:

$$C = \frac{L}{6.6\sqrt{\frac{\Phi}{100000}}}$$

By substituting this value of C in equation (1), transforming and reducing, we obtain.

$$\Phi = \frac{1.8\,E^2.\,10^{12}}{n^2\,L^2}$$

We may now substitute as follows: $E = 99.4$, $n = 60$, $L = 4{,}248$ (inches). This gives $\Phi = 275{,}000$.

The magnetic cross section of the field core is $\frac{\Phi}{50000} = 5.5$ sq. in., and the length of one turn, $6.6\sqrt{\frac{\Phi}{100000}} = 11$ ins.

The number of conductors is $\frac{4248}{11} = 386$. As we have four poles, the number of conductors should be divisible by four, and we will, therefore, take 384 conductors.

These conductors may be distributed in 32 slots, giving 12 conductors per slot. We will wind these conductors 2 wide and 6 deep. The diameter of No. 16 double cotton-covered magnet wire is 61 mils. For insulation we allow 30 mils on each side of the slot and 30 mils at the bottom; also a clearance space of 30 mils below the surface of the teeth. This gives for the dimensions of the slots a depth of 426 mils and a width of 182 mils.

The cross sectional area of the core was 5.5 sq. ins., and the ratio of axial length to radial depth, 100. We will make the core 3½ ins. long by 1 9-16 ins. deep, which approximately satisfies the above two conditions.

The mean magnetic density in the teeth may be taken at 32,000 lines per sq. in. There are in all 550,000 lines, each passing through the teeth twice, which is equivalent to 1,100,000 passing once. The magnetic cross section of the teeth should, therefore, be $\frac{1,100,000}{32,000} = 34.4$ sq. ins. Dividing this area by 3.5 ins. the length of the teeth, we obtain 9.83 ins., the circumferential space taken up by the teeth. The circumferential space taken up by the slots is $32 + .182 = 5.82$ ins. The total inner circumference of the stator is, therefore, 15.65 ins., and the diameter practically 5 ins.

Some of the lines of force that pass through the primary circuit do not enter the secondary circuit, but leak around it. The leakage co-efficient is about 1.2. The total lines of force crossing the gap number, therefore, $\frac{1,100,000}{1.2} = 916,600$. By dividing this number by the surface of the gap we obtain the mean magnetic density in it, $\frac{916,600}{55} = 16,600$.

The output of our motor is 186 watts, and the allowance for friction 30 watts. The total energy transformed from the electrical to the mechanical state is, therefore, 216 watts. The C^2R loss in the secondary is 15 watts. It is a well-known fact that the ratio of the motor speed to the speed of synchronism is the same as the ratio between the energy transformed in the secondary to the energy absorbed by it. The speed of our motor at load will be, therefore, $\frac{216}{231}\ 1800 = 1680$ r. p. m. When a rotor of 5 ins. diameter, running at 1,680 r. p. m. develops mechanical energy at the rate of 216 watts, the tangential force on its circumference is 4.34 lbs. Now a conductor a inches in length, carrying an alternating current whose $\sqrt{\text{mean square}}$ value is I in a sinusoidal field of mean intensity, B, has a mean force of $\frac{I\ a\ B}{10,180,000}$ lbs. exerted on it. If their be C conductors, the force is $\frac{C\ I\ a\ B.}{10,180,000}$ lbs. We may substitute the values of a and B and equate this expression to our circumferential force, $\frac{C\ I \times 3.5 \times 16,600}{10,180,000} = 4.34$,

$$\text{Hence, } \frac{4.34 \times 10,180,000}{3.5 \times 16,600} = C\ I = 760.$$

The number of secondary conductors, multiplied by the amperes per conductor is, therefore, 760. We will put 15 conductors on the secondary, which will give a current of a little over 50 amps. per conductor. The loss in each conductor is one watt, and the resistance per conductor, therefore, $\frac{I}{2500}$ ohm. This includes, of course, the resistance of the soldered joints and the return on the ends, which cannot be exactly calculated, but which may be taken at one-half of the total, so that the actual resistance of the conductor is only $\frac{I}{5000}$ ohm. The size of wire of which a length of 3½ ins. has a resistance of $\frac{I}{5000}$ ohm is No. 8 B. & S., which we will use for the cage winding of the motor. The maximum density of our equivalent rotating field of constant strength was 25,000 lines per square inch. We have to make a small allowance for the insulation of the discs, and also take into account the fact that our actual rotating field is fluctuating, which subjects parts of the core to a considerably higher magnetic density. At 35,000 lines per square inch the hysteresis loss per cubic foot at 60 cycles is 360 watts. There are 129 cubic inches of iron in the stator core below the teeth, and the hysteresis loss would, therefore, be 28 watts. In the teeth the magnetic density will reach 60,000 lines per square inch. The hysteresis loss at this density and frequency is 840 watts per cu. ft. There are 17 cu. ins. of iron in the teeth, and the hysteresis loss is, therefore, 8.5 watts. In the rotor the hysteresis loss is practically nil, as the frequency of the reversal of magnetism is proportional to the slip of the rotor, which at full load is only one-fifteenth the frequency of reversal of the magnetism in the stator core. Our total hysteresis loss is, therefore, 36.5 watts, which is well within the limit of our allowance for it.

CONSTRUCTION OF MOTOR.

The construction of the motor will now be explained, reference being had to the accompanying drawings. As the motor is symmetrical in both vertical planes, the drawings of the stator show it part in full view and part in section.

The stator is built up of discs stamped from No. 27 transformer iron. These discs have an internal diameter of 5 ins. and an external diameter of 9 ins. They should be varnished on one side with an insulating varnish, thinned down so as to form a thin, uniform coating on the surface of the discs. The discs must be dried before being assembled. Two disc-shaped brass castings of the same internal and external diameter as the discs and a thickness of 3-16 ins. serve as end plates. These castings have four lugs on the outer edge, strengthened by ribs. Through these lugs the clamping rods pass.

A round piece of sheet iron, 10 ins. in diameter, should be procured, from which to make a templet. Find the center of the sheet iron and lay out a circle of 4⅝ ins. radius. Divide the circumference into four equal parts and centermark the division points. Drill the center and the division joints on the circle with a small drill. (About No. 40.) This templet is clamped on the brass castings so that the holes on the circle come about over the center of the lugs. With the same small drill used before a hole should now be drilled through the brass casting. These holes are then enlarged with a ¼-in. drill. The clamping rods are ¼-in., either Bessemer or cold-rolled steel rods, cut off to 6¼-ins. With a ¼-in. standard die a thread is cut on each end of the rod to a distance of 1¼ ins.

The disc may now be assembled. The clamping rods are not strong enough to properly compress the discs, and this should be done under a drill press or in a vice while the nuts are tightened up. Enough discs must be put on to make the length 3½ ins. when tightened up. After

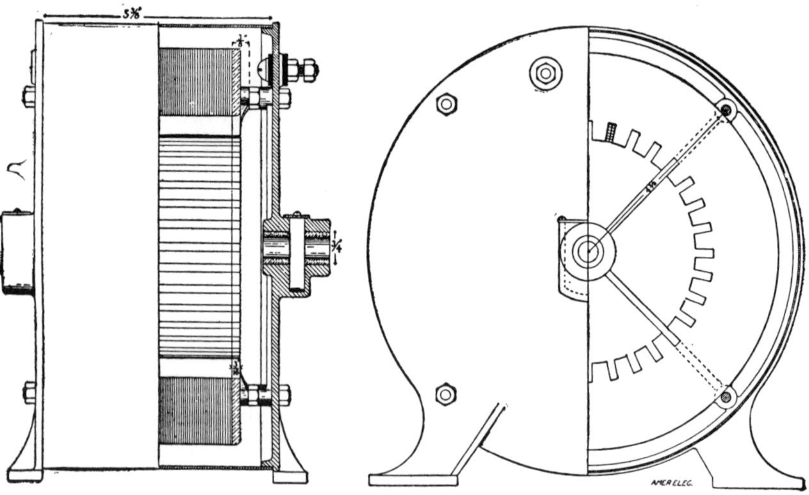

FIGS. 156 AND 157.—END AND SIDE ELEVATION, HALF IN SECTION.

the core has been put together it should be chucked in a lathe and a light cut taken out of it to make the inner diameter 5 1-64 ins. Care must be taken not to make the bore too large, as this would much reduce the efficiency and capacity of the motor.

The rotor is built up of discs of the same material as the stator, ⅝ in. internal diameter and 5 ins. external diameter. Two disc-shaped brass castings serve as end plates. The hole in the center of these castings should be finished to ⅝ ins.

The shaft is turned up from a piece of cold-rolled steel of ¾-in.

diameter. The middle part of the shaft is turned to such a diameter that the discs fit over it and the ends so as to be a running fit in a ½-in. hole. A ⅝-in. thread is cut on each end of the middle part of the shaft.

The discs for the rotor need not be insulated but can be built right up on the shaft and clamped by the two hexagon nuts shown in the drawing. A hole is drilled into the shaft through the end plates, as shown, and a steel rod or round spike is driven into the hole and sawed off. The rotor is now put into a milling machine, and with a ⅛-in.

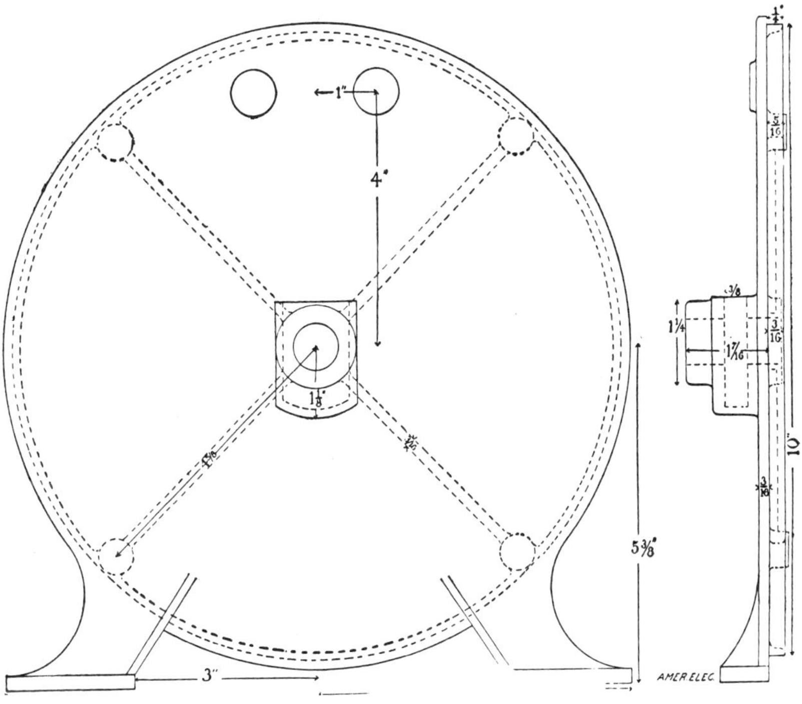

FIG. 158.—FACE AND EDGE VIEWS OF AN END DISC.

milling cutter, 15 slots are cut 3-16 in. deep. The No. 8 copper wires have to be driven into these slots. They are soldered at the ends to the brass end plates, the solder being applied liberally and made to fill up all around the wire. The rotor should now be put in a lathe and turned down to a diameter of 4 63-64 ins. Care must also be exercised here to avoid taking too large a cut. While the rotor is in the lathe, the nuts on the shaft are turned to ⅜ in. in length.

We now take the bearing supports, and by means of our templet drill the holes for the clamping rods. In fastening the templet to the castings a center should be made to coincide. In one of the castings two ⅜-in. holes for the binding screws are drilled through the center of the bosses provided for this purpose. A piece of soft wood is cut that will fit into the oil chamber. A ½-in. hole is drilled through this piece of wood, through which the shaft may pass.

The rotor is now wrapped with paper until it fits tightly in the stator. It is put into place and the bearing supports are slipped on with the wood in place in the oil chambers. The bearing supports are fastened down by means of nuts on the clamping rods. Some babbitt metal should now be melted in a ladle, the motor set on end and the outer parts of the bearings filled with babbitt. The bearing supports should now be marked so that they can be put on the same way again after they have been taken off. Take off the bearing supports, reverse the shaft through them and fill up the remaining end with babbitt. The ends of the babbitt lining should now be trimmed up, the bearings put back on the core and a ½-in. reamer run through them.

We now take the stator core and put it in a shaper to cut the slots. The slots are 182 mils wide and 426 mils deep, and there are 32 of them

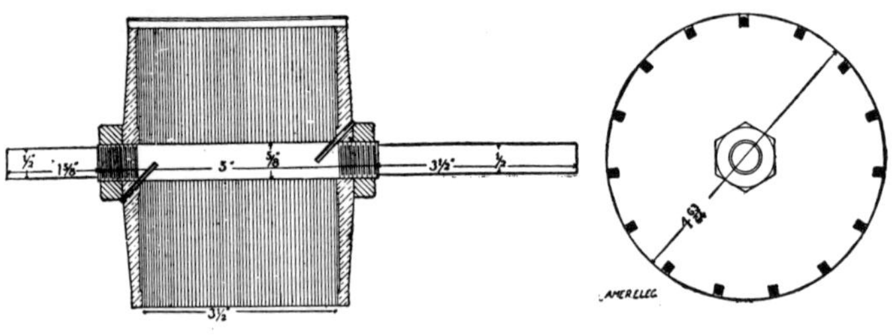

FIG. 159.—DETAILS OF ROTOR.

equally spaced around the inner circumference of the stator core. After the slots have been cut, all the sharp edges are rounded off with a file and the core is cleaned of all iron dust and grease. The insulation is next put on. Shellac dissolved in alcohol is the best insulating substance to stick the insulating material to the core. Two thicknesses of pressboard, 15 mils thick should be used in the slots, the troughs being made about ⅛ in. longer than the core. The ends and outside of the core are insulated in the same manner as ordinary direct-current armatures, and it will not be necessary to describe this specially. Any one not familiar with the method of insulating armature cores may refer to the descriptions of small motors in Chapters I to IX.

The wire is wound in the slots two wide and six deep, while on the outside of the core it is wound only two deep. The coils are connected as shown in the diagram. In the first eight coils the ending of one coil is connected to the beginning of the next. The ending of the eighth coil is connected to the ending of the ninth, the beginning of the ninth to the end of the tenth and so on until the sixteenth. The beginning of the sixteenth is connected to the beginning of the seventeenth, the end of the seventeenth to the beginning of the eighteenth, and so on to the twenty-fourth, where the connections are changed again. The beginning and ending of the whole winding are brought out to the binding posts. These consist of 1/4-in. round-head machine screws, 1 1/4 inches long, passing through the bearing support, being insulated from it by fiber washers and bushings. A tube of sheet iron is made, 10 ins. diameter and 5 ins. long which will just fit over the circular offset on the bearing supports and serve to protect the windings of the stator. This completes every part of our motor, and it may now be assembled.

The motor is not self-starting, and has to be brought up to speed by some external means. If the bearings are well aligned, as they should be, a vigorous start by hand on the pulley will be sufficient to make the motor pick up. The motor should be started immediately the current is turned on, and for this reason a switch should be placed convenient to the motor.

CHAPTER XVI.

SIMPLE TRANSFORMER IN FOUR SIZES.

The transformers here described can be built by any amateur without the use of machine tools; some form of winding machine being the most important piece of constructive apparatus. In order to eliminate the bugbear of stampings, the core is made of an unusual type, involving the use of simple rectangular strips of transformer sheet iron, No. 27

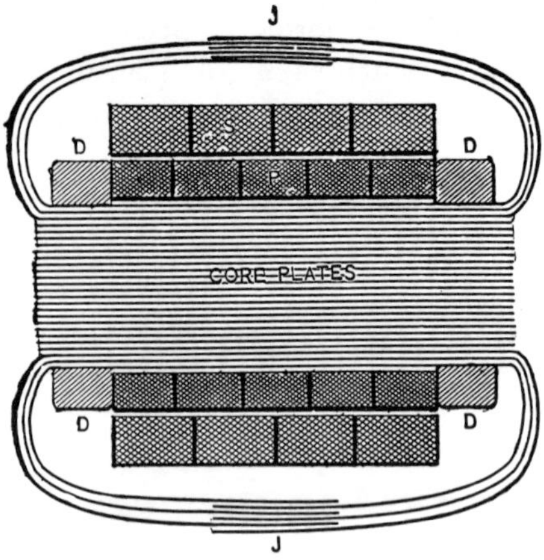

FIG. 160.—SECTIONAL PLAN.

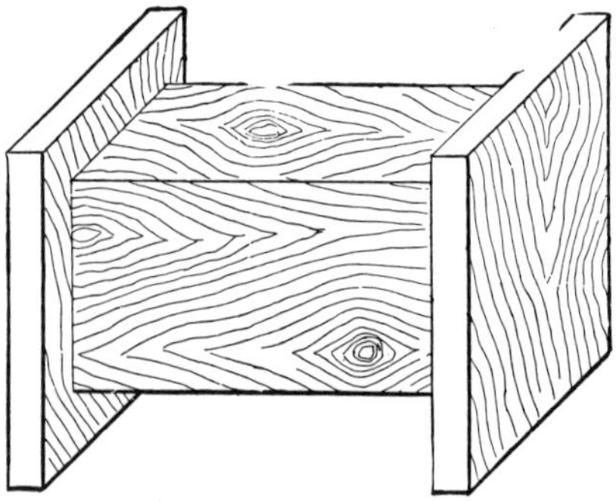

FIG. 161.—WINDING CORE BLOCK.

gauge, all of one size. The coils surround the core exactly like the windings of an induction coil, and the core sheets are bent back around the outside of the coils and lapped, as indicated in Fig. 160, which is a horizontal section through the center of the transformer, with most of the core sheets omitted beyond the ends of the winding. All four sizes for which data are given are of identical construction, the only difference being dimensional.

The first step is to make two wooden core blocks, like Fig. 161; one on which to mount the primary bobbin and one for the secondary bobbin.

One head may be put on permanently, but the other must be removable. The dimensions are given in Table I. A spindle of ¾-in. round iron should be put through the center of the core lengthwise, and if a lathe is not available for winding purposes, one end of the spindle may be bent into a crank and the whole structure mounted between two simple upright posts, to form a winding machine.

The next step is to make the retaining bobbin for the primary winding. Take a sheet of heavy fuller board, of the size specified in the table of dimensions, and cut four slits in each long edge, as shown in Fig. 162 at *a* and *b*; these slits are ¼ in. long in all cases. Bend the sheet into the shape shown at Fig. 163, forming a sort of box with open ends, and slip

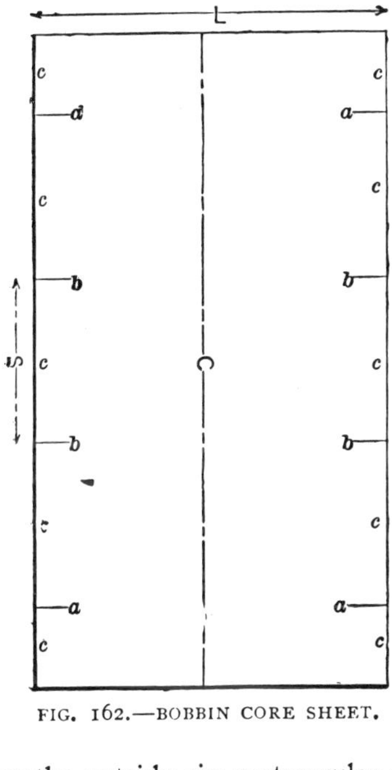

FIG. 162.—BOBBIN CORE SHEET.

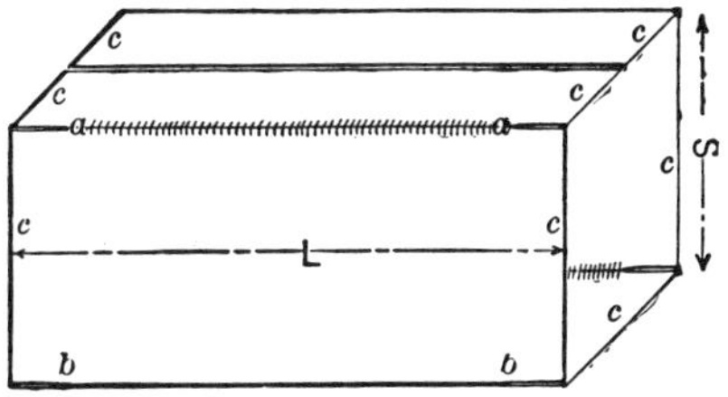

FIG. 163.—BOBBIN CORE.

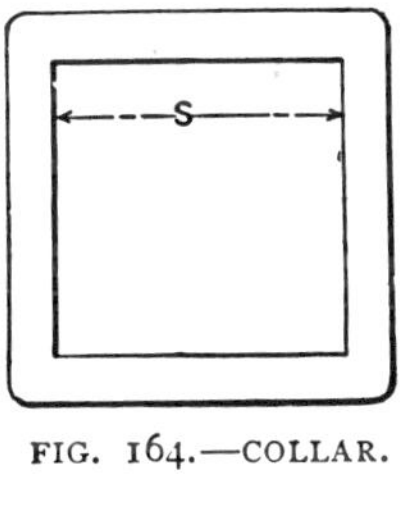

FIG. 164.—COLLAR.

over the outside six rectangular collars of vulcanized fibre, like Fig. 164; these are 1-16 in. thick. Bend the flaps, *c*, Fig. 163, outwardly at right angles with the wall of the "box," so that when the collars are finally adjusted into place along the outside, each end one will be held on by four flaps, as indicated in Fig. 165. The surfaces of the flaps may be coated with thick shellac varnish in order to keep them in place against the faces of the end collars.

Next, mount the complete bobbin, Fig. 165, on its wooden core block, Fig. 161, and prepare it for the primary winding. The partitions or collars must be adjusted at equal distances apart, as in Fig. 165, and the spacing maintained temporarily by means of wooden blocks. Before applying the winding, the seam where the edges of the fuller board meet must be covered by a strip of the same material laid clear across the side of the box in each of the compartments and secured in place by varnish. Then the coils may be wound on, care being taken to observe rigidly the prescription of Table II as to number of turns per section. The starting and finishing ends of each coil or section must be of heavier wire than that of the coil itself, and rubber-covered with an outer braid; No. 18 wire is a good size for the two smaller sizes of transformer, and No. 14 for the two larger ones.

After winding and securing the ends with heavy linen thread, tag the ends, marking the inner or starting ends, "B," and the outer or final ends, "F." Then tape each section thoroughly and lead the terminals of the various sections lengthwise along the outside of the whole winding to one

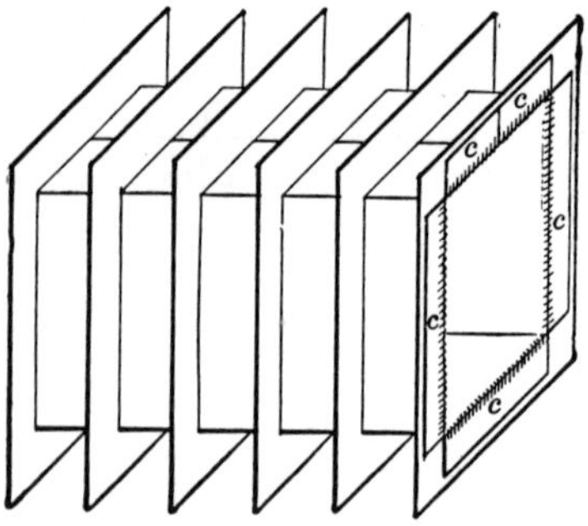

FIG. 165.—COMPLETE BOBBIN.

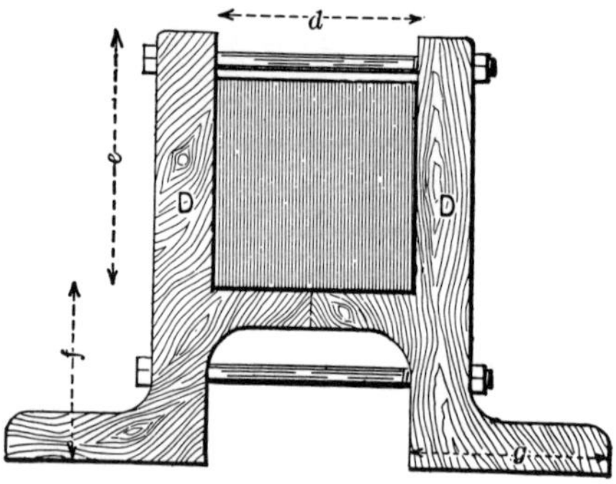

FIG. 166.—CORE-CLAMPING DOGS.

end, securing these terminal wires to the surface of the structure by means of a few extra turns of tape. All of the terminals should project from one end of the complete winding, and they should be laid side by side in regular order.

The secondary bobbin or box is made in exactly the same way as the primary, but has only five collars instead of six, and is larger in size, as Table I shows. After winding the secondary, tape it on the outside and tag the ends, like the primary; varnish both heavily with either P. & B. or shellac varnish, and set them aside to dry.

Next mount between two pairs of wooden dogs the requisite number of core plates to make the proper thickness, as indicated in Figs. 166 and 167, drawing the dogs up snug, and wrap the core plates tightly with three layers of plain linen tape along the portion between the dogs.

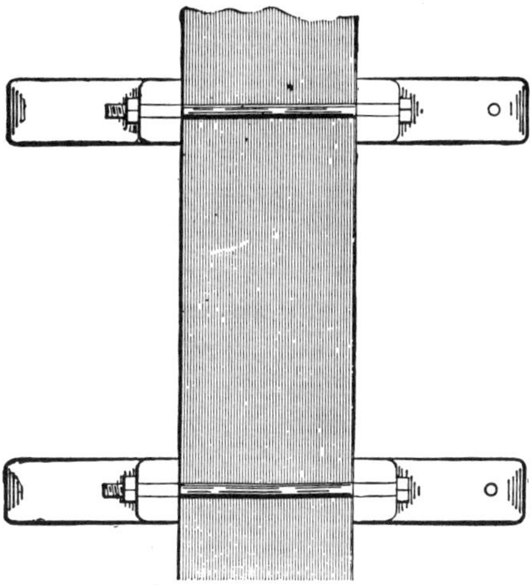

FIG. 167.—CORE-CLAMPING DOGS.

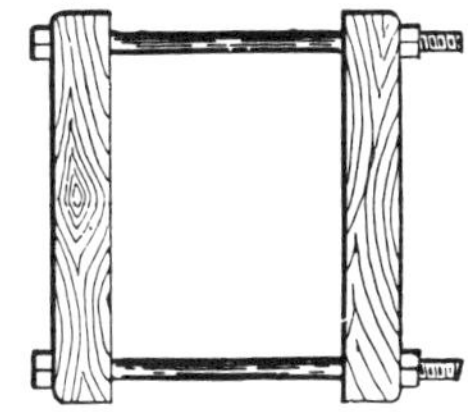

FIG. 168.—JOINT CLAMP.

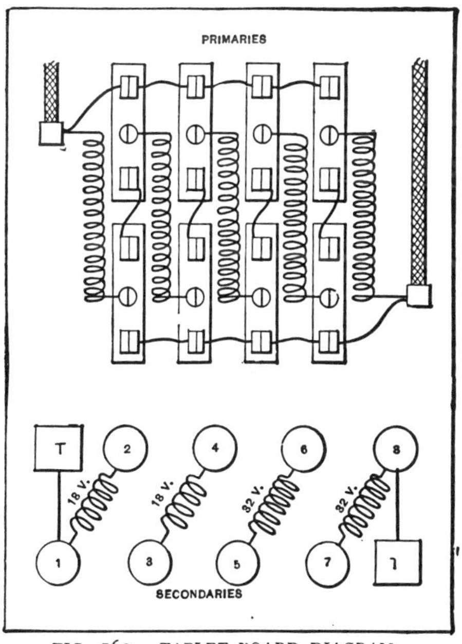

FIG. 169.—TABLET BOARD DIAGRAM.

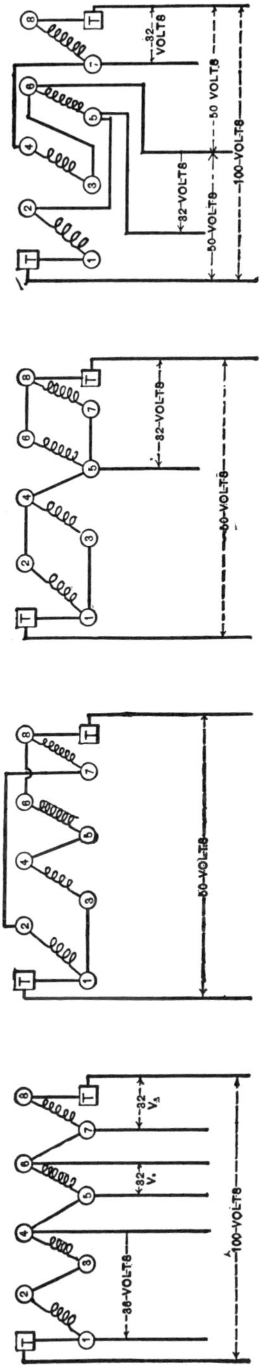

FIG. 170.—FOUR OF THE AVAILABLE COMBINATIONS OF SECONDARY WINDINGS.

Varnish the outer layer of tape lightly; remove the pair of dogs, and slip on the primary and secondary windings separately. The windings should be so disposed that the sides out of which the terminals lead are on top. When the windings are in place, put back the pair of dogs previously removed and bend the core plates around the ends, lapping them as indicated in Fig. 160; one-half of the plates should be carried around one side and one-half around the other. The lapped joints must be tightly clamped together by means of two wooden strips 1½ ins. wide and 2 ins. thick, drawn together with ⅜-in. iron bolts, as shown by Fig. 168.

To the upper ends of the four dogs which hold the core, screw a tablet board to which the terminals of the windings are led and from which the main transformer connections are made. The tablet board is most easily made of "soapstone slate." which is merely a very soft grade of light gray slate. Wood or fiber will not do and hard rubber would be expensive. On it mount eight single-pole double-throw "baby" knife-blade switches, two primary terminal blocks, eight heavy binding posts. and two secondary terminal blocks, as indicated in Fig. 169. The sketch also shows how the coils are connected to the switches and binding posts.

The transformers are all designed for 1,000 volts primary e.m.f., with all of the primary coils in series, and 100 volts secondary e.m.f. with all of the secondary coils in series. The primary winding is divided into five equal sections of 200 volts each, so that it can be grouped for 200, 400, 600, 800 or 1,000 volts. In grouping for 400 and 800 volts one section must be left open, and in grouping for 600 volts two sections must be left disconnected from the main primary terminals. The original object of the writer in dividing the primary into 200-volt sections was to permit the transformer to be supplied from either an ordinary 1,000-volt primary circuit or a 200-volt motor circuit. The reader will readily understand that this arrangement is not compulsory; the primary may be wound in a single coil without any partitions, if desired, although it will be found more reliable if two or three partitions be used to reduce the voltage per section.

The secondary winding as designed is divided into two sections of about 18 volts each and two sections of about 32 volts each. No switches are used for making the various combinations because too much space and complication would be required. The connections are to be made between the various binding posts by means of short lengths of heavy iron—No. 6 or No. 8 gauge. In order to avoid excessive ohmic loss the wire should fit the hole in the binding post snugly. All of the binding posts should have two holes and binding screws each.

With primary switches all thrown inward, the primary sections are in series for 1,000 volts between the terminal posts. With all of them thrown outward the primaries are in multiple for 200 volts. Other combinations may be easily traced out. At the secondary end several com-

TABLE I.—MECHANICAL DIMENSIONS.

Size of transformer, watts	200	500	750	1000
Length of core plates	20 ins.	22 ins.	30½ ins.	31½ ins.
Width " "	1¾	2½	2¾	3
Thickness of compressed core, d	1¾	2½	2¾	3
Primary Bobbin.				
Length of sheet, C	7½	10½	11½	12½
Width " L	4⅞	5¼	5 9/16	5⅜
Length of finished bobbin	4⅜	4¾	5 1/16	4⅞
Width of one side, S	1⅞	2⅝	2⅞	3⅛
Depth of flanges	⅝	½	½	½
Secondary Bobbin.				
Length of sheet, C	12½	15½	16½	17½
Width " L	4⅞	4¾	5 9/16	5⅜
Length of finished bobbin	4⅜	4¼	5 1/16	4⅞
Width of one side, S	3⅛	3⅞	4⅛	4⅜
Depth of flanges	½	½	½	½
Core Dogs, D				
Depth of core clamp, e	3½	4¼	4½	4¾
Thickness parallel with the core	1	1	1	1
Thickness parallel with the bolts	1½	2	2	2½
Height, foot to core, f	1½	1½	1¾	2
Length of foot, g	2½	3	3	3½

TABLE II.—ELECTRICAL AND MAGNETIC DATA.

At 1000 volts primary and 100 volts secondary; 133 cycles.

Output of transformer watts	200	500	750	1000
Primary Winding.				
Size of wire, B. & S.	No. 26	No. 22	No. 20	No. 19
Turns per section	640	300	250	220
Total turns	3,200	1,500	1,250	1,100
Depth of winding, layers	20	12	11	11
Resistance, hot	128	30	18	13¼
C^2R loss, full load	5⅛	7½	10⅛	13¼
Secondary Winding.				
Size of wire, B. & S.	No. 16	No. 11	No. 10	No. 9
Turns per small section	57	27	23	20
" " large "	103	48	40	35
Total secondary turns	320	150	126	110
Number of layers	5	4	3	3
Resistance, hot	2	0.34	0.2	0.175
C^2R loss, full load	8	8½	11⅜	17½
Losses, Full Load.				
C^2R, both windings	13⅛	16	21.5	30¾
Hysteresis	13	8	10.7	8¼
Eddy currents	1⅞	1	0.8	1
Total losses	28	25	33	40
Full load efficiency	86%	95%	95.6%	96%

binations are obtainable, as shown by Fig. 170; the most serviceable will doubtless be found to be the one on the left of the sketch.

The tablet board must be kept enclosed by a box cover when the transformer is in use. The fuse blocks and master switches should be located at a little distance from the transformer, and the cover should never be removed except when the primary switch is open.

CHAPTER XVII.

CONSTRUCTION OF A REACTIVE COIL.

It is well known, of course, that reactive or "choking" coils are used in alternating-current work instead of ordinary resistance coils for the purpose of reducing the e.m.f. in a portion of a circuit because they are much less wasteful than resistance coils or rheostats. As the uses of reactive coils are so diverse no one design can be given which will fit all cases; hence only one form will be described here in detail and rules will be given by means of which anyone can modify the design to fit any case.

The reactive coil here described is designed for use in series with one, two or three open arcs on a constant-potential circuit of 100 or 110

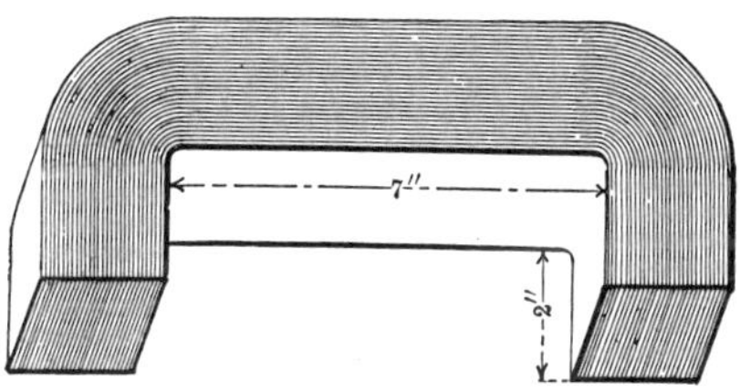

FIG. 171.—THE CORE.

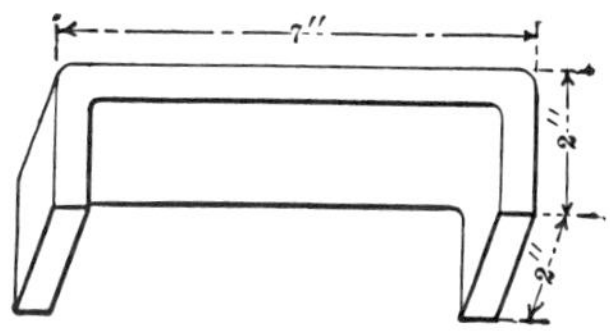

FIG. 172.—TEMPLATE.

volts, or with one lamp on a 50-55 volt circuit, or with three enclosed arc lamps fed from a 200 volt circuit. Its dimensions and windings are based upon a magnetic density of 30,000 lines of force per square inch in the core and it may be adjusted to pass any current from ½ ampere to 15 amperes. The apparatus is of the adjustable core type, and if desired, the winding can be tapped at various points and the coil can be used as an auto-transformer.

The core consists of rectangular plates of No. 27 transformer iron, 14 ins. long and 2 ins. wide, bent into the form shown by Fig. 171. The thickness of the core must be 2 ins. The easiest way to assemble it is to make an iron template like Fig. 172, of ½-inch strap iron and bend the successive plates of the core into shape over the template, one at a time, leaving them in position as they are bent. Clamp the strips to the template as shown in Fig. 173 and bend the ends down without resorting to any hammering whatever. The core must measure 2 ins. in thickness when clamped tightly. As each strip is bent down into shape its ends should be squared off with a pair of tinners' snips, so that when all are

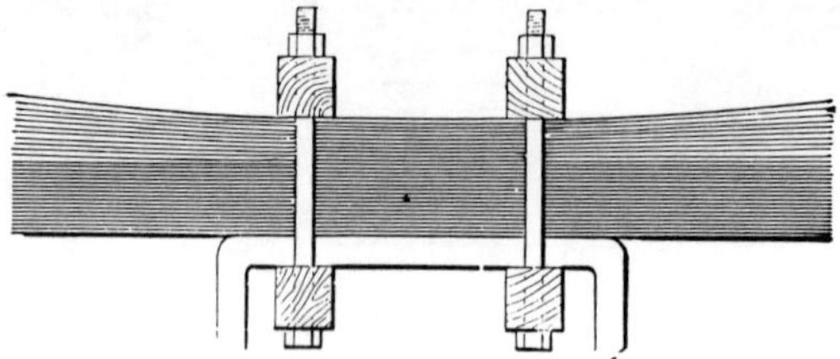

FIG. 173.—CORE STRIPS ON TEMPLATE.

bent the ends will all be flush, forming a laminated pole-face at each end of the core, as indicated in Fig. 171.

When the last strip is in place, remove the template, replace the clamps near the corners of the core and bind the core strips tightly together with heavy cord (at least 1-16 in. in diameter), winding a full layer from bend to bend, and pulling each turn just as tight as the cord will

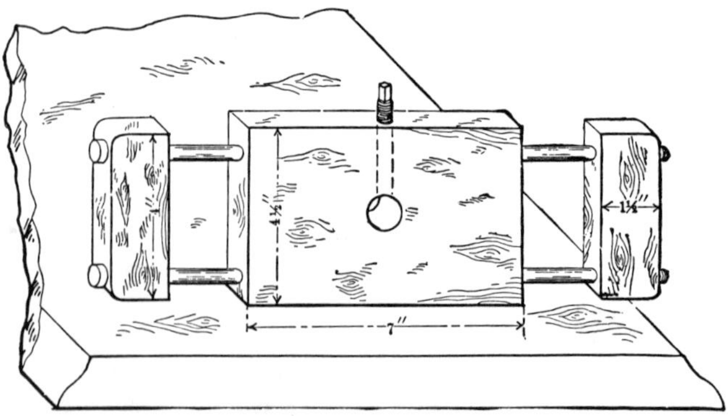

FIG. 174.—WOODEN CLAMPING BLOCKS.

stand it. The best way is to take a couple of turns at one corner and tie the cord; then loosen the clamp and move it an inch away, setting it up tight again; wind the cord over this inch of space and move the clamp another inch, continuing this procedure until the whole core is covered between the two bends. After binding the core with cord in this manner, cover it with two layers of insulating tape, carrying the tape around the bend and out almost to the end of the right-angle poles. Then wind on that portion of the core between the bends 200 turns of No. 8 double cotton-covered magnet wire in four layers. When this is done secure the ends of the core between three clamping blocks mounted on a base board, as shown in Fig. 174.

The yoke which completes the magnetic circuit is somewhat similar in form to the core just described, as Fig. 175 indicates, but the right-angle projections at each end of the yoke are much shorter than those of the main core. The exact length of these projections is immaterial except that it should be not less than an inch and not more than two. The yoke will preferably be built up in the same way as above described in connection with the main core, and after it is bound together with twine it should be mounted in the clamp shown in Fig. 176 in such a way that the center of the spindle projecting from one side of the clamp will coincide with the center of the yoke structure. The clamp jaw should be made a snug fit for the yoke so that the bolts will not need to be drawn up very tightly. The bolts must be insulated from the metal of the clamp

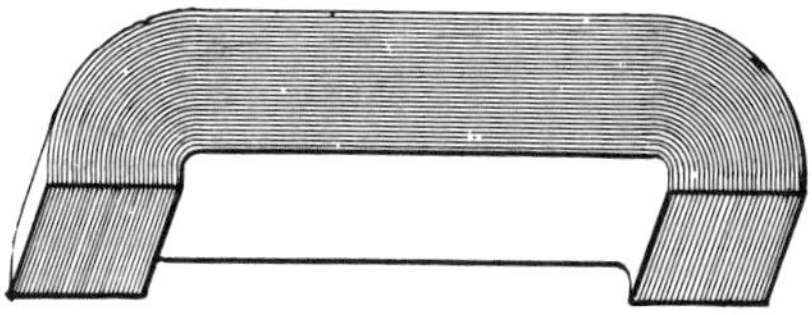

FIG. 175.—THE YOKE.

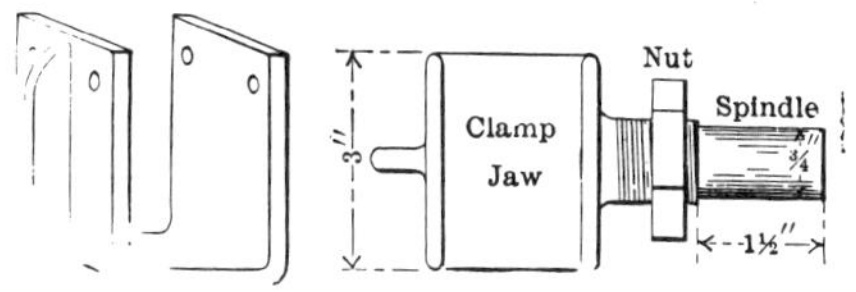

FIG. 176.—CLAMP FOR YOKE.

by bushings of either hard fiber or rubber in order to avoid forming a closed circuit around the yoke, which would result in a heavy flow of current through the clamp and bolts.

The spindle is seated in a bushed hole through the center of the long clamping block shown in Fig. 174, so that the ends of the yoke can be brought into alignment with the pole-faces of the main core. A round-nose set screw through the upper edge of the clamping block will serve to hold the spindle in any position to which it may be adjusted. Two adjustments are available, one in a rotary direction about the center of the spindle, and the other in a straight line toward and away from the pole-faces of the main core. This latter adjustment is preferably made by means of a thin nut fitted to a fine screw thread on the spindle, as indicated in Fig. 176, the set-screw in the clamping block being used merely to secure the core in any position to which it may be adjusted. To use the apparatus as a choking coil, connect the winding in series with the lamp or lamps, adjust the regulating nut on the spindle so as to secure the length of air gap specified in the accompanying table and secure finer gradations by twisting the yoke into or out of alignment with the pole-faces of the main core until the exact choking effect is secured, when the set-screw may be used to hold the yoke in that position. The length of air-gap given in the body of the tables refers to each of the two gaps, not the sum of the two.

If it should be desired to use the apparatus as an auto-transformer the yoke should be brought into accurate alignment with the pole faces, pushed up solidly against them and held in this position permanently by means of the set-screw. Used in this manner, the winding will have to

LENGTH OF AIR GAP; OPEN ARC LAMPS.

Ampere.	1 Lamp; 52-volt circuit.	Lamps on 104-volt circuit.		
		1	2	3
4	$\frac{9}{64}$	$\frac{1}{32}$	$\frac{1}{16}$	$\frac{1}{2}$
6.6	$\frac{1}{4}$	$\frac{1}{20}$	$\frac{1}{10}$	$\frac{7}{8}$
7.5	$\frac{17}{64}$	$\frac{1}{16}$	$\frac{7}{64}$	$\frac{15}{16}$
10	$\frac{3}{8}$	$\frac{5}{64}$	$\frac{9}{64}$	$1\frac{1}{4}$
12	$\frac{7}{16}$	$\frac{3}{32}$	$\frac{3}{16}$	$1\frac{1}{4}$
14	$\frac{1}{2}$	$\frac{7}{64}$	$\frac{7}{32}$	$1\frac{1}{4}$

LENGTH OF AIR GAP; ENCLOSED ARCS.

Ampere.	1 Lamp; 100-volt circuit.	Lamps on 200-volt circuit.		
		1	2	3
4	$\frac{5}{64}$	$\frac{1}{64}$	$\frac{1}{32}$	$\frac{1}{4}$
6.6	$\frac{1}{8}$	$\frac{1}{40}$	$\frac{1}{20}$	$\frac{7}{16}$
7.5	$\frac{9}{64}$	$\frac{1}{32}$	$\frac{3}{64}$	$\frac{1}{2}$
10	$\frac{3}{16}$	$\frac{3}{64}$	$\frac{1}{16}$	$\frac{3}{4}$
13	$\frac{1}{4}$	$\frac{1}{16}$	$\frac{3}{32}$	1

be tapped. Each turn of the winding will represent 1-200 of the e.m.f. of the circuit, so that to operate a single 33-volt lamp on a 50-volt circuit one terminal of the lamp must be tapped into the winding of the auto-transformer 133 turns distant from the other terminal, or preferably at 2-3 the distance from one end, as indicated in Fig. 177. Fig. 178 indicates the arrangement of the taps for three enclosed arc lamps on a 200-volt circuit. Taps would be taken out at exactly the same points to supply three open arc lamps on a 100-volt circuit.

Should the reader desire to construct a reactive coil to suit any other conditions, the following simple formulas will give the required dimensions and winding. For work on a 133-cycle circuit the choking effect in volts for a coil built for the average conditions of practice is given by the formula:

$$0.18 \times T \times A = E \quad \ldots \ldots \quad (1)$$

In this formula A is the area of the core cross-section and T is the number of turns of wire. In order to make this formula hold good the number of turns of wire must agree with the formula:

$$\frac{3.4\,L}{C} = T \quad \ldots \ldots \quad (2)$$

In this formula L is the length of the magnetic circuit within the

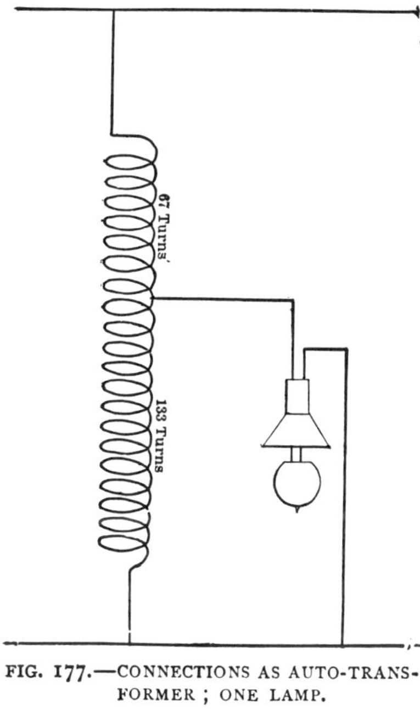

FIG. 177.—CONNECTIONS AS AUTO-TRANSFORMER; ONE LAMP.

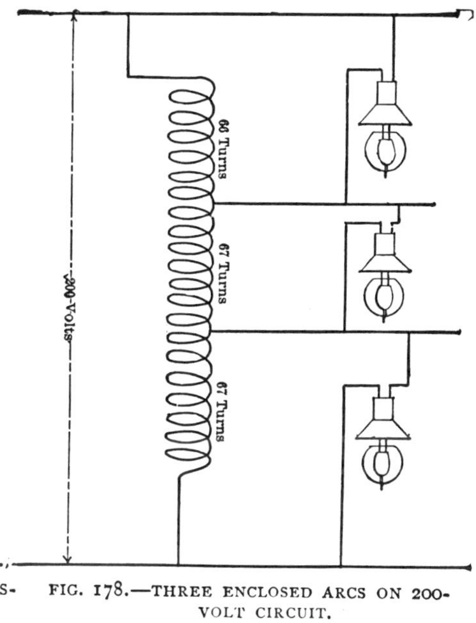

FIG. 178.—THREE ENCLOSED ARCS ON 200-VOLT CIRCUIT.

iron, including the yoke, and C is the current in the winding. For 60-cycle circuits

$$0.11 \times A \times T = E \quad \ldots \ldots \quad (3)$$

in which the number of turns must agree with the formula:

$$\frac{4.5\,L}{C} = T \quad \ldots \ldots \quad (4)$$

The two joints between the yoke and the main core when they are in actual contact are equivalent to about 36 ins. of iron under the conditions assumed in formulas (1) and (2), and to about 52 inches under the conditions upon which formulas (3) and ('4) are based; therefore, in figuring the closed circuit either 36 or 52, as the case may be, must be added to the actual length of the magnetic circuit in the iron.

The size of the wire to be used on a coil may be ascertained by allowing 1,000 circular mils of cross section per ampere of current. In designing a reactive coil the dimensions for maximum output should be calculated, first assuming the yoke in contact with the main core, because anything below the maximum effect can be obtained by adjusting the distance between the yoke and the main core.

As an example, suppose it were desired to build a coil to regulate the impressed e.m.f. of a circuit of 6.6-ampere lamps fed from a 200-volt transformer, on a 133-cycle circuit, the range of regulation or choking effect being from 25 volts to 150. Transposing formula (1) into:

$$\frac{E}{0.18 \times T} = A$$

and assuming temporarily 50 as the number of turns, the area of the core will be for the maximum reactive effect of 150 volts, 20 sq. ins. Making the core 4 x 5 ins. will give this area.

Now transposing formula (2) to read:

$$\frac{C \times T}{3.4} = L$$

and substituting for C and T the values of 6.6 and 50 respectively, we find that the length of the magnetic circuit (if it were all iron) would need to be 97 ins., which, of course, would be a ridiculous dimension. It must be remembered, however, that an air-gap is equal to practically 1,800 times its length of good sheet iron under the conditions assumed in formulas (1) and (2); therefore it is only necessary to make the iron core long enough to accommodate the winding and then insert an air-gap of such a length as to bring the total reluctance equal to that of 97 ins. of iron.

No. 12 wire will be large enough to carry the current, and the diameter of this over the insulation is 0.092 in.; 50 turns side by side, therefore, will make a coil 4⅝ ins. long, which is not excessive. The total length of the iron part of the magnetic circuit may be made about 20 ins., so that the two air-gaps must be made equivalent to 97 — 20 = 77 ins. of iron. As each inch of air-gap is equal to 1,800 ins. of iron, the total length of air-gap required will be:

$$\frac{77}{1,800} = 0.043 \text{ ins.},$$

so that each air-gap will be about 7-64 inch long in order to bring the choking effect of the coil down to 150 volts with 6.6 amperes flowing through the winding.

The foregoing formulas and instructions are based on a magnetic density of 30,000 lines per square inch in the core of a coil to operate on a 133-cycle circuit, and 52,000 lines per square inch in the core of a coil to work on a 60-cycle circuit. While the density can be carried somewhat higher than this and the size of the core correspondingly decreased, more satisfactory results will usually be obtained by employing the densities here given, because with higher densities the hysteresis loss in the core will cause it to overheat and jeopardize the coil. The two densities above specified give the same core-loss, to wit: 0.68 watt per cubic inch of core iron.

CHAPTER XVIII.

THE CONSTRUCTION AND CALCULATION OF RHEOSTATS.

The resistance material of rheostats for the regulation of current or potential in electrical circuits may be metallic wire, carbon or graphite, or acidulated water. In the present article rheostats in which the first named material is employed will only be considered.

The different conductor materials used in the construction of commercial rheostats are iron, German silver and copper. Each of these materials has advantages in particular cases. The advantages of iron are cheapness and the ability to withstand high temperatures. German silver has a high resistivity or specific resistance and a low temperature co-efficient. Copper is only used where large currents have to be carried, as, for instance, in electro-plating work, where one dynamo supplies several tanks requiring different voltages, and regulation is effected by inserting resistance into the circuits requiring the lower pressure. In this case, copper, by virtue of its higher conductivity, makes it possible to use smaller conductors, thus facilitating the construction of the rheostat.

The table on page 155 gives the carrying capacity of tinned iron wire under different conditions. The last column gives the length of wire having a resistance of one ohm.

In designing motor-starting rheostats, the values given under the heading "Safe current for one minute" should be used, while the carrying capacities given in the other two columns apply to dynamo field rheostats, motor regulators and such other rheostats as have to carry current continuously. No definite resistivity and carrying capacity can be assigned to German silver, as it is an alloy, and different makers use different proportions of the elements. In the tables given by Matthiesen the resistivity of German silver is given as 2.2 times that of iron. For the same rise of temperature a German silver wire would, therefore, carry about two-thirds the current of an iron wire of the same size. Most commercial German silver has, however, a specific resistance higher than that indicated by the above ratio.

The wires of rheostats are mounted in a number of different ways. They may be embedded in enamel or some other refractory insulating material; they may be wound on a plate or slate; they may be wound on a framework of iron rods insulated with asbestos, or on

Size of Wire, B. & S.	Safe Current in Wood Frame.	Safe Current in Iron Frame.	Safe Current for One Minute.	Feet per Ohm.
8	17.4	20.3	43.6	250
9	14.6	17.1	36.6	173
10	12.3	14.3	30.8	137
11	10.3	12.0	25.8	108
12	8.7	10.1	21.7	86.4
13	7.3	8.5	18.3	68.5
14	6.1	7.1	15.3	54.3
15	5.1	6.0	12.9	43.1
16	4.3	5.0	10.8	34.1
17	3.6	4.2	9.1	27.1
18	3.00	3.5	7.6	24.3
19	2.52	2.9	6.3	16.5
20	2.17	2.5	5.4	13.5
21	1.82	2.1	4.5	10.7
22	1.53	1.77	3.8	8.49
23	1.28	1.49	3.2	6.73
24	1.08	1.20	2.3	5.34

insulated metallic spools with layers of asbestos between the layers of wire. Finally, the wire may be wound into coils which are stretched between insulators on an iron frame or in a frame of insulating material. When the wires are embedded in enamel, they are placed on the surface of and in close proximity to a cast-iron base plate which assists in radiating the heat. Slate is also quite a good conductor of heat, and plates of slate are often used for smaller rheostats. Spool-wound coils of wire sometimes present an advantage where the rheostat is only used for a short period at a time, as, for instance, in motor-starting rheostats, as this method of winding permits of getting a large amount of wire into a small space, and the capacity of the rheostat under such conditions depends more on its capacity for taking up heat than on the radiation. When spiral coils are employed, they are generally placed in a case with openings to facilitate the circulation of air.

The diameter to which spiral coils of wire are wound varies with the size of the wire. If for a given size of wire the diameter is taken too large, the coils must be stretched considerably to obtain the necessary stiffness. No. 24 (B. & S.) iron wire may be wound into coils of ½ in. diameter, while No. 16 may be wound into coils of from ¾ in. to ⅞ in. diameter, and other sizes proportionally. The wires are wound close on a mandrel in a lathe and are stretched as they are put in position. For the larger sizes of wire a stretching of 20 per cent. is sufficient, while coils of No. 24 of 6 ins. or more in length must be stretched to about double their length. Some manufacturers place asbestos tubes inside coils of smaller wire, which, as they stiffen the coils, permit coils of larger diameter and reduce the stretching required.

Dynamo Field Rheostats.—Shunt and compound-wound generators are generally regulated by means of a rheostat in the shunt field circuit. In Figs. 179 and 180 are shown two dynamo field rheostats, both of which are of fire-proof construction. · The form shown in Fig 180 is intended for small machines, while that at Fig. 179 is adaptable to any size.

FIG. 179.—DYNAMO FIELD RHEOSTAT.

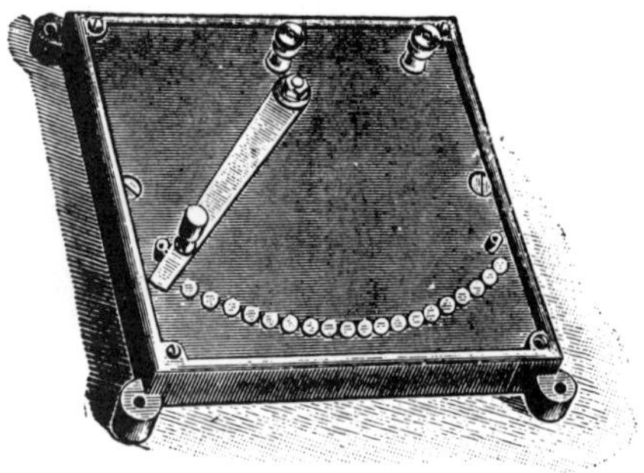

FIF. 180.—DYNAMO FIELD RHEOSTAT.

In the factory it is generally easy to experimentally determine the resistance required to cut down the voltage of a machine to the desired lower limit. Cases may, however, arise, where this is not handy, and the resistance can then be calculated, provided the excitation-voltage curve of the dynamo and the resistance of its field are known. The calculation may be illustrated by a practical example. The main curve in Fig. 181 is the excitation-voltage curve of a 1.5-kw 55-volt generator.

The machine is run at such a speed that without any load and without any extra resistance in the field circuit, it generates 65 volts. A rheostat is required which will cut down the voltage to 40. The field resistance is 36.4 ohms.

From the curve we see that at 65 volts the exciting ampere-turns are 6,200, while at 40 volts they are only 2,300. The ampere-turns are proportional to the voltage applied to the shunt. When 65 volts are being generated, the voltage at the terminals of the shunt is 65. At 40 volts it must, therefore, be $65 \times \frac{23}{62} = 24.15$. The rest of the e. m. f.

(40 — 24.15 = 15.85 volts) must be taken care of by the drop in the rheostat. As the same current goes through field coil and rheostat, their resistance must be to each other as the drop of potential in them. Thus we get for the resistance of the rheostat $36.4 \times \frac{15.85}{24.15} = 24$ ohms. The

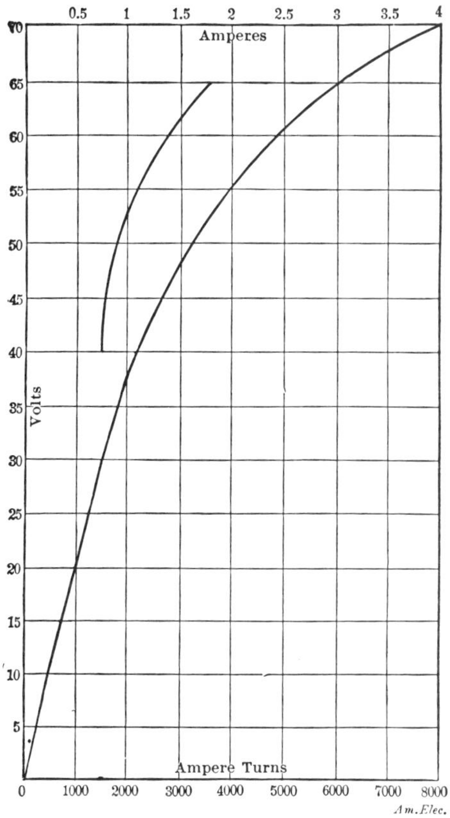

FIG. 181.—EXCITATION CURVE OF 1.5 KW, 55-VOLT DYNAMO.

largest current that any part of the rheostat ever has to carry is a little less than $\frac{65}{36.4} = 1.78$ amperes, and the smallest current $\frac{40}{36.4 + 24}$ $= .65$ amperes. After finding a few intermediate points in the same manner as we found the smallest current, we can draw a curve showing the field current for the different voltages. This curve is also shown in Fig. 181. For small rheostats, like the one under consideration, but one size of wire is generally used. In the present case an iron wire No. 22, or a

German silver wire No. 20 would have the required current-carrying capacity. The total length of wire required would be for iron 204 ft., for German silver of the resistivity given above, 148 ft. In all large rheostats, however, the size of wire decreases from the "out" terminal to the "in" terminal. The calculation of the different portions may be illustrated by the present example. Supposing that twenty-five steps of about 1 volt each are desired. From the field-current curve and the table of iron wire we see that the largest wire that would be used is No. 22. Of this we must make the first three sections of the rheostat. When the fourth section is inserted the field current is reduced so much that No. 23 will carry it. The resistance of these three sections is to be calculated in the same manner as the total resistance of the rheostat was calculated above. We then calculate the sections requiring No. 23, No. 24, etc., successively.

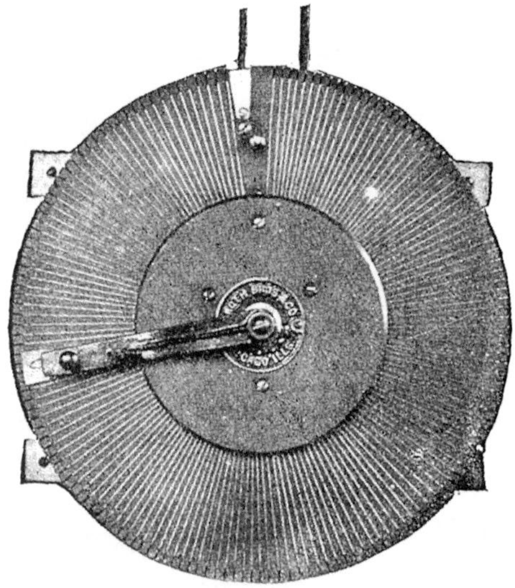

FIG. 182.—MULTIPLE-CONTACT RHEOSTAT.

In some lines of work it is desirable to be able to change the e. m. f. of a generator very gradually; that is, by very small steps. This requires a large number of contact points, and as a rheostat with a large number of contacts as generally made (Figs. 179 and 180), is quite expensive to manufacture, several types have lately been brought out in which an attempt is made to simplify the construction. One of these is illustrated in Fig. 182. It consists of a plate of slate in the form of a concentric ring, the inside and outside edges of which are grooved to receive the resistance wire. The wire is wound on a plate in a continuous winding nearly all around the ring, as seen in the illustration. The plate carrying the wire is clamped between two other plates of slate. The front plate carries the contact lever, while to the back plate are fastened two strips of brass by means of which the rheostat is fastened to the switch-board. The sliding contact piece bears directly on the wire. The rheostat illustrated has 180 steps.

Fig. 183 shows diagrammatically a rheostat in which the number of steps is equal to twice the number of contacts less one. It consists of an ordinary rheostat with a slightly different contact arrangement. The contact lever carries, in addition to the regular contact piece, another

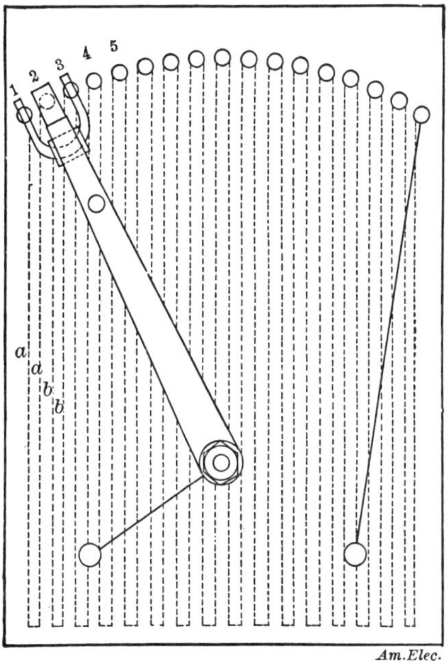

FIG. 183.—MULTIPLE CONTACT RHEOSTAT.

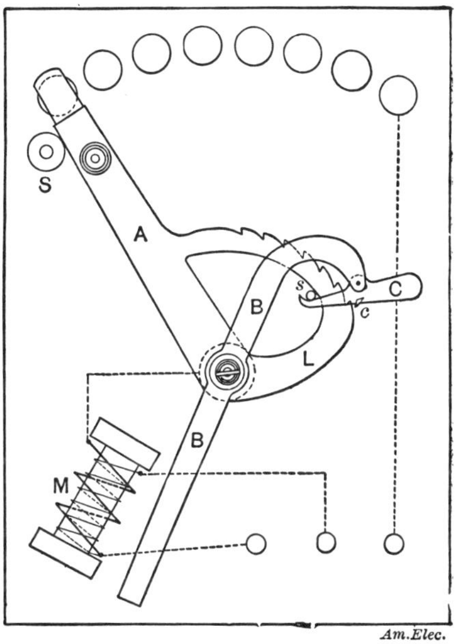

FIG. 185.—MOTOR-REGULATING RHEOSTAT.

double contact piece which is, however, insulated from it. The main contact piece is wider than the distance between two neighboring contact points, while the two prongs of the double contact piece are narrower than this distance. Suppose that the current enters through the contact lever. It will then pass from contact 2 to contact 3 through the sections, a^1 a and b^1 b in parallel, and from 3 through the rest of the sections in series. The resistance of two sections in parallel is, of course, equal to one-half the resistance of a single section. When the lever is turned to the right, the main contact piece comes in contact with 3 and the prongs of the extra contact piece are between contacts 1 and 2, and 2 and 3 respectively. The parallel resistance is then cut out. This type was suggested by Vedovelli.

Motor-Starting Rheostats.—When a shunt motor is started up, a resistance must be placed in its armature circuit, to prevent an abnormal rush of current. This resistance is cut out stepwise as the motor gains speed. Motor starters have generally but one size of wire all through, of sufficient cross section to carry the full load current for one minute. (See table below.) The resistance of the rheostat should be such that when it is connected across the mains, a current equal to the full load current of the motor will pass. From this condition the resistances of starting rheostats for motors of different outputs and voltages, compiled in the following table, have been calculated. (Ten per cent. is allowed for armature and friction loss in the motor.)

RESISTANCE IN OHMS OF MOTOR-STARTING RHEOSTATS.

HP.	1	3	5	7	10	15	20	30	40	50
110V	15	5	3	2.1	1.5	1	.75	.5	.37	.3
220	60	20	12	8.4	6	4	3	2	1.5	1.2
500	300	100	60	42	30	20	15	10	7.5	6

Motor-starting rheostats are nearly always automatic; that is, they

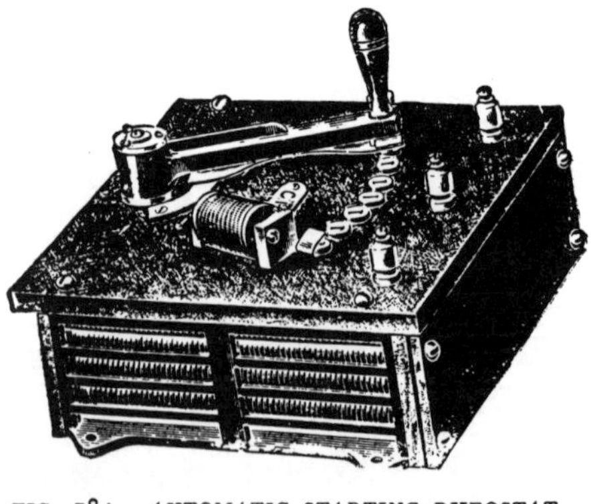

FIG. 184.—AUTOMATIC STARTING RHEOSTAT.

have an electro-magnetic attachment by means of which the armature is automatically cut out of circuit whenever the main current fails for any reason. This protects the motor from injury when the current in the mains is established again. Fig. 184 shows a much-used type of automatic starting rheostat. The rheostat has an electromagnet on its face plate; the coil of this magnet is in series with the field winding of the motor. The contact lever is of iron and has a spiral spring inside its hub. The magnet holds the lever in position when the resistance of the rheostat is cut out of circuit, but when the current in the field circuit ceases the lever is brought back to the dead button by the action of the spring.

Motor-Regulating Rheostats.—The speed of motors may be regulated by means of a rheostat in the armature circuit. For the special case of a constant torque on the motor, the speed is proportional to the counter e. m. f. and the current remains constant, both for shunt and series motors. The resistance necessary to reduce the speed to a certain fraction of its original value may be found by the following rule:

Multiply the e. m. f. of the mains minus the drop in armature (and field in case of series motors), by the difference of unity and the given fraction, and divide the product by the current. The quotient obtained is equal to the required resistance in ohms.

Motor-regulating rheostats are also made in which the contact lever is automatically held in any position and automatically released in case of overload or when the main current is interrupted. Fig. 185 shows a front view of such a rheostat. Two levers are rotatable on a stud fastened to the base plate. One of these, the contact lever, *A*, has an elbow-shaped side projection, *L*, with a number of notches on its outer edge, corresponding to the contact points. The other lever, *B B*, is of iron, and is two-armed. The lower arm serves as armature to the electromagnet, *M*. The other arm has pivoted to it at its extremity a third lever, *C*, provided with a catch, *c*.

The two arms of lever *C*, are of unequal moment and one of the arms rests continuously against a stud, *s*, fastened to the base plate. A helical spring (not seen in diagram) coiled around the hub of the lever *A*, holds the lever against the stop, *S*, when the rheostat is not in use. When the lever, *A*, is turned to the right, lever *B* also turns on account of the friction between the two levers, and is thus brought near the poles of the magnet, *M*. The catch, *c*, engages into the notch corresponding to the contact point, on which the contact lever is left. The catch is locked by the electromagnet through the armature, *B*. The magnet, *M*, is wound differentially, one winding being in the field and one being in the armature circuit of the motor. The turns of the windings are so proportioned that under ordinary conditions the effect of the field current predominates. When the current in the armature circuit rises above a certain value, the magnet is so much weakened that it cannot resist the action of the spring on the hub of lever *A*. It leaves go its armature, *B*, the catch, *c*, disengages, and lever *A*, returns to the "off" button. The same happens if the main or the field circuit should be opened.

CHAPTER XIX.

SIMPLE VOLTMETERS, AMMETERS, AND WATTMETERS.

Nearly all forms of meters depend upon the magnetic effects of the current for their action. These may be divided into solenoid instruments, magnetic vane or needle instruments, and moving coil instruments. Another class of instruments of great importance uses the heating effect of the current, which produces expansion in a strip of metal or a wire, as the source of their indications.

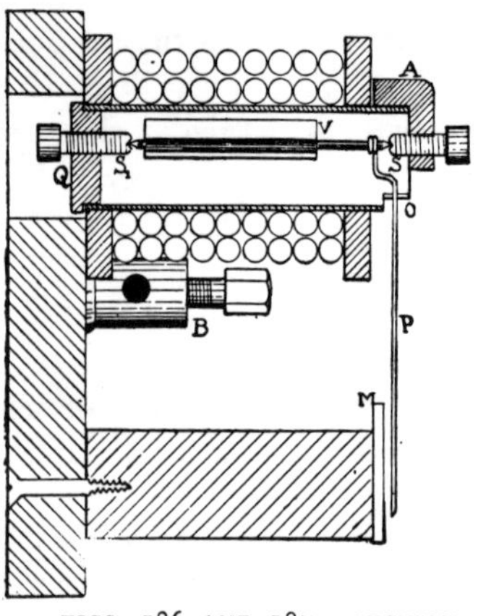

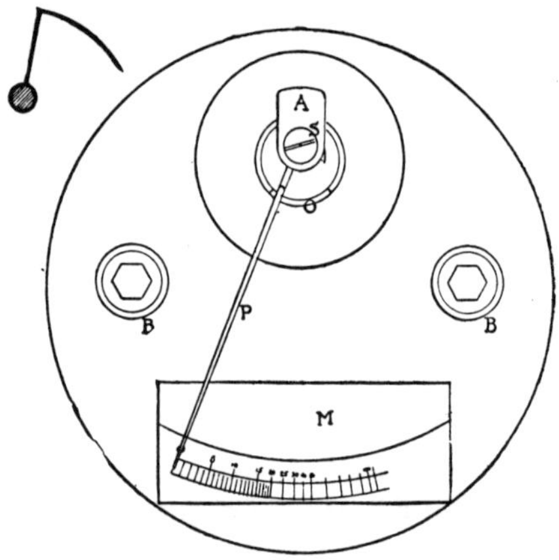

FIGS. 186 AND 187.—SECTION AND ELEVATION, MAGNETIC VANE AMMETER.

A very satisfactory instrument is shown by Figs. 186 and 187. Here the fact that the lines of magnetic force crowd close together along the inner sides of a solenoid is used as the principle of action. The coil of large wire is wound on a brass tube with wooden or fiber heads, one end of the tube being closed with a brass plug, *Q*. A piece of brass, *A*, is soldered to the other end of the tube, and through this and the plug are screws, *S* and S^1, coned out for the reception of the pointed ends of the pivot, as shown. The screws must be of brass or other non-magnetic material. They are not arranged in the center of the tube, but are a little above it, say 5-32 in., if the tube is an inch in diameter. On the pivot, which is of hard steel, is mounted a little vane or wing of thin sheet iron (the sort used by photographers for the basis of "tin-types" is best) bent to the shape shown in the small detail drawing. This should have about the relative size shown in the illustration. The needle, *P*, is of aluminum wire, for the sake of lightness, and the whole is so balanced, by soldering on bits of copper wire if necessary, that it hangs normally as shown in Fig. 187. The vane being eccentric to the tube carrying the coil tends to approach its inner surface when current passes. It must be so attached to the pivot that this tendency causes the needle to sweep over the scale, *M*. As in the instrument just described this scale is of paper mounted on a scrap of looking-glass. The whole is attached to a circular wooden base and forms a convenient wall or switchboard instrument. A cover to exclude dust and keep off stray air-currents would be a valuable addition.

This ammeter may be made very sensitive and accurate if care is taken in its construction. The lighter the needle the more sensitive the instrument, other things being equal. To decrease its sensitiveness the lower part of the needle system should be loaded so as to bring the center of gravity of the whole lower and thus cause a greater tendency for the needle to return to its zero position.

The ammeter may be converted into a voltmeter by the use of a fine wire high-resistance coil of many turns in place of the coarse coil shown. For such an instrument, used as an ammeter, and measuring currents up to 100 amperes, about eighteen or twenty turns of wire ¼-in. in diameter will be found sufficient.

There are many cases where it is desirable to know the direction of the current as well as its volume. Neither of the instruments described indicates this. The next ammeter to be described not only indicates the direction and amount of the current, but also possesses two valuable qualities not shared by the cruder forms described—portability and freedom from vibration of the needle. In other words, it is a "dead-beat" instrument, the needle going promptly to its place on the scale and stopping without vibration. It is a very satisfactory and useful instrument and will be described at some length on account of its various good qualities.

The ring, *R* (Figs. 188 and 189), is made of good quality tool steel, 1 in. x ½ in., and bent around a diameter of 4½ ins. The ends do not meet, but are rounded off as shown by dotted lines in Fig. 189, leaving an opening about 1 in. between them. After this ring has been forged into shape and finished by the rounding of the two ends and the boring of the two holes for the screws, *m* and *n*, it is hardened by heating it

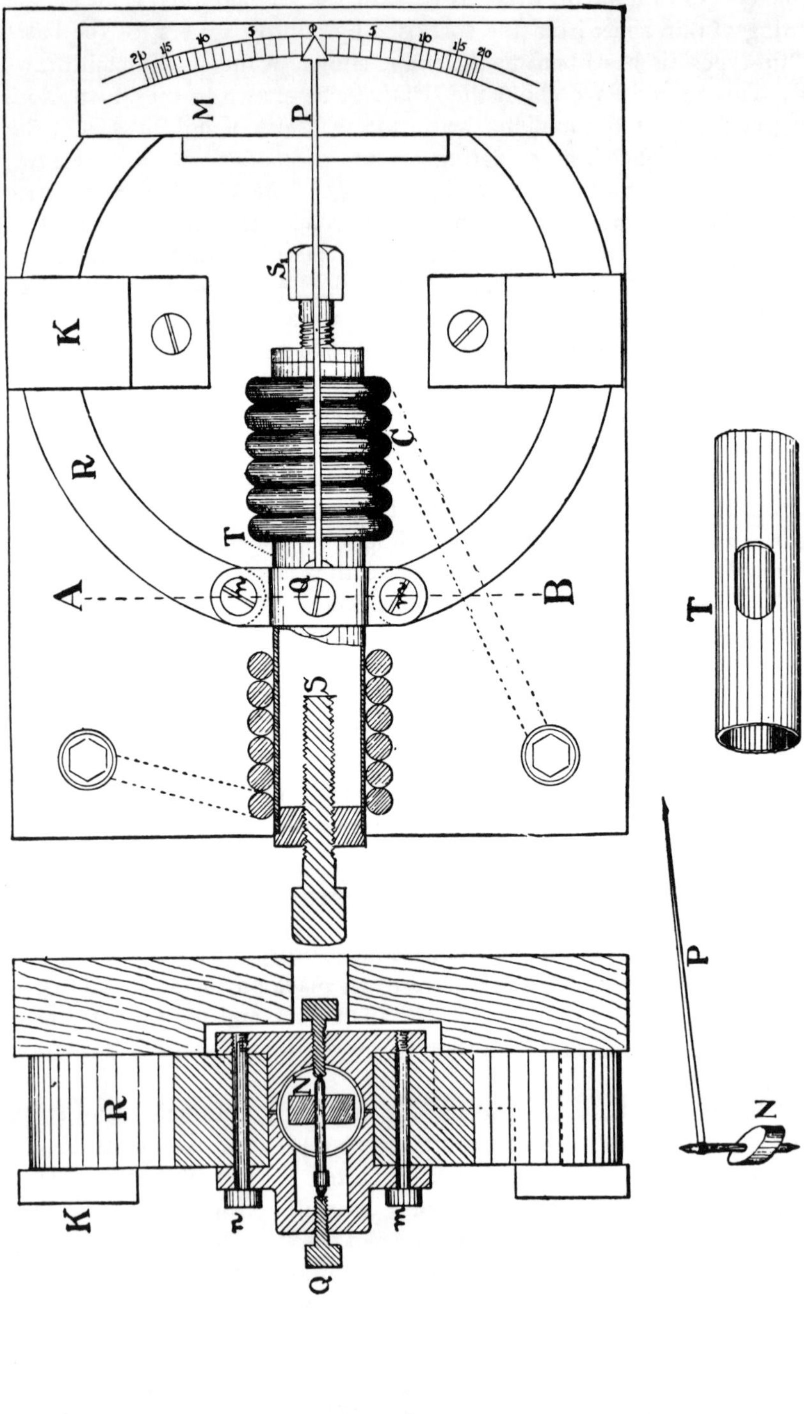

FIGS. 188 AND 189.—PERMANENT MAGNET PORTABLE AMMETER.

red hot and suddenly cooling it in water, or better, a solution of sal ammoniac. It is then magnetized by wrapping it with about seventy-five turns of wire and passing a strong current, or by rubbing it on the poles of a dynamo, and after magnetization it is boiled for an hour in water. Then it should be magnetized again, and again boiled, this being repeated several times, the magnetizing always being done in the same way and never reversed. By this method of magnetizing and boiling the ring is brought to a permanent state and does not lose its magnetism, as would be the case if no such precaution were taken.

The tube, *T*, shown also in one of the small detail drawings, is of brass, 1 in. outside diameter and about 1-16 in. thick and 4¼ ins. long. In the middle of it is cut an opening ½ in. wide by ¾ in. long, as shown in the small drawing. This is for the withdrawal of the iron needle described below. Referring to Fig. 188, which shows a cross section of the instrument on the line *A B* (Fig. 189), it is seen that the tube is supported between small castings of brass which are clamped between the poles of the permanent ring magnet by the screws *m* and *n*. To the lower of these castings the tube is soldered at its middle, the opening already described being on the upper side, as shown at *Q*, Fig. 189. The upper casting clamps the tube solidly in place when the two screws, *m* and *n*, are drawn up tight.

Through these two castings are tapped brass screws coned out for the reception of the pivot carrying the needle, *N*. This pivot is of hard steel wire, the ends being coned in a small lathe or ground off to shape on an emery wheel. On it is mounted the soft iron needle, *N*, shown in perspective in one of the small drawings. This should be of the softest and purest iron obtainable, ⅜ in. long over all, about ⅜ in. wide in the middle, and about the same thickness, measured along the pivot. It must be filed carefully into perfectly symmetrical shape. Attached to the pivot is the aluminum pointer, *P*, which sweeps over a scale and mirror as described above. It will be noticed that the zero point of this scale is at the middle, the needle being deflected in either direction, according to the direction of the current.

In the ends of the brass tube, *T*, are soldered two plugs carrying the soft iron screws, *S* and S^1. These are intended for the regulation of the scale of the instrument, and once adjusted are to be left alone. By screwing them in nearer the needle, *N*, the instrument becomes more sensitive, that is, it gives a larger deflection for the same current. Hence, when the instrument is assembled, the maximum current it is intended to register should be sent through it and screws adjusted until the needle is at its extreme deflection.

To soften the iron screws, *S* and *S*', put them in a small sand crucible and cover them with powdered lime. Then heat the crucible to a cherry red heat in a charcoal or anthracite fire, leaving it to cool very slowly as the fire dies down. The lime prevents the formation of a scale of oxide on the screws, which may be cleaned after they are cool by dropping them for a moment in weak muriatic acid and washing in water containing a little ammonia.

The coil is wound as shown on the brass tube, its ends being attached to appropriate binding posts. The figure shows a coil of only twelve turns, intended for the measurement of fairly large currents, but for small currents the wire may be smaller and the turns more in proportion. For currents up to 15 amperes use No. 10 wire and put on sixty turns, thirty on each end of the tube.

The ring magnet, *R*, is fastened down to the wooden base by clamps, *K*. As shown in the illustration the instrument is a table form, adapted for use in a horizontal position only. If the needle system shown in the small drawing is balanced perfectly for all positions (which would require a small counterweight to compensate for the pointer, *P*), the instrument may be used in any position.

For use as a voltmeter the coarse coils shown should be replaced by coils of fine wire wound on insulating spools and slipped over the ends of the spool. Indeed, a combination instrument may be made by slipping these spools on over the ammeter winding as shown. For the best results these spools should contain the largest possible amount of the finest possible wire, and should be connected, in addition, through a resistance in the base of the instrument, care being taken to so wind the latter that it does not produce any magnetic effects. The total resistance of the coil and the additional resistance for measurements up to 150 volts should not be less than 9,000 or 10,000 ohms. For higher voltage additional resistance or a shunt must be used. Unless the resistance is very high so much current will flow through the coils that they will heat or even burn out if the instrument is left in circuit, hence the necessity for care in providing enough resistance. It is also to be noted that a voltmeter is exposed to the full pressure of the current that it is measuring, and that its insulation cannot be too careful. In an ammeter the fall of potential is utterly negligible and insulation is not a feature of particular importance, but with a voltmeter this is different, a short circuit leading instantly to disastrous results.

The above instrument, which is similar to those made by Carpentier and Ayrton & Perry, is a thoroughly satisfactory shop meter, if carefully and accurately made. Unfortunately, like the other instruments described above, it is useless for alternating currents.

The instrument shown in Figs. 190 and 191 belongs to the moving coil type and is adapted to either alternating or direct currents. On a wooden base are mounted two wooden rings, *R R*, about 6 ins. in diameter, upon which are wound about 20 turns of No. 10 wire. The blocks, *W* and *W* carry on their upper surfaces small copper cups, *A* and *A*1, which are soldered to the copper strips, *m* and *n*, and so connected that the current circulating in the coils, *R R*, includes in its circuit the two cups and the coil, *C*, which is suspended between them. This coil is wound on a thin hardwood frame of the shape shown in the drawings, the upper of the two small drawings showing its upper surface. The coil has 20 turns of No. 10 wire, its ends being soldered to the copper pins, *K K*. One of these is shown on a larger scale in the small drawings. At their bottoms they are filed into knife-edges, so that when they are placed in the copper cups, *A* and *A*1, the coil, *C*, rocks on

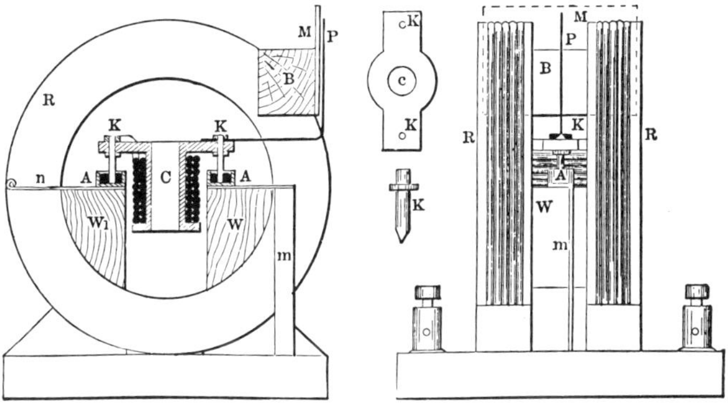

FIGS. 190 AND 191.

them. The bottoms of the copper cups are slightly hollowed to keep these knife-edges in place at the center, and the whole system is balanced so that it rocks very easily on the knife-edges. The cups, *A*, and, *A*1, are filled about half full of mercury, on the top of which is placed a drop of kerosene, or, better, of a mixture of about 4 parts of kerosene and 1 of alcohol in which is dissolved 5 per cent. of cyanide of potash. This preserves the mercury surface from oxidation and also relieves the surface tension. A pointer, *P*, is attached to the rocking coil and plays in front of a scale as shown. The instrument should be covered with a bell-glass, or a wooden box having a glass front, to keep out dust and air currents.

While this is by no means an ideal instrument it is a satisfactory one for many purposes. It should be mounted on a shelf on a wall or in some other place where it will not be disturbed or subject to vibrations, and it should be taken down and cleaned occasionally. The nearer the knife-edges come to the center of gravity of the suspended system the more sensitive the instrument will be. It is advisable to make the coil, *C*, as light as possible, and for this reason aluminum wire should be used in place of copper for winding it if this can be obtained. This instrument is not "dead beat," but it has a fairly constant scale through a range of about 30 degs., which is a great advantage.

For alternating current work the "hot-wire" instruments possess some important advantages, among them that of not affecting the circuit in any way, as the coil instruments do by their self-induction. The instrument next to be described is somewhat similar to the Cardew hot-wire volt-meter, and, while rather delicate, is an excellent instrument in careful hands. Figs. 192 and 193 are front and side elevations of it, partly in section.

The principle upon which it depends is the expansion of a small wire heated by the current to be measured. A wooden box, 8 ins. in diameter, is turned to form the head of the instrument, and to this is attached as shown a piece of iron gas-pipe, *T*, 1 1-4 ins. inside diameter and about 3 ft. long. At the bottom of this pipe (the instrument is intended to be attached to a wall with the wooden box part uppermost) is a brass plug, insulated from the pipe by an insulating bushing, *B*, and carrying a threaded rod and nut, *N*. This threaded rod is grooved, a small pin engaging in the groove so that when the nut, *N*, is turned the rod is screwed in or out, but does not turn around. There should not be the least lost motion about this fitting, as the motion of the threaded rod must be accurate and exact. In the upper extremity of the threaded rod is soldered a small bit of copper wire split at its upper end. This is to clamp the platinum wire, *w*, whose expansion is recorded by the instrument.

At the upper end of the tube, *T*, is arranged a small drum, *R*, having coned ends resting in the screws as shown, or the screws may be pointed and the drum coned out for their reception. The drum should be not more than 3-16 in. in diameter, and is best made of steel. It is shown relatively too large in the illustrations. Around this drum the platinum wire, *w*, is wrapped two or three times, the surface of the drum being first enameled with a mixture of finely powdered asbestos and water-glass (soda silicate) painted on and dried with gentle heat. It is not absolutely necessary to enamel the surface, but it is well to do so, as this prevents any current from flowing through the cone bearings of the drum and heating them.

In the tube, *T*, is secured a small piece of 1-2 in. iron pipe, closed at

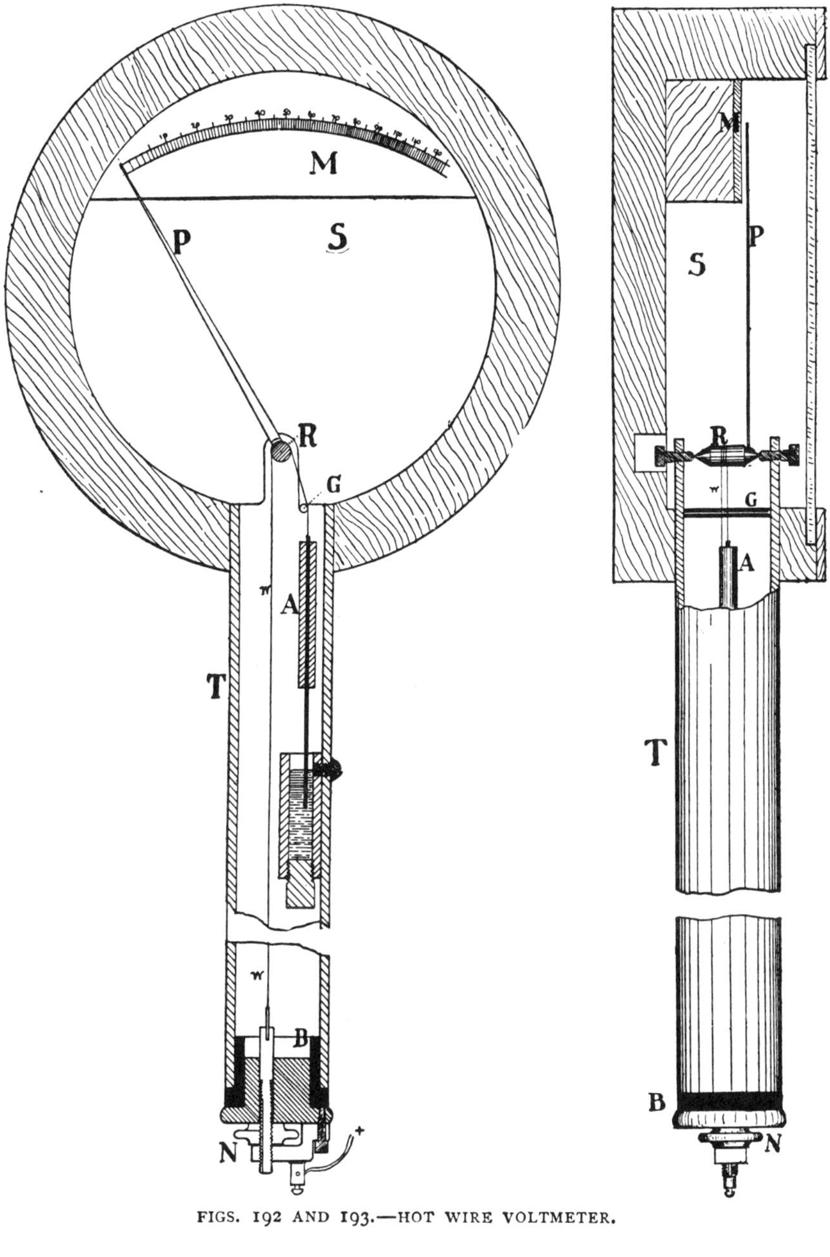

FIGS. 192 AND 193.—HOT WIRE VOLTMETER.

the bottom with a plug and partly filled with mercury. The platinum wire, after passing around the drum, is deflected by the glass rod, *G*, so that the wire attached to the weight, *A*, suspended from the platinum wire, dips in the mercury. This wire should be of copper and the weight should be of lead or brass, its weight depending upon the size of the platinum wire, but enough to keep it taut. For a No. 38 platinum wire the weight should be about 1½ ozs.

A resistance is absolutely necessary with this type of voltmeter, and the otherwise empty space, *S*, will contain it comfortably. For voltage up to 150 a resistance of about 4,000 ohms should be used and the platinum wire should be about No. 38 or 40. The resistance can be wound on a wooden spool and is best of German silver wire, about No. 32, preferably not smaller.

As the expansion of the iron pipe and the platinum wire is not the same, the daily changes of temperature, the stretch of the fine wire, etc., will make the zero point somewhat uncertain. For this reason the nut, *N*, and its adjuncts are provided, so that the needle can always be adjusted to zero before the reading is made. The instrument is quite sensitive, absolutely dead-beat, and entirely unaffected by external magnetism. The best way to increase its accuracy without making it more delicate is to lengthen the pipe, *T*, but this has practical disadvantages. It should be mounted by means of wooden cleats against a wall and left there. Care should be taken in winding the resistance coil not to make it inductive, as this will vitiate the accuracy of the instrument and introduce complications in the circuit. To wind the resistance non-inductively two wires should be wound at once on the spool, and then connected together so that the current circulates in different directions through each.

A sensitive and excellent wattmeter may be very simply made as follows: On a wooden base build up with brass screws or glue a wooden box, as shown in Figs. 194 and 195, about 7 ins. × 8 ins., and 7½ ins. high, having a glass front, *G*. Against the back of this box is mounted a wooden spool, *A*, carrying a coil of a few turns of coarse wire. For currents up to about 20 amperes No. 8 wire and 8 or 10 turns are advisable. In the center of the top of the box is cut a hole, and mounted above this in a wooden ring, *R*, is an ordinary glass lamp chimney of the shape shown (such as is used with "student lamps"). In the top of this is a plug, *H* fitting so that it can be turned with the fingers, and carrying two wires, *m*, *n*, fitting rather tightly in the fibre or wooden plug. If wood is used it should be boiled in paraffine, as the voltage measured is in full force between the two pins, *m* and *n*. These pins should be ½ in. apart, and should be split at the bottom so that the two thin suspension wires, *S* and *S*, may be pinched in the split part.

Suspended from these two wires is the coil of fine wire, *K*. This is wound on a fibre spool, 2 ins. outside diameter, 1 in. inside diameter, and having a space for the winding ½ in. wide measured along the axis of the spool. It should be as thin as possible to reduce its weight, and should contain about 300 turns of No. 32 wire. In winding this the greatest care should be used to insulate it thoroughly, and the wire

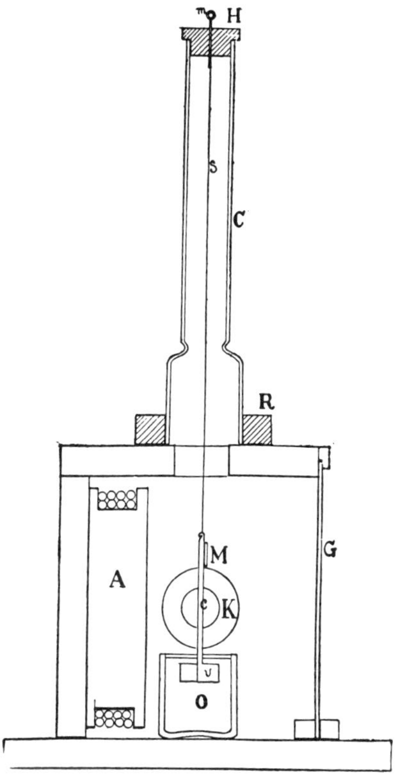

FIG. 194.—SIDE ELEVATION, WATTMETER.

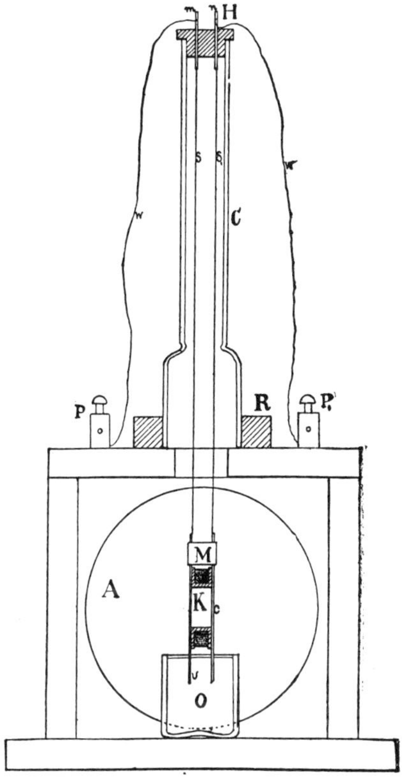

FIG. 195.—END ELEVATION, WATTMETER.

should be copiously shellacked and the coil thoroughly baked after completion. The ends of this coil are connected to the two thin sheet copper or aluminum strips, c, which are shellacked to the sides of the spool. At the lower end of each of these strips is a vane or projection, *v*, the two being arranged to point in opposite directions. These are intended to dip into kerosene oil contained in the cup, *O*, and to dampen the swinging of the suspended coil. Across the two strips, *C*, is fastened with gummy shellac a small mirror, *M*, best made by silvering the surface of a bit of glass, such as is used for microscope slide covers, though a piece of ordinary thin looking glass will answer. The two suspension wires, which should be of exactly equal length, are fastened to the strips, *c*, with small drops of solder. These suspension wires should be of copper, not larger than No. 36 gauge and preferably smaller. It is not necessary to remove the insulation from them.

Two binding posts, *P* and *P*1, are provided for attaching the fine wires, *w*, leading from the pins, *m* and *n*. To these posts are connected the terminals from the two sides of the circuit to be measured, while the main current is led through the coil, *A*. The deflections of the swinging coil are read off by means of a lamp and scale, a simple and excellent arrangement for this purpose being described by Mr. J. F. Hobart in Chapter XXI. For small deflections the scale readings are proportional to the watts. This instrument is equally good for alternation and direct current work, and in careful hands is surprisingly accurate. It is quite good enough for incandescent lamp measurements.

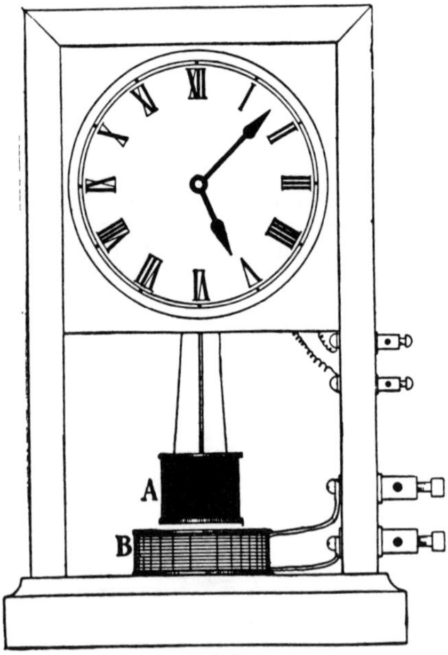

FIG. 196.—RECORDING WATTMETER.

Of course, a resistance must be used in series with the fine coil, the amount being proportional to the voltage of the circuit. Up to 150 volts, using No. 32 wire on the coil, the resistance should be about 2,500 ohms, and it should be wound non-inductively as described above. If very accurate determinations are to be made the self-induction wire circuit may be compensated by a condenser consisting of glass plate with a strip of tin foil on each side, but this is an unnecessary refinement for ordinary work.

A watt-hour meter may be made from any pendulum clock, a good one being, of course, preferable. The arrangement is shown in Fig. 196. The bob of the pendulum is replaced by a fine wire coil of many turns, small wires being led up to points near the center of motion of the pendulum and thence to the binding posts on the case. Below the pendulum and as near it as possible is a coarse wire coil, *B*. The main current flows through this while the fine wire swinging coil is bridged across the circuit. The clock is first adjusted carefully to keep correct time with no current in the coils. The mutual attraction between the coils causes either an acceleration or retardation of the rate of the clock in proportion to the watts passing, and all that is necessary is to know the constant of the apparatus and note the gain or loss of the clock. Up to a gain or loss of about 3 minutes per hour this meter is very accurate. Of course, to get the constant, it is necessary to calibrate the meter by working it for several hours on a known load. It is equally applicable to direct or alternating currents.

On a constant pressure supply system, direct current, the swinging coil may be replaced by a bar magnet, and the instrument then becomes a coulomb-hour meter, or, if the voltage is constant, a watt-hour meter. The well known Aron meter, much used in England, is constructed on this principle.

CHAPTER XX.

D'ARSONVAL GALVANOMETER.

Of the several types of galvanometers, there is, perhaps, no other which covers so wide a range of usefulness as the D'Arsonval, and certainly there is no other which can take its place in the dynamo room, for, owing to its intense magnetic field, it can be used in close proximity to dynamos, and it is not affected by wires carrying heavy currents, as are other types. It is a dead-beat instrument, enabling readings to be taken very rapidly, and it has not the delicate suspension of other forms, making it a convenient portable instrument. The form which it is the purpose of this article to describe will be found to meet the requirements of all but the most delicate tests, when, of course, a high-priced instrument is essential.

The magnet for this instrument is built up from sheets 1-16 in. steel of the form shown in Fig. 197. Six of these plates are bolted together, making the complete magnet 3-8 in. thick. The plates should be cut or forged from the best steel and bolted together; while in that position they should be carefully finished, after which they may be taken apart and tempered, but care should be taken to mark them with a prick punch so that they may be reassembled in the same relative positions. To harden them, heat a large flat piece of iron and lay one of the pieces on this. When it becomes a cherry red, quickly plunge it, points first, into a pail of water, repeating the operation with each of the remaining pieces. After being hardened they may be magnetized by means of the usual coils about their poles. Each should be magnetized separately. They may then be reassembled and polished.

A base should be provided of hard rubber or well seasoned hard wood, and should be turned, as shown in Figs. 197 and 199. Cut a mortise in the center which shall be a snug fit for the magnet and allow it to go ¼ in. deep. Cut two strips of 1-16 in. brass, 3-8 in. by 5 3-8 ins., and at one end of each turn up 3-8 ins., forming a right angle. Drill through the short limb of each for a screw to fasten it to the base; at

2 7-8 ins. from the base of these L-shaped pieces drill a 3-16 in. hole and bolt one to the back of each limb of the magnet. Place the magnet in position and fasten by screws to the short limbs of the brass strips.

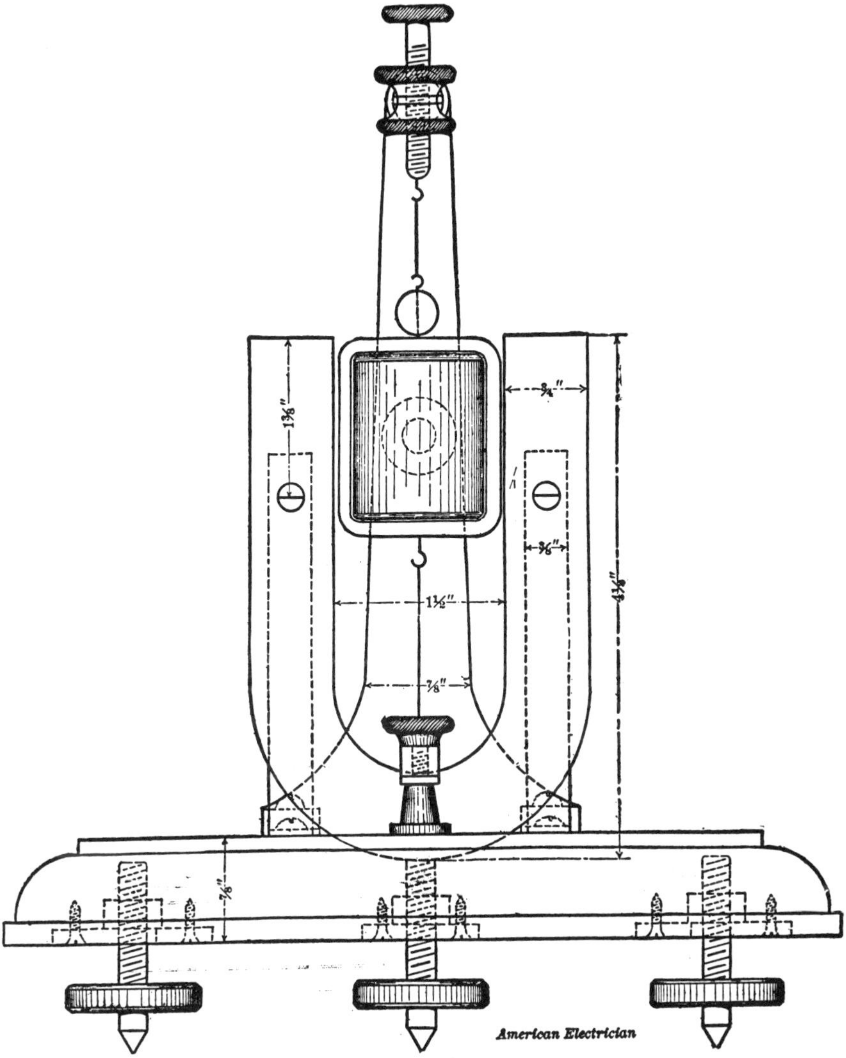

FIG. 197.

A pattern must be made for the standard shown in Fig. 198, and one cast from brass. It will be well to have this ½ in. longer than shown, as experience indicates that the suspension is improved by being made slightly longer. The casting should be finished, all holes drilled and secured to the base, as shown in Fig. 199. A brass button should be turned up and a slot cut therein to receive the end of a piece of 1-16 in. sheet brass, 7-16 in. wide, which should be soldered into this slot. A 3-16 in. hole is drilled through the end of this strip, 1½ in. from the

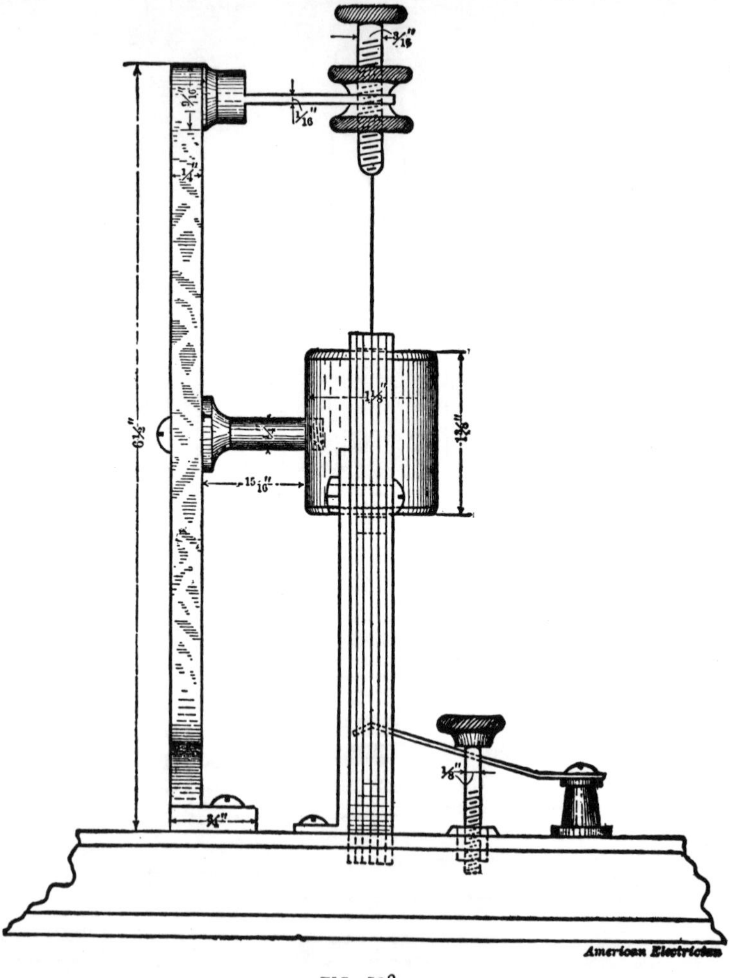

FIG. 198.

face of the casting, to the top of which this button is screwed. A milled head screw, 3-16 in. in diameter and 1 5-16 in. long, is provided with two milled nuts, and has a small hook, made from No. 20 hard brass wire, soldered to its end. This screw is passed through the hole in the brass strip; screw and nuts are shown in Figs. 197 and 198.

The cylinder shown in Figs. 197, 198 and 199 is turned from soft iron, and is supported by a brass arm from the standard. Its position is plainly shown in Fig. 198. A spring of 1-32 in. brass, 5-16 in. wide, is supported from a small brass pillar in the position shown in Fig. 198. A small milled-head screw passes through this spring and into a threaded

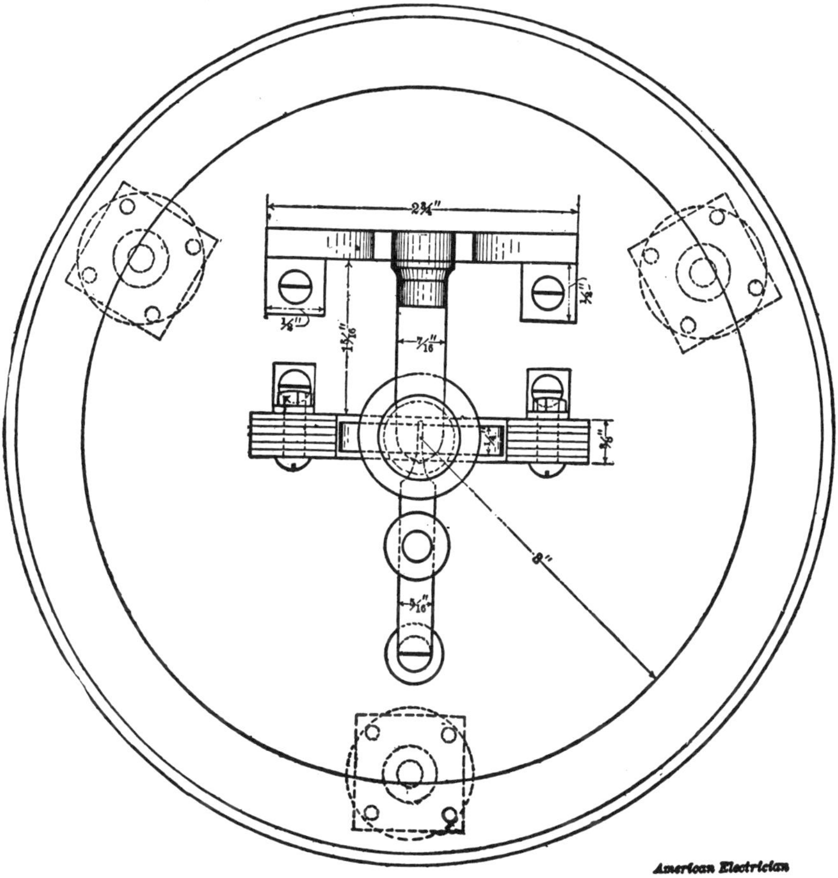

FIG. 199.

plate set into the base. This screw serves to adjust the tension on the spring to which the lower suspension is attached. A small hook, made from No. 18 or 20 brass wire, is soldered to the end of the spring in such a manner that it will be exactly under the center of the iron cylinder. One of the leveling-screws is shown in Fig. 200 and their positions are indicated in Figs. 197 and 199.

Two small binding posts are fastened on the extreme edge of the base so as to be outside of the glass globe, which must cover the completed instrument and rest upon the base. These posts are to be connected beneath the base, one to the small pillar on the front of the instruments and the other to the standard.

To wind the coil a form is necessary; this is made from a piece of brass ¼ in. by 1 3-6 in. by 1½ in. with a plate 1½ in. by 1 15-16 in. fastened by screws, to each side. The corners, over which the wire bends, should be slightly rounded. The wire to be used should be silk-covered and should not be larger than No. 36 B. & S., and the finer it is the more sensitive will be the instrument. Wind it carefully and in even layers, using no paper or other insulation aside from that on the wire. The completed coil should be 1 7-16 in. wide, but the exact length is immaterial. Place the form with the wire still on it in an oven and heat until as hot as the hand can bear; then with a clean soldering copper, drop on paraffine until the wire is completely saturated with it. After it has cooled, carefully remove the coil from the form and trim off the superfluous paraffine.

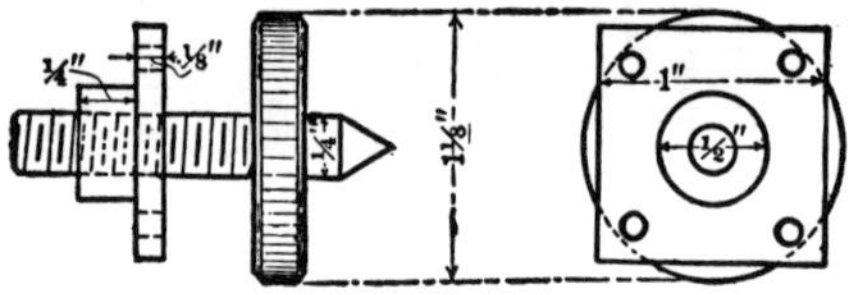

FIG. 200.

Cut from very thin copper, such as a leaf from a dynamo brush, two plates ¼ in. by 5-8 in. and to the center of one solder a small hook made from No. 20 or 22 hard brass wire; to the other one solder a similar hook, but with a shank 7-16 in. long. With a fine silk thread bind one of these plates to each end of the coil, being careful to place a thin piece of mica between the plate and the coil, and to have the hooks exactly in the center. To each plate solder one terminal of the coil and shellac all but the hooks.

A mirror 3-8 in. in diameter is now to be cemented to the shank of the longer hook by means of a little thick shellac varnish. This mirror may be made from a microscope cover glass, and silvered by the following formula: Take 100 parts by volume of a 10 per cent. solution of nitrate of silver and add, drop by drop, a quantity of ammonia just sufficient to dissolve the precipitate formed. Make up the volume to ten times the amount by adding distilled water. Dilute a 40 per cent. solution of formaldehyde to a 1 per cent. solution. Dip the glass, previously cleaned with chamois, in a mixture of two parts of silver solution to one of formaldehyde. After ten to fifteen minutes wash in running water and varnish the back. The silver will adhere to both sides and must be removed from the face.

The coil, with its mirror, is now to be suspended between the hook on the screw at the top and the one on the spring at the bottom. The suspension is a very fine phosphor bronze wire or strip, and can be best obtained from the makers of such instruments. For the suspensions take two pieces of the proper length and form a small loop in each end of each. These loops must fit snugly over the hooks. Suspend the coil by these, having the mirror at the top. Adjust the instrument so that the suspensions are taut with some little strain on them from the lower spring. See that the coil swings freely between the magnet limbs and the iron cylinder and is parallel with the front of the magnet.

The instrument is then complete, but should be provided with a reading telescope or a scale and lamp, which is not quite as convenient, but is simpler. It consists of a board 2 ft. long attached to a base and carrying a scale at about the same height as the mirror on the galvanometer. Just below the center of the scale is a 3-4 in. hole with a fine wire stretched perpendicularly across it. A lamp is placed with its flame opposite the hole and behind it and, by means of a suitable lens, the image of this wire is thrown on the mirror and reflected back to the scale, thus acting as a pointer. An ordinary magnifying glass will answer in the absence of a better lens. The scale should be about 2 ft. from the galvanometer and the lens should be between the two, rather nearer the scale. The exact positions must be left to experiment in each individual case.

CHAPTER XXI.

SENSITIVE MIRROR GALVANOMETER.

The instrument described herewith is intended to obviate almost entirely the necessity for skilful manipulations, upon the principle which pays so well in the machine shop, viz., that the whole be so designed in its several parts, that the machine work shall be reduced to a minimum, or even dispensed with altogether, save a little drilling, etc.

The above scheme has been adopted in making the galvanometer, which, after having been turned out "with jack-knife and pliers," will

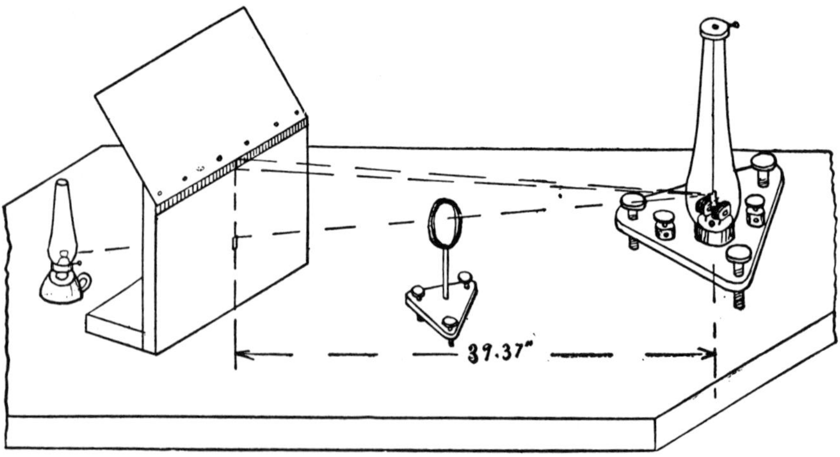

FIG. 201.—SENSITIVE "HOME-HADE" GALVANOMETER.

give results closely approaching those received from a more elaborate and costly instrument. Fig. 201 gives a view of the instrument complete. It consists of five parts—the lamp, the screen, the lens, the coils and the needles.

For the lamp, a bicycle lamp leaves nothing to be desired, though a common kerosene hand lamp, as shown in the engraving, answers every purpose. The vertical board is as high as the lamp, and the scale is attached to the top edge of the board. The scale may be an ordinary yardstick, or ruler fastened to the board, or it may be a strip of paper ruled to millimeters, and shellacked to the board.

The tin shade is simply to cut off some of the light which otherwise would be reflected over the top of the scale, and dim the bar of light. A clean, sharp slit may be made by cutting a somewhat large hole in the board, and covering it with a bit of cardboard or brass, in which a slit of the size found by experience to be best has been cut.

The lens may be an ordinary reading glass, or it may be one of the cheap lenses to be obtained in almost any shop for a few cents. Almost any form of lens can be made to answer, but preferably it should be a double convex of very long focus—16 ins. to 18 ins. If a reading glass is

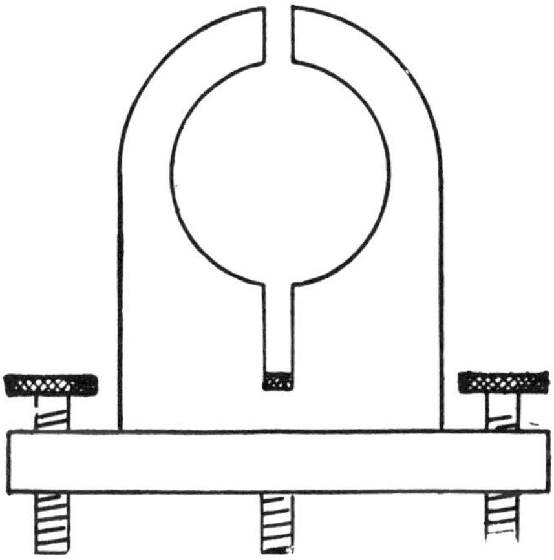

FIG. 202.—METHOD OF MOUNTING PLAIN LENS.

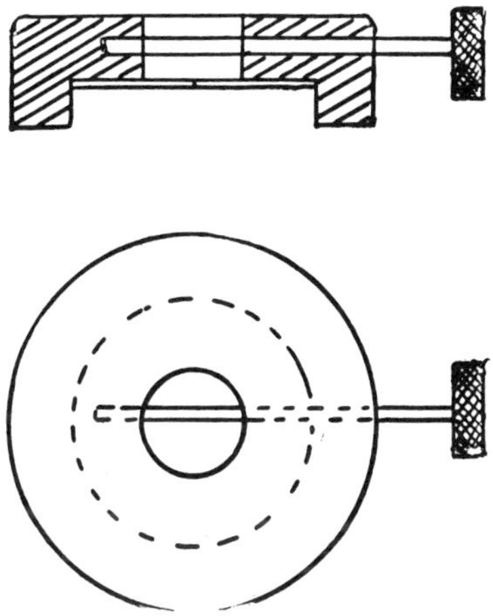

FIG. 203.—CAP FOR TOP OF GLASS CHIMNEY.

used, it may be mounted by placing the handle through a hole in the base-board as shown. If a plain lens is to be used, a cheap mount is shown by Fig. 202. A bit of board is cut out as shown, and the hole through it is just a trifle smaller than the lens. A narrow V-shaped groove is then cut around the center of the inside of the hole, and a saw kerf run into the board as shown. This allows the lens to be pressed into the groove, and the spring of the wood holds it there.

The six leveling screws are common brass wood-screws, 4 ins. long, about ¼ in. in diameter, with the top of the head filed off flat. The edges of the disc thus formed may be milled in pretty good shape by rolling the edge of the head under a single-cut file of the required degree of fineness. Place the screw on a hardwood board, or better yet, on a sheet of lead, and by rolling under a file, the milling can be quickly done. By all means use a lathe if you have one, in preference to the file method.

The third member is built on a bit of board cut about 8 ins. on a side of triangular shape, as shown. Three leveling screws are let in, and two binding posts are placed in connection with the coil. These posts are shown in the engraving. A common medium sized lamp chimney is procured and fitted to a circular piece of wood ¾ in. thick. The wood is screwed to the base, and the coils are fastened to the wood; the mirror must be placed one meter (39.37 ins.) from the scale.

Another circular piece of wood is fitted to the top of the chimney, as shown in Fig. 201. A detail plan and section of this piece is shown by Fig. 203. It is bored out to fit on the chimney, and a ½ inch hole is bored in the center, completely through the wood. A wire with a sort of thumb-head is bored into the wooden cap so as to pass through the center of the ½ in. hole. A bit of cardboard is glued into the bottom of the large hole, and a pin-hole punched through the exact center, permits the suspension fibre of the needle system to be carried to the wire and wound up by turning the thumb-head above described.

The coils may be made according to the work to be done; the writer has three sets of coils with his own instrument, two in each set, and uses whichever set the character of the work requires. The first is made of about 50 ft. of single-silk-covered copper magnet wire, the size being No. 19, B. & S. gauge. The second set of two coils is wound of No. 33 or No. 34 wire. The third coil is wound with No. 36 wire. Nearly ¼ lb. was put on the two coils, and the combined resistance of the complete coils is about 1,000 ohms—500 ohms each.

A form for winding the coils is shown by Fig. 204. It is made of wood, held together with two screws. A couple of binding wires are laid in before the coil is wound. About six layers of the wire above mentioned can be made out of one-half the 25 ft. mentioned. Two of these coils are used, connected in series and to the binding posts. After winding, the binding wires are fastened, the coil is drenched with shellac and placed in the cook-stove oven for an hour. The core is then removed, additional binding placed on the coil if found necessary, and again baked at low heat for two or three hours. This holds the coil permanently. Two coils are to be used, and the needle system suspended between the coils, which are placed 3-8 inches apart.

For the needles with the low resistance coils the writer used a common sewing needle. The temper was drawn, the eye and point filed off, leaving a bit of wore 1¼ ins. long. A nick was filed in the center, then the needle was hardened and magnetized, and broken through the nick, thus giving two needles magnetized pretty near alike. A piece of cardboard 2 ins. × ½ in. was pierced, and the needle stuck through it, as in Fig. 205, and held by a drop of hot sealing wax.

A bit of mirror, *m*, was waxed to the top of the cardboard, and the suspension fibre fastened between the mirror and the cardboard, as shown. The upper end of the fibre is carried to the cap on top of the chimney, attached to the thumb-head wire, and wound up until the lower needle hangs in the middle of the coil, and the upper needle clears the top of the coil about ¼ in. The instrument is now ready for setting up and adjusting in the usual manner.

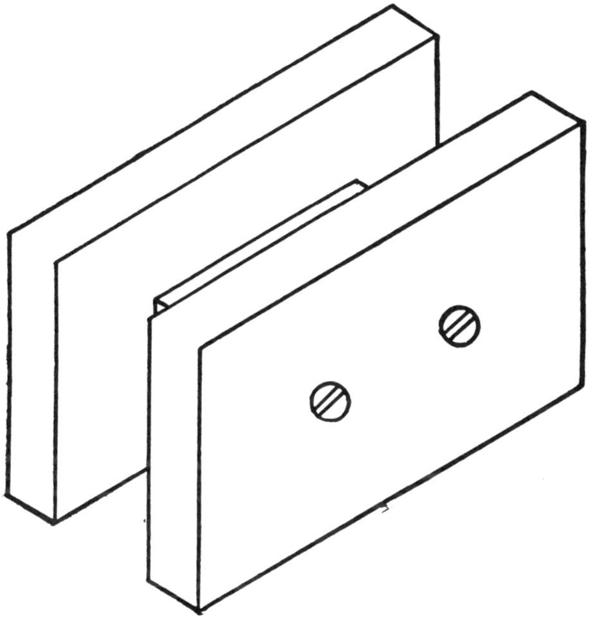

FIG. 204.—FORM FOR WINDING COILS.

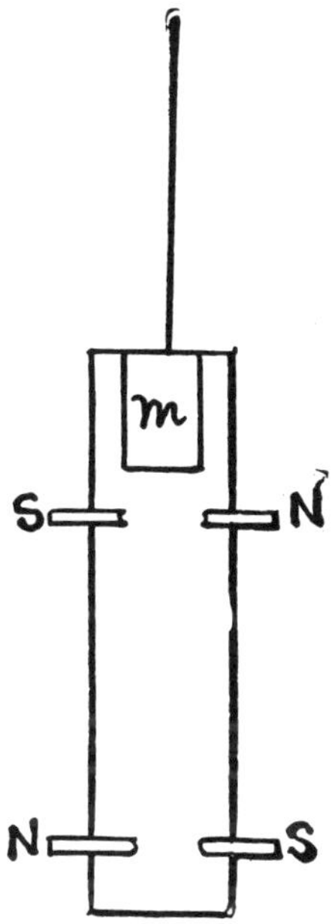

FIG. 205.—NEEDLE SUSPENSION.

The second set of needles is made in the same manner, except that pieces of fine watch spring less than ⅛ in. wide, may be used preferably, three pieces being placed together with a single thickness of paper befor each needle. The pieces should be file-marked, hardened, magnetized and broken in pieces, the same as the needles.

Finding that the light needles and the low resistance coils gave an instrument readily affected by thermal currents, the writer made the third set of needles of steel tape about ⅜ in. wide, and used five pieces in each needle, separating each with a paper. All the needles in the three systems were ⅝ in. long. The third set was rather heavy, but in connection with the 1,000-ohm coils proved very sensitive, although slow-moving.

Different effects were secured by using either set of the needles with the other coils, making six possible combinations. Where extreme sensitiveness is not required, the writer found it desirable to use a directing magnet, and not depend upon the torsion of the suspension, or over-strength of one of the needles, to return the beam of light to zero.

With 1,000 ohms in each arm of the bridge, and 6 volts from the battery, a considerable deflection is obtained by changing *R* a single ohm, and with the bridge adjusted at 1,000 to 1 at *a* and *b*, the galvanometer readily deflects beyond the capacity of the bridge, which is .001 ohm, with 1,000 ohms galvanometer resistance.

For the suspension in this instrument the reader may use a hair, quite fine, say about .002 in. in diameter. From the needles to the point of suspension there should be about 8 in. of effective hair. Just how much better the instrument would be with a raw silk fibre the writer has no means of knowing at present, but it was as delicate as will be required for any ordinary work. The "efficiency" of the low-resistence instrument is rather greater than that of the high-resistance form, while the "figure of merit" is greater the more turns of wire are placed on. For measuring very low resistances, the low-resistance coils will give perhaps the best results.

CHAPTER XXII.

A THOMSON ASTATIC GALVANOMETER.

While the Thomson astatic galvanometer has been more or less superseded by the D'Arsonval instrument, it is, nevertheless, an excellent instrument for the detection of feeble currents where great sensitiveness is required. It is, therefore, used for nul or zero methods, such as measuring resistance by the Wheatstone bridge; Rayleigh's and Bosscha's

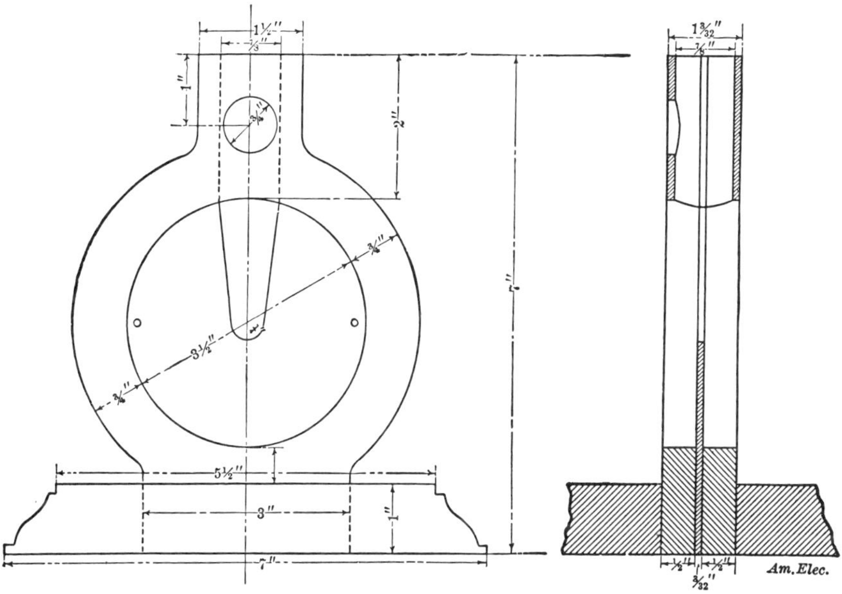

FIG. 206.—FRAME OF GALVANOMETER.

methods of comparing the electro-motive force of primary cells; Thomson's method of comparing the electrostatic capacities of two condensers, etc. It requires, perhaps, more skill and patience on the part of the user than the D'Arsonval, and cannot, like the latter, be so readily used in the neighborhood of dynamos or wires carrying strong and variable electric currents.

The galvanometer about to be described has proved a very useful instrument for laboratory work and may be made very sensitive if desired.

The woodwork, as shown in Figs. 206 and 207, is made up of four pieces. Into the base is mortised and glued the upright portion, consisting of a thin central sheet or diaphragm, 3-32 in. thick, glued between the two cup-carrying supports. This thin partition holds the two coils apart and is cut away, as indicated in Figs. 206 and 208, to give sufficient room for the suspended system. A hole, 7-16 in. in diameter and 2 ins. deep, should be drilled from the top, and another hole in the front, 3-4 in. in diameter, as shown in Figs. 206 and 208.

This front opening is to be covered with a clear piece of perfectly plane glass, ground square or round, as shown in Fig. 215. To make this

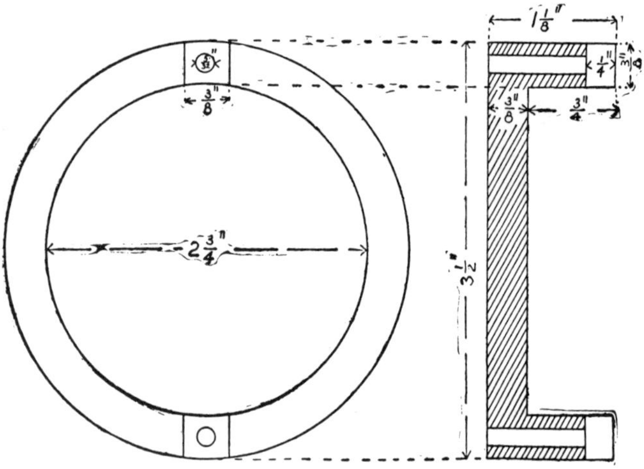

FIG. 207.—COIL CUPS.

glass fit air-tight and yet not break when the screws are tightened, put a washer of rubber, felt, or chamois skin under the edge. Fig. 208 is a half-tone picture of a similar instrument. After the two main binding posts, three leveling screws, two spirals of No. 26 silk-covered copper wire, soldered to the under nuts or washers of the binding posts, and two brass rods for holding the coil cups, have been put in place, set the binding posts well apart and near the edge of the base to allow plenty of room for subsequently putting in place and removing the front coil-cup. Fig. 207 is a scale drawing of the wooden cups for holding the coils. These cups should fit snugly into the uprights in order to be as air-tight as possible.

Well seasoned, hard maple is a very good wood to use and polishes nicely. The inside surfaces of the woodwork should be treated to a coat of shellac. To give the outside a fine polish proceed as follows: After making the outside surface as smooth as possible with very fine sand-paper, coat it with hard oil. When perfectly dry, rub it down well with powdered pumice stone, using a soft rag and boiled linseed oil thinned with kerosene. Repeat this several times and the last time or two use powdered rotten stone in place of the pumice stone.

Fig. 209 shows one of the leveling screws and Fig. 213 the tapped brass rod and nuts for holding the cups in place. There are two of these brass rods, threaded their entire length, eight brass washers, four small square brass nuts and four larger ones. Two of the latter are shown on

FIG. 208.

FIG. 215.

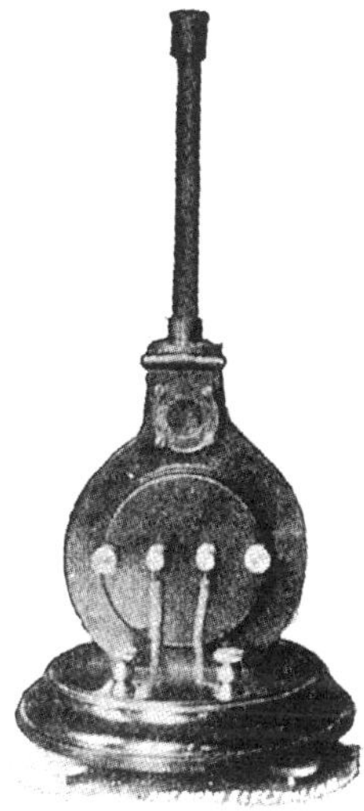

FIG. 216.

the front of the finished galvanometer (Fig. 216). Fig. 207 shows how these rods are fastened in place by the small square nuts, one nut and washer on each side of the thin partition.

Fig. 210 shows a form on which the coils are wound. The one from which this drawing was made was constructed of iron because a large number of coils were wound upon it, but, where only two coils, as in the present case, are to be wound, it could be made of some hard wood by slight modification. Before winding the coil cover that part of the form which is to be filled up with wire, with two or three thicknesses of paraffined paper to facilitate the removal of the coil when finished. The wire coil should fill the form to within ¼ in. of the edge. To make a coil with a resistance of 200 ohms, making a 400-ohm instrument when the two coils are connected in series, will require about 13 ozs. of No. 30 B. & S. double-silk-covered copper wire. By connecting the two coils in parallel, instead of in series, the galvanometer will have 100 ohms resistance. In winding these coils it is well to first wind on a single layer of No. 26 wire, to which the No. 30 is soldered, and to terminate the winding with a layer of No. 26. This protects the finer wire and furnishes a more suitable wire for making soldered connections to the binding post washers. In soldering connections use resin and not soldering fluid, as the latter, unless completely neutralized and dried off, may later corrode the metal and cause trouble.

The coil, when wound, should be well treated with shellac and allowed to dry before attempting to remove it from the form. To do this, it is necessary to heat the form until the paraffined paper softens enough to allow the removal of the coil from the form. The shellac should hold the coil in shape. The ends of the coil are soldered to washers which are firmly fastened in counter-sunk holes in the inside of the wooden cup by the binding post screws. After placing the coil in the center of the cup pour a mixture of melted paraffine and resin into the cup, around the coil, to hold it there. Paraffine alone is too soft in warm weather to sustain the coil in place. Fig. 211 is a horizontal cross section of cup and coil complete. Fig. 215 gives an inside and outside view of the same. When the coils are in place the hollows at the centers of the coils should come exactly opposite each other and just above the bottom of the cut-away portion of the separating wooden partition in the middle of the upright supports, and the surface of both coils should press against this separating partition.

A more sensitive instrument, having the same resistance, can be made by winding the two coils as follows: Over a single layer of No. 26, wind 136 ohms of No. 36, heaping up the wire slightly toward the face of the coil which is to be nearest the needle—that is, on the side of the form where the conical spindle has the smallest diameter. Over this wind 52 ohms of No. 31, and finally, put on 12 ohms of No. 26.

The two coils are to be connected either in series or parallel, by spiral wires running through and under the wooden base, so that the current, flowing through both, will tend to deflect the needle system in one and the same direction; that is, one coil must not oppose the other in its action upon the magnetized needle. The ends of these spiral wires should be soldered to washers under the nuts of the main binding posts. Two main binding posts, similar to Fig. 217 and four smaller ones, for the coil cups, similar to Fig. 218 will be needed.

Fig. 212 shows two brass pieces which are connected together by a stout piece of glass tubing, 6 ins. in length. The outside diameter of

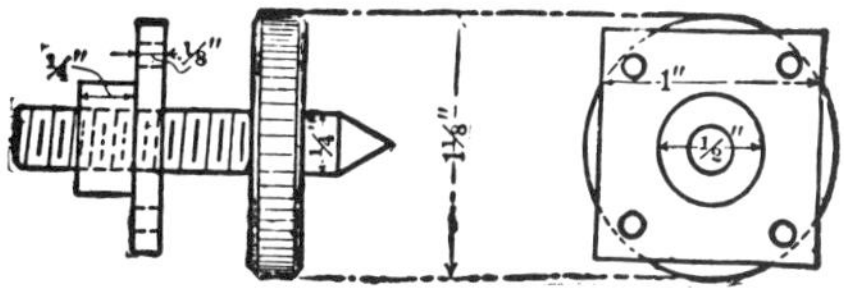

FIG. 209.

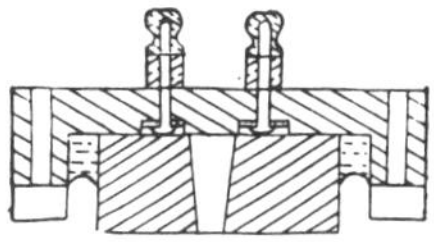

FIG. 211.

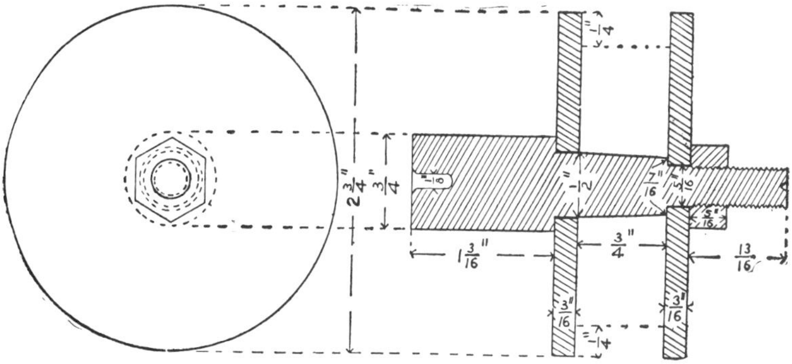

FIG. 210.—FORM FOR WINDING COIL.

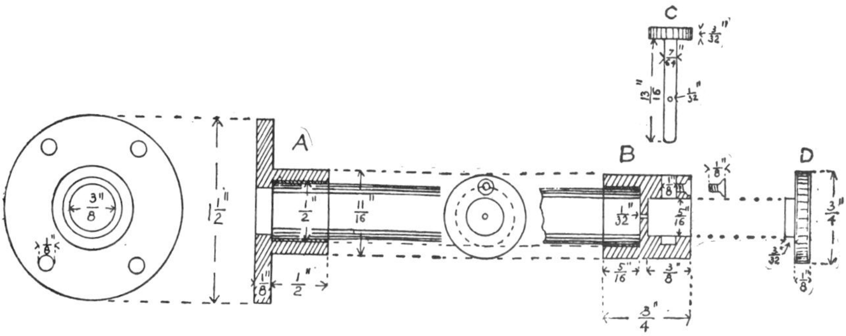

FIG. 212.—DETAILS OF SUSPENSION TUBE.

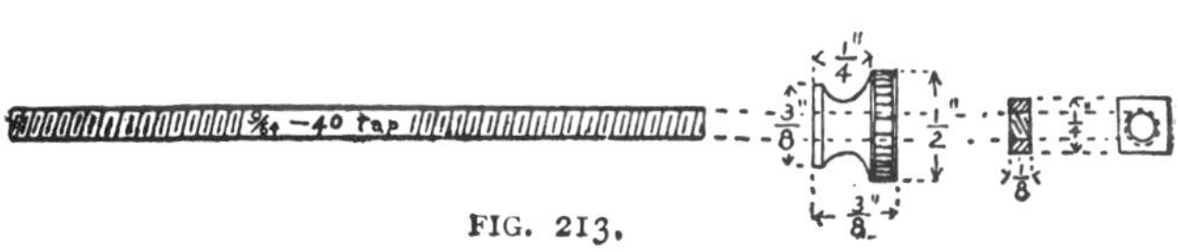

FIG. 213.

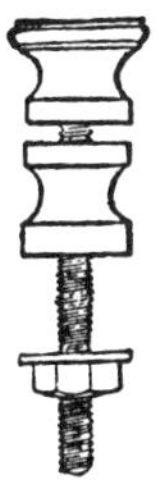

FIG. 217.

FIG. 218.

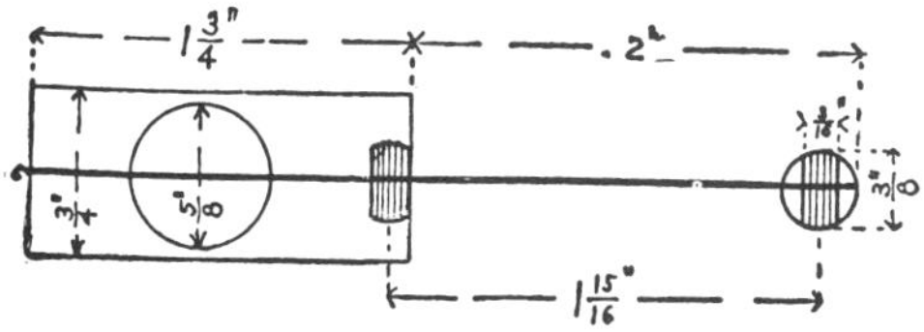

FIG. 214.

this glass tube should be enough smaller than ½ in. (the diameter of the hole into which it is to go) to allow a piece of chamois skin to be wrapped once around the tube and thus make the two ends fit snugly into the two brass pieces, *A* and *B*. In order that it may not turn, the bottom one is to be firmly secured with shellac or LePage's glue on both sides of the chamois skin; but the top piece of the skin is glued on only one side to allow the top brass piece to be revolved around the glass tube so that the torsion may be removed from the suspending fibre, whenever necessary, in adjusting the instrument to give zero reading. By using a longer glass tube and, consequently, making the suspending silk fibre longer, the sensitiveness of the galvanometer may be increased.

The bottom brass piece, *A* (Fig. 212), is fastened to the top of the wooden framework by four brass screws; but, between the wood and metal, a piece of felt is interposed. This is in the form of a circular disc of 1 3-8 ins. external diameter and cut away at the center to allow passage for the suspending fibre. This felt serves two purposes: First, it prevents air currents passing through and disturbing the action of the needle system; secondly, it affords a means of adjusting the suspension tube into perfect alignment with the rest of the instrument by tightening the proper screws.

It is perhaps needless to remark that no iron in any form, except for the needles, must be used in the construction of this form of galvanometer.

The astatic needle system is shown in Fig. 214. It consists of a very thin, rectangular piece of mica, ¾ in. × 1¾ ins., to which are fastened, by shellac, a plane glass mirror from ½ in. to 5-8 in. in diameter, the upper set of steel needles, and a thin glass fibre about 1-50 in. in diameter. The object of the mica vane is to dampen the vibrations and help bring the needle system to rest and, for this purpose, it should be as large as the space in which it swings will allow. The space which contains the needle system and the silk suspending fibre should be made as air-tight as possible to prevent external gusts of wind or currents of air from entering and causing the needle system to vibrate and shake. When finished and set up ready for use, the mirror should come opposite the glass covered hole in the front of the instrument. It is difficult to make a good mirror and it is more satisfactory to purchase one from an electrical instrument maker. However, if the reader desires to make one he will find the necessary directions on page 178.

The glass fibre used to connect the lower set of needles to the mica vane on which the upper set of needles is fastened, must be perfectly straight and very light. By a little practice a good one can be made from a small glass tube by heating it over a fish-tailed gas burner and drawing it out. Make a number of these and select the best one. To the top of this secure, by shellac, a minute hook made of No. 36 or No. 38 bare copper wire, and to the lower end glue a very thin circular piece of mica upon which is glued one set of steel needles. The two pieces of mica must be in the same plane. This needle system should be as light as possible (not over ten grains), to make its moment of inertia small. Other things being equal, the smaller this moment of inertia the quicker it will come to rest.

The steel needles may be made from steel piano wire about 1-40 to 1-50 in. in diameter, or No. 8 guitar string. The temper should first be drawn and the wire straightened and cut into convenient lengths—about

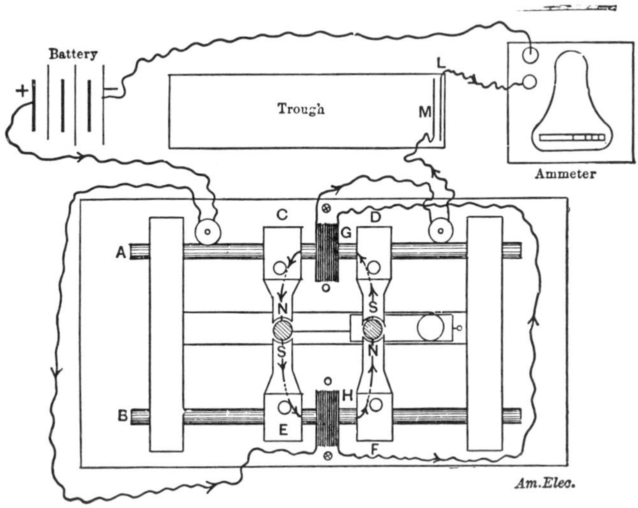

FIG. 219.—METHOD OF MAGNETIZING NEEDLES.

6 ins. long. These pieces must then be tempered glass hard. This should be carefully done, as the entire piece should be of the same hardness. From this wire cut or break off twelve pieces of the proper length (as indicated in Fig. 214) for the needles and secure them in place with shellac, six above and six below, being careful to have a thin air space, about 1-50 in. between each needle, so that they will not touch each other.

The best way to magnetize the needles and obtain the astatic system—that is one set magnetized equally but oppositely to the other—is to magnetize the set by an apparatus shown in Fig. 219. *A* and *B* are soft-iron rods, 7-16 in. in diameter. *C*, *D*, *E* and *F* are soft-iron pole pieces between which the needles to be magnetized are placed, as shown. *G* and *H* are two coils, each consisting of 350 turns of No. 18 B. & S. copper wire. The coils must be connected in series so that the magnetic potential of each will produce a magnetic flux and poles as indicated in the figure. The trough is merely a variable liquid resistance containing some liquid, such as salt water or a solution of washing soda, and two metal plates or electrodes, one of which can be moved from one end to the other, so that the current may be varied from 0 to about 5 amperes. Any other convenient variable resistance which will do this may, of course, be used.

After making connections and adjusting the strength of the solution so that the proper current can be obtained, place the needles between the pole pieces, as in Fig. 219, and put on the full current of 5 amperes and then gradually and slowly reduce the current to zero. Repeating this process several times should magnetize the needle sufficiently, and if the steel were properly hardened, it may never need remagnetizing.

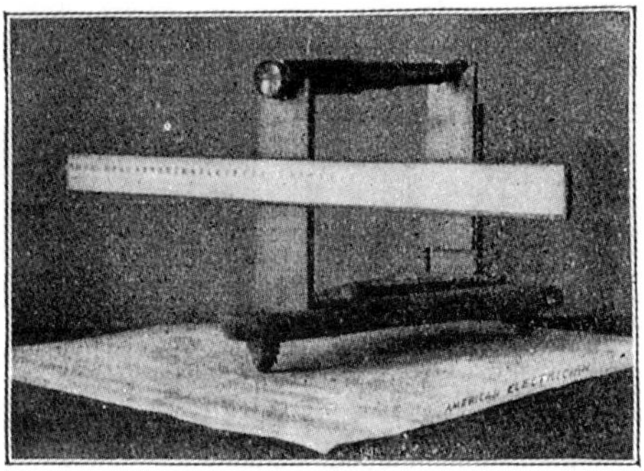

FIG. 220.—TELESCOPE AND SCALE.

For suspending the needle system, get a single fibre of unspun silk at least 10 ins. long. To remove all initial torsion, hang this up for a day or so, having fastened to its lower end a small weight of non-magnetizable material, such as a brass screw. Pass this silk fibre through the 1-32 in. hole in *B*, Fig. 212, and secure one end to *C*, by tying it through the 1-32 in. hole in that piece. One or both of the coil cups being removed, work the free end of the silk fibre down through the glass tube. Fasten this free end to the small hook on the upper end of the needle system with shellac. When the shellac is dry, wind up the fibre on *C*, until the lower needles hang in the center of the coils. Then level the galvanometer until the lower needles hang midway between the two coils and turn perfectly free. With both coil cups in place and the galvanometer properly leveled, the needle system should still vibrate free and smooth. It will depend upon the resultant polarity of the astatic system whether the instrument should be set facing East or West. This can best be determined by trial.

If a plane mirror is used, a telescope and scale for observing the deflection of the needle system is more convenient than a lamp, scale and lens. A telescope, suitable for this purpose, can be purchased for $1.80 or less, and a one-half meter scale on cardboard can be obtained for twenty-five or fifty cents. If a terrestrial telescope is used, it may be greatly improved, for this purpose, by removing the set of rectifying lenses, in which case the figures on the scale must be reversed and inverted. Fig. 220 shows a suitable stand for this purpose, which may be easily made. It is customary to place the telescope and scale just one meter from the mirror. The rear support for the telescope and the board upon which the paper scale is fastened are arranged to slide up and down for adjusting. The scale should be as much below the level of the mirror as the object glass of the telescope is above it.

Quite a number of galvanometers of the above type with the telescope and scale have been in regular use in the electrical laboratory at Lehigh University for the past four or five years. There are also several four-coil galvanometers of a similar design, whose resistances run up as high as 5,000 ohms, but for general use the two-coil instruments are sensitive enough. Considerable of the work on these galvanometers was done by students taking the electrical course during their third year.

CHAPTER XXIII.

A CHEAP TESTING SET.

The Wheatstone bridge is one of the most valuable instruments that can be devised, not only for its primary office of measuring resistances, but for testing faults of any kind. Most beginners are not aware that a convenient substitute can be made that will answer in many cases, and the expense of which need not exceed two or three dollars. Following a description of such an easily made apparatus is given, and it may be of interest to those anxious to possess a testing set and who do not feel that they can afford the large sum usually asked for such an outfit.

Select a board of some dry wood and shape it nicely to the dimensions of 42 ins. long by 8 ins. wide. Procure some flat copper rod which is at least ½ by ¼ of an inch in sectional area. Of this there should be one continuous bar 36 ins. long, and two shorter pieces about 3 ins. long. Then secure a good straight piece of German silver wire, about No. 14 B. & S. gauge. Go over this with a micrometer, testing it at every inch in its length, and be sure that the piece selected is of uniform diameter and has no nicks or marks of any kind.

With a suitable size of drill bore holes ¼ of an inch from the ends of the short pieces of copper rod, and ⅛ of an inch from the flat sides, as shown in Fig. 221. The wire should now be soldered into the copper block, care being taken that it comes exactly flush with the block, no drops or beads of solder appearing around the joint. To do this, the better way is to heat the copper block very hot and thoroughly tin the inside of the hole and fill it with solder, which, if the block is hot enough, will remain melted. Having previously tinned the outside of the wire, insert it in the hole thus prepared, pushing it straight in until the proper distance has been reached. Be very careful not to pull it outward unless the solder does not fill well around the hole where it enters, because any outward pull of the wire would draw a meniscus of solder around it which would be undesirable. The length of wire between the two connecting blocks, as shown in Fig. 222, should be exactly one meter long. This should be most carefully adjusted for accuracy.

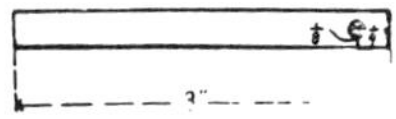

FIG. 221.—COPPER BAR.

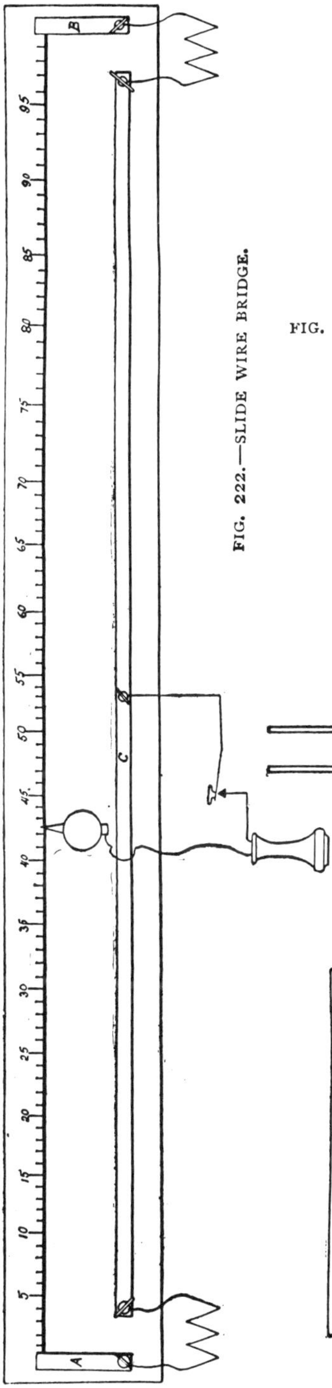

FIG. 222.—SLIDE WIRE BRIDGE.

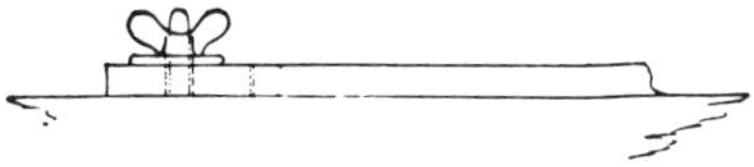

FIG. 223.—METHOD OF FASTENING TERMINAL BLOCKS.

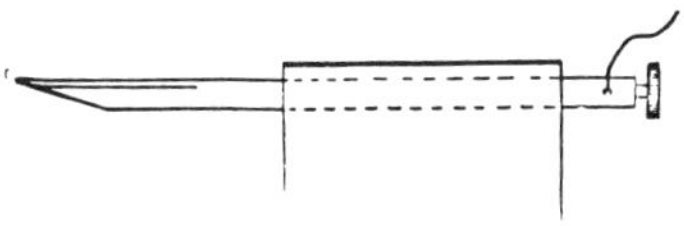

FIG. 224.—WEIGHTED INDEX.

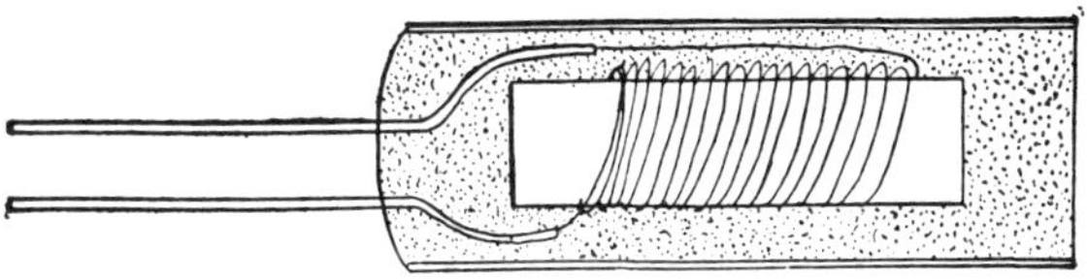

FIG. 225.—RESISTANCE COIL.

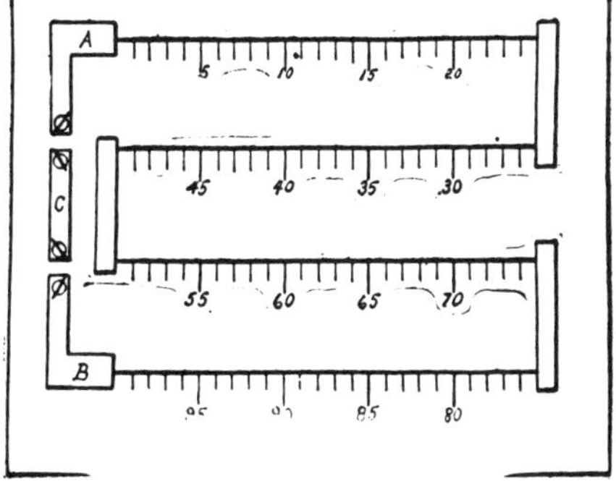

FIG. 226.—IMPROVED BRIDGE.

The copper blocks are provided with binding screws at their terminals for the convenient insertion of known and unknown resistances, and are secured to the board as shown in Fig. 223 by screws passing up from beneath. The blocks carrying the wires are arranged to hold it so that it will be free from all kinks and twists, but not to put any mechanical strain upon it, which might change its resistance; in other words, the wire should be straight, not stretched. A little platform of thin wood or pasteboard should now be built up underneath the wire, so that it will support it throughout its length, and on the top of this strip of pasteboard should be pasted a meter scale; if the wire has been accurately adjusted, it will exactly fit between the terminal blocks. If these directions and the drawings are carefully followed, the result will be a bridge of considerable range.

To use this instrument, a source of e. m. f. is connected to the end blocks and a galvanometer, or, as shown, a telephone receiver, is connected to the long middle block, and its other terminal to a weighted index, which can be moved along the board between the wire and the long copper rod and make contact with the former at any point desired. This index is preferably made of a brass rod, one end of which has been filed up to a V-shape and drawn out to a sharp point, as shown in Fig. 224. The other end of the brass rod is conveniently shaped up into a binding post, to which the detector terminal may be atttached. The weight may be a piece of lead cast about it and suitably shaped. The sharp edge of the brass should be very soft in order not to mar the wire at the point where it rests upon it, and for that purpose should be annealed by heating it in a flame and allowing it to cool slowly. A few standard resistances should be provided; one, ten and one hundred ohms will be sufficient. These may be conveniently made by measuring off the proper amount of insulated wire with a testing set, coiling it about a small wooden block and inserting it in a short piece of large brass tubing, which is then filled with paraffine. These coils may be made of moderately fine wire and stout, projecting terminals of as nearly no resistance as may be, led out through the paraffine for purposes of connection. Fig. 225 shows a section of such a coil prepared for use.

The known resistance is selected as nearly as possible equal to the unknown resistance to be measured. For instance, if we are measuring a series of field coil we know its resistance to be at most but a fraction of an ohm, consequently we should use the lowest resistance coil that we have, which in this case would be one ohm. Similarly, in measuring a shunt field coil whose resistance may be 16 or 20 ohms, we should use the ten-ohm coil, and so on. The known resistance is connected across the binding posts at one end of the bridge, and the unknown resistance is similarly connected at the other. A source of e. m. f., two or three dry cells will answer, should be connected from *A* to *B*. The weight index is adjusted along the wire for a position where no deflection of the galvanometer occurs, and the point is read on the meter scale.

If a telephone is used instead of a galvanometer, it is best to interpose a key in the circuit, as shown in Fig. 222. The procedure is to then seek a point on the scale where the opening and closing of the key produces no click in the receiver.

Multiply the known resistance by the length of the wire on the unknown side and divide it by the length of wire on the known side, and the result will be the value of the unknown resistance.

The bridge that has just been described is rather a clumsy construction, being nearly 4 ft. long, but it can be modified as follows, and the result will be more convenient:

A board 12 ft. x 14 ins. is provided with blocks of copper, as shown in Fig. 226. Four wires, each 25 centimeters long (about 10 ins.), should be soldered into them, as shown in the diagram, the blocks at the end being shaped as shown so as to be brought up conveniently near a third one. Assuming the blocks to be of negligible resistance, the wire must then be divided into four lengths of 25 scale divisions (one meter) each, and thus a more compact and satisfactory instrument will be obtained.

CHAPTER XXIV.

CONSTRUCTION AND USE OF A PHOTOMETER.

In the photometer here described, and for the construction and use of which directions are given, an attempt has been made to gather into one instrument the good points of a number, and at the same time to avoid their defects. In this an especial indebtedness is owed to a portable photometer made by the Electric Motor and Equipment Company, of Newark, N. J., which must be here acknowledged.

For the construction shown no materials are required which are not readily procurable at small cost, nor are there any processes of manufacture involved which are beyond the capabilities of a man with very simple tools and ordinary mechanical ability. In many things some simple change might make a more finished instrument, but more machine work would be involved. Such modifications will readily suggest themselves to anyone undertaking the construction.

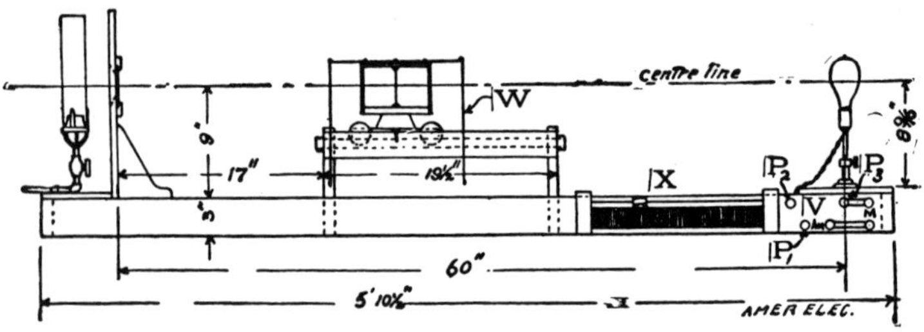

FIG. 227.—THE PHOTOMETER COMPLETE.

A general drawing of the instrument is given in Fig. 227. The frame, which supports everything, is of ⅞-in. white pine 3 ins. wide, set together in an open rectangle 5 ft. 10½ in. long by 6¾ ins. wide over all. This frame may be supported at a convenient height by brackets on any side wall. At the left is an argand gas burner, in front of which is a board with an adjustable slot cut in it, the slot being directly in front of the argand flame. At the other end, on a suitable stand, is the incandescent lamp whose candle-power is to be measured. Between them is a frame on which travels a car containing a piece of paper set in a plane at right angles to a line joining the two lights. This paper has a grease spot on it, and in use the car is moved backward and forward until a place is found where the grease spot disappears. The candle-power of the lamp is then read off on a scale immediately below the car. For adjusting the e. m. f. on the lamp terminals a rheostat is provided, shown between the incandescent lamp and the track for the car at X in the drawing. The instrument is in principle a Bunsen photometer.

In Fig. 228 the car is shown in detail. The car proper is of wood, open in front, closed at the back, with ends cut away in openings 2½ ins. in diameter. Half way between these two ends is the screen—seen in

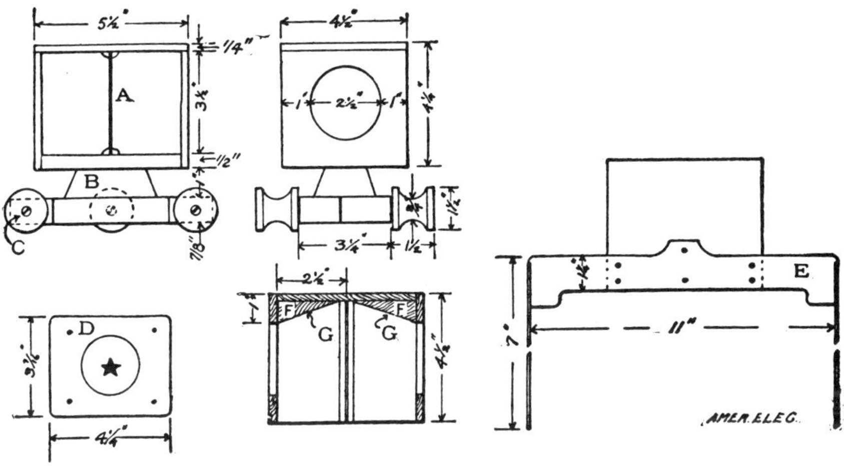

FIG. 228.—THE CAR IN DETAIL.

edge view at *A*. This is of two pieces of sheet metal (see *D*), about No. 14 or No. 15 B. & S. gauge, in both of which there is cut a circular hole 2 ins. in diameter. Between these pieces is put the paper with grease spot, as shown, and the two plates held together with small screws so the paper and grease spot can readily be renewed when necessary. The compound plate, *D*, slides in place from the front of the car in small wooden guides as at *A*.

To the right of the frame *D*, in the figure, is a section of the car taken half way between the top and bottom. The two blocks, *F*, *F*, are wooden ones, extending from bottom to top of the car on which pieces of mirror are mounted. These mirrors should be of good quality, as near alike as possible, and covering a large part of the surfaces *G*, so that an observer sees in each of them the reflection of the paper and grease spot—one side of the paper in one and the other side in the other. The exact angle at which these mirrors should be placed will best be determined in each case by trial. As shown, the results should be perfectly satisfactory. The piece *B*, between car proper and base, on which the wheels are mounted, is to be of lead. It may be cored out some, but is to give weight and stability to the car on the track and so must be heavy. The base to which wheels *C* are secured is of wood ⅞-in. thick and shaped to carry the lead block and three wheels as shown, two running on the forward track and one on the rear track. These wheels are simply porcelain insulators or knobs, which any central station will be likely to have in stock, or which may readily be procured from any dealer in electrical supplies. They run on wood screws as axles. If they are turned from iron or brass, *B* may be made of wood. The three parts making up the car are put together with two slender bolts passing down through all, from the car. *E*, in another part of the figure, shows a brass plate of about No. 12 or 14 gauge, screwed to the top of the car and with light stiff wires (about No. 10) soldered to it and projecting forward. To each of these wires is hung a curtain of black drilling, 4½ or 5 ins. wide and 10 or 12 ins. long, hung as shown at *W* (Fig. 227). These curtains are to screen the eyes of the observer from the lights at the end of the bar. When this car is finished, it must be painted black all over. Not a shiny black, but a dull, dead black, such as may be had at least cost probably by cutting a little lamp black with turpentine and adding just enough shellac varnish to make the black stick, but not get glossy.

Before taking up another part of the photometer, a few words as to methods of making the screen will be well put in. The sort of paper used is not of great importance. It must be white and quite nearly opaque. Preferably both sides should be as nearly alike as possible. The grease spot must be as transparent as possible, have a very clearly defined edge and the more edge the better. In Fig. 231 there is shown a star which is a good size to use and which has plenty of edge. To make the grease spot, cut the star out of brass plate (say ⅛-in. thick) and mount it on a rod set perpendicular to the plane of the star. Melt some paraffine over a water bath, put the star into the paraffine and let it warm some; then take it out, let it drain and set it down on the paper selected. A number of grease spots having been made, pick out the best for use. I have found a typewriter's paper, imitation linen, smooth, and rather thin, quite satisfactory—care being taken not to use any part containing a water mark.

The car runs on a track of which a satisfactory idea may be had from Figs. 227 and 232. The two tracks are of common ½-in. iron pipe, shown at *P* (Fig. 232), slipped through end pieces, whose form and size are shown in the same figure; and finished by iron caps screwed on each end. It will be necessary to pick out good smooth pieces of pipe for the purpose and perhaps do a little filing, so that in use one will not unconsciously tell by the "feel of the track" when the car is in some certain position along it. *Q*, in Fig. 232, is a strip of 7-16-in. wood which is in front of the pipe tracks in Fig. 227, and is to have the scale for reading candle-power mounted on it.

The framework must be mounted rigidly in place and the distances between it and the end fixtures, as well as the distances between them (the

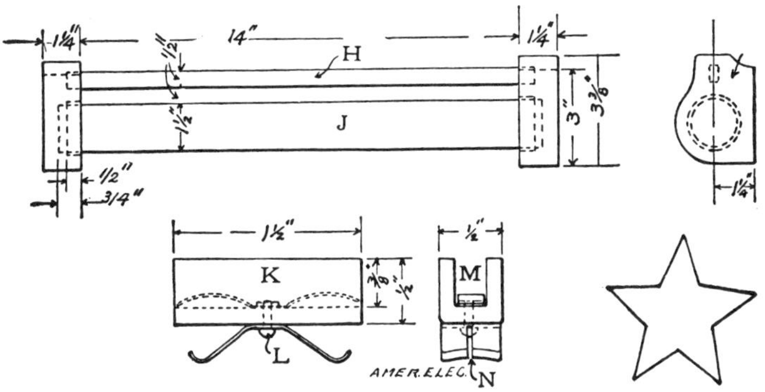

FIGS. 229 AND 230.—DETAILS OF RHEOSTAT AND CONTACT MAKER.

FIG. 231.—STAR OF SIZE FOR GREASE SPOT.

screen and the incandescent lamp) made exactly as shown. It is on the accuracy of these measurements that the value of the table given on page 207, determining the candle-power scale, depends. The scale of candle-power, to be mounted on the strip in front of the iron tracks as before mentioned, may be put on a piece of cardboard tacked to the board, the divisions laid out as per table on page 207 with waterproof India ink and then the whole shellacked over to preserve it and prevent weather changes from warping the cardboard out of shape.

The table is to read candle-power from eight to thirty. This is an abundant range for 16-c.p. lamps. Half candle-powers may also be marked off by divided distances between marks on the scale, except for values when these are given in the table. A pointer is attached to the car directly under the greased paper (as per Fig. 227) by which to read car position and so candle-power.

At the left end of the frame is an argand gas burner (See Fig. 227). An oil lamp might replace the argand burner, but if gas is available it will be much more convenient to use. In the figure the argand burner is shown quite near the slot board, so near indeed that it is probable the board will have to be covered with asbestos paper on the side next it. To procure the results mentioned in Part II., it will be necessary to keep the burner up close. The exact position of the gas burner is not of importance except that it must be directly behind the slot when viewed from the car. The screen is shown in detail in Fig. 233. It is arranged with a slider to control the width of the slot S, which is 1 in. x 2 ins. at largest opening. If facilities are at hand it will be well to provide a screw and

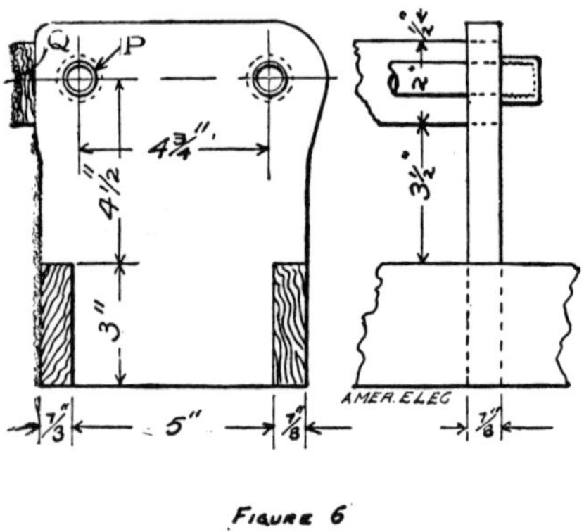

FIG. 232.—DETAIL OF TRACK AND SCALE ARRANGEMENT.

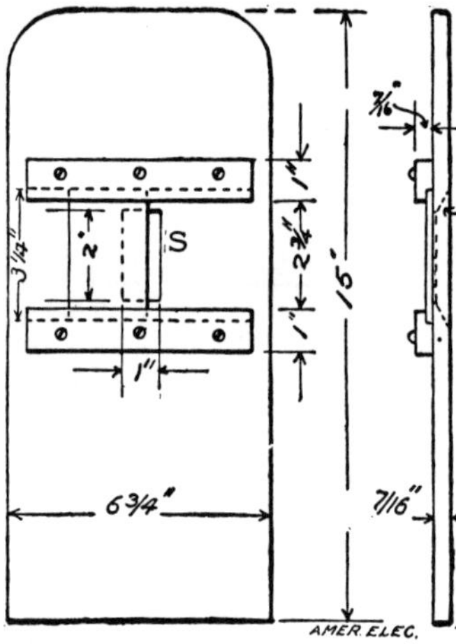

FIG. 233.—DETAILS OF SCREEN.

nut on the slider to control the width of the slot within small limits and readily. Bevel off the slot in fashion shown at R.

At the right hand end of the frame is the standard to hold the incandescent lamp, connection board and rheostat. An ordinary lamp socket has a ⅛-in. pipe screwed into it and this drops into a ⅜-in. pipe in a cast iron base through a ⅜-in. cap drilled at the end and with set-screw put through to hold the lamp rod at any height. The lamp cord is brought from the socket as shown and its ends carried to the two binding posts, P_2 and P_3, respectively. The mains are connected to the binding posts at M (Fig. 227), voltmeter at posts between which the letter V is put, and ammeter, if used, at posts on each side of Am, as marked. Otherwise this ammeter gap is bridged by a copper link. Two connecting straps are shown on the face of the frame and other connections are completed inside, so that the circuit is from one post at M to P_3, then through the lamp to P_2, to brass rod marked *H* (Fig. 229), through the contact into the rheostat coil, then to *P*, and so out at the other side of connection for the mains. The voltmeter connected thus reads e. m. f. on the lamp terminals and the ammeter, both the current through the lamp and voltmeter, so that if this latter takes appreciable current, as it usually does, allowance must be made, unless the voltmeter is always off circuit when the ammeter reading is taken.

For the rheostat construction see Figs. 229 and 230. *J* is a round wooden core which, with end pieces, is preferably made of hard wood. Over *J* is a thin covering of asbestos paper, put on smooth and tight, and shellacked to place. Over 14¾ inches of length at the middle of the rod is wound a tight spiral of No. 20 German silver double cotton covered wire also shellacked to place. This coil may be secured at its ends by tieing it down as with an armature coil or by soldering together the last two or three turns at each end. Along one side of the cylinder so formed, after the shellac has dried, and under the brass rod *H* (¼ in. thick x ½ in. wide), the insulation on the outside is to be rubbed off with sandpaper, making a path over the bared wire, ⅝ or ¾ ins. wide. This leaves each turn insulated from its neighbors, but makes it possible to connect readily with any turn of the solenoid. Bridging the distance from *H* to the wire on *J*, and making the contact just mentioned, is the contact device of Fig. 230. This is a brass body *K*, cut from ½-in. square rod, with an opening *M* which rides easily on the rod *H*. Secured to it are German silver springs, as shown within, to make contact with the rod, and outside to make contact with the turns of wire on the coil. A bolt like that at *L* can be had from almost any old lamp socket. It had better be very securely put in place. As the success of the rheostat depends on this contact device, it must be carefully made and the lower spring piece divided on each side as at *N*, so there may be four contacts on the rheostat wires. With this rheostat the volts on a 16-candle-power 110-volt lamp can be controlled over a range of ten volts by almost imperceptible steps and one ampere can be put through it continuously without dangerous heating. The whole thing is held together by wood screws from the ends, *O*, into the rod, *J*, and then screwed in place at the side of the frame.

After the photometer is all put together and mounted on a side wall, the whole thing must be given a couple of coats of a dead-black paint, and the wall immediately behind the instrument painted the same way. A dark room need not be provided. All very bright lights, however, must be absent from the vicinity of the instrument or turned out when it is in use.

HOW TO USE THE PHOTOMETER.

Two needs must be supplied before any work can be done with the finished instrument. There must be standard incandescent lamps and a thoroughly reliable voltmeter. The incandescent lamps may be purchased from any of the larger lamp companies, marked with the voltage at which they give 16 candle-power, and from what direction they must be viewed to give this candle-power. Sixteen-candle-power standards must be used. By a thoroughly reliable voltmeter is meant one which has 110 volts at its terminals when the needle points to this, that always comes to a 110-volt reading when 110 volts is applied, and which can be read to one-fifth of a volt. Unless such a voltmeter is used, only very rough candle-power measurements can be made. Errors as much as ½ candle-power to 1 candle-power in 16 will come in, due to this cause alone, with a voltmeter in use whose readings are slightly in doubt. The circuit used must be one on which the e. m. f. at the lamp can be held steady for a similar reason. An ordinary 16-c.p. lamp changes candle-power about one unit per volt change in applied e. m. f. at the normal or rated e. m. f.

Having the standards and the voltmeter for the work, proceed to measurements this way. Two men are required, one to watch the voltmeter and change lamps, the other to work the car. Connect the circuit, put one of the standards in the socket, set it at its marked voltage and turned so one views it from the photometer car in a direction which makes its candle-power 16, light the argand burner, put the car so its index is at mark 16, and adjust the slot at the argand until the grease spot disappears. The argand lamp has now become the standard and is assumed to stay at fixed candle-power for further work. Next remove the standard lamp, put in one whose candle-power is to be determined, bring it to a proper applied e. m. f., and while one man reads the voltmeter and keeps the lamp at a constant terminal potential difference, the other one moves the car backward and forward, seeking a place where the grease spot cannot be distinguished from the surrounding paper, and when the place is found he reads the candle-power of the lamp now in the socket. So work is proceeded with, a return being made periodically to the standard to check the argand burner, which is the working standard, and perhaps also to other lamps as well to procure check readings of their candle-power. A check on the argand by the incandescent lamp standard is of course made by putting the car at 16 candle-power, adjusting the incandescent lamp in position and applied e. m. f., and observing whether the grease spot is still invisible. The one handling the car must keep his eye solely for observations of the grease spot and will not be able to do good work until he has gotten accustomed to it by ten minutes' preliminary work. Also a person without experience in photometric measurements will not be able to get good, that is, concordant and correct results. A practiced observer will have results agree with this photometer so that individual observations never differ by more than a half-candle-power in sixteen.

It will be very unusual for the grease spot to simultaneously disappear from both of the images in the mirrors for any one position of the car. This is mainly due to difference in the color of the lights used. Old incandescent lamps are less troublesome than new ones in this respect. In making a reading one should set the car to a position where the difference in tint of spot and field is the same, whichever mirror he views.

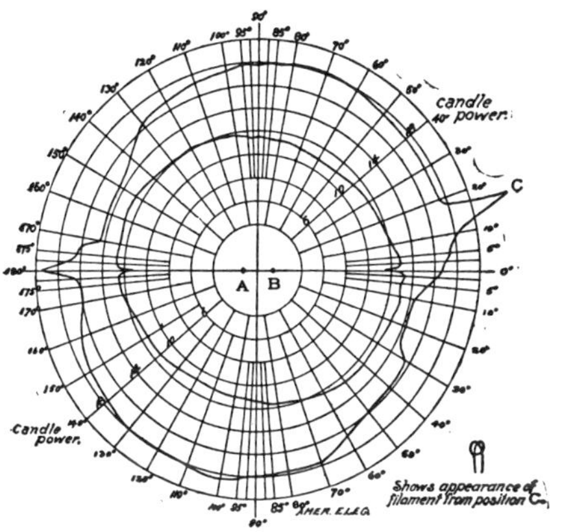

FIG. 234.—DIAGRAM OF HORIZONTAL CANDLE POWERS ON TWO LAMPS, ONE WITH COILED, THE OTHER WITH PLAIN HORSESHOE FILAMENT.

It must be observed that when making measurements no reversing of car is to be done, nor is there necessarily a dark room. Simply avoid lights or white surfaces before the eyes of the observer and have the lights in the vicinity of the instrument as few as possible, of as small candle-power as possible, and constant in position and candle-power. If one can just manage to read newspaper print from the surrounding lights it will be dark enough. Avoid light most carefully at the incandescent lamp end.

If 110-volt lamps are to have their candle-power determined it commonly happens that the dynamos run at 110 volts. Hence it is well nigh impossible to get 110 volts at the photometer and in any event the rheostat would be useless. One should have a few small storage cells to put in series in the circuit on this account, to raise the e. m. f. to about 120, and then there is abundant opportunity to procure the 110 volts steadily, even under considerable fluctuations in the e. m. f. on the circuit.

A 16-c.p. lamp does not measure 16 candle-power when viewed from any direction. There is a good deal of variation in candle-power according to the direction from which one views the lamp, and especially is this true for coiled or looped filaments. Even the candle-power in a plane at right angles to the lamp axis, the horizontal candle-power is irregular. One can readily see this for himself by holding a piece of white paper near a lighted lamp and observing the bright streaks of light in certain directions. The kind of variation which exists is shown by two curves of horizontal candle-power plotted together in Fig. 234. These were taken from two lamps, one of which had a coiled filament and the other a plain U-form filament.

Most lamp manufacturers are now rating their product by the candle-power measured when the lamps are rotating on a vertical axis at 180 r. p. m.—a value decided on at the National Electric Light Convention a year or two since. A rotating socket is not included in the photometer design here given. Fig. 235 shows, however, a rotating socket recently built for use in the Electrical Laboratory at Drexel Institute here shown, which might replace the incandescent lamp stand in the design given in this article. The one shown in the cut has embodied in it certain features worth noting. There are four brushes rubbing on four contact rings. Two carry current into and out of the lamp through a socket so arranged that the voltmeter connected through the two other brushes is actually connected to the lamp terminals, and so the true e. m. f. on the lamps is known. To maintain a steady rate of rotation the little motor shown is run by two storage cells, given up to this duty alone. The speed of the rotating socket is also under control and can be varied through all necessary limits by shifting the little rubber-covered pulley to different radii under the large rotating disc, which is mounted on the stem of the lamp whose candle-power (that is, mean horizontal candle-power) is to be determined. With proper tools and facilities such a rotating device, or one equivalent, can be put on the photometer here described and so a very complete instrument be had.

FIG. 235.—A VIEW OF A ROTATING SOCKET.

The standard lamp is better if not rotated. Care must be taken that the standard is not viewed from a position near to one like *C*, in the figure of horizontal distribution, however, where the turning of the lamp through only five degrees varied the candle-power by more than five units. A plain U-shaped filament is much the best for a standard lamp.

TABLE FOR CONSTRUCTION OF CANDLE-POWER SCALE.

Distance from Slot Board in Inches.		Candle-power to be Marked on Scale.
Exact Value.	Nearest 64th.	
21.961	$21\frac{31}{32}$	30
22.198	$22\frac{13}{64}$	29
22.444	$22\frac{7}{16}$	28
22.7	$22\frac{45}{64}$	27
22.967	$22\frac{31}{32}$	26
23.246	$23\frac{1}{4}$	25
23.537	$23\frac{17}{32}$	24
23.842	$23\frac{27}{32}$	23
24.162	$24\frac{5}{32}$	22
24.498	$24\frac{1}{2}$	21
24.8525	$24\frac{55}{64}$	20
25.227	$25\frac{15}{64}$	19
25.623	$25\frac{5}{8}$	18
26.0435	$26\frac{3}{64}$	17
26.491	$26\frac{31}{64}$	16
26.97	$26\frac{31}{32}$	15
27.4825	$27\frac{31}{64}$	14
28.035	$28\frac{1}{32}$	13
28.634	$28\frac{41}{64}$	12
29.285	$29\frac{9}{32}$	11
29.634	$29\frac{5}{8}$	10.5
30.000	30	10
30.384	$30\frac{25}{64}$	9.5
30.790	$30\frac{51}{64}$	9
31.219	$31\frac{7}{32}$	8.5
31.672	$31\frac{43}{64}$	8

TABLE FOR CONSTRUCTION OF NO. 2 CANDLE-POWER SCALE.

Distance from Slot Board in Inches.		Candle-power to be Marked on Scale.
Exact Value.	Nearest 64th.	
24.027	$24\frac{1}{32}$	45
24.212	$24\frac{7}{32}$	44
24.400	$24\frac{13}{32}$	43
24.593	$24\frac{19}{32}$	42
24.794	$24\frac{51}{64}$	41
24.997	25	40
25.210	$25\frac{7}{32}$	39
25.429	$25\frac{27}{64}$	38
25.651	$25\frac{21}{32}$	37
25.884	$25\frac{57}{64}$	36
26.124	$26\frac{1}{8}$	35
26.370	$26\frac{3}{8}$	34
26.625	$26\frac{5}{8}$	33
26.890	$26\frac{57}{64}$	32
27.163	$27\frac{5}{32}$	31
27.451	$27\frac{29}{64}$	30
27.746	$27\frac{3}{4}$	29
28.052	$28\frac{3}{64}$	28
28.374	$28\frac{3}{8}$	27
28.707	$28\frac{45}{64}$	26
29.055	$29\frac{3}{64}$	25

The candle-power of 8-c.p. lamps can readily be obtained with this photometer also. Put a standard 8-c.p. lamp, at some marked voltage, in the lamp socket; put the car at 16 candle-power and adjust the slot at the argand burner until the slot disappears. Then proceed to measure the 8-c.p. lamps as though they were 16's, halving the value of candle-power on the scale in each case for their real candle-power. A similar method may be used also in determining the value of the candle-power procured from 32-c.p. lamps. Use a standard 32-c.p. lamp as in case of the standard 8-c.p. above, and when readings are made on the candle-power scale double the value obtained in each case. Unless the gas used is intrinsically of very good candle-power and the argand burner is put very close to the slot board, this last cannot be done since the gas burner will not give enough light. If it is a possible plan carefully avoid any flickering edges of flame showing through the slot when it is viewed from the car position. The following alternative method of measuring 32-c.p. lamps will always be satisfactory.

Set up a second lamp socket (number 2) fifteen (15) inches to the right of the one already provided in the regular construction, or seventy-five (75) inches from the slot. Construct a second scale (number 2) under the one already made (number 1), to be used only when the lamp under test is in the more distant socket.

This scale will be constructed by measurements given in the table on page 207, all measurements being from the slot before the argand lamp.

Use 16-c.p. standard in socket number 1, put the car at 16 candle-power on scale number 1, and set the argand lamp by it so the spot disappears. Then put the 32-c.p. lamp, whose candle-power is desired in number 2 socket, and measure its candle-power in the usual way, but reading its candle-power on number 2 scale. If the two sockets are connected in parallel they will be ready for operation alternately at any time, the same instruments and rheostat being used without any change in connections.

CHAPTER XXV.

CONSTRUCTION OF A SIMPLE STORAGE BATTERY.

The accompanying engravings show the construction of a plate and a single cell of storage battery of the Faure type, which may be built with no tools beyond a pair of heavy tinners' shears, a small punch and a slitting saw of the sort used for cutting thin metals.

Each plate is made up of twelve strips of lead cut to the shape shown in Fig. 236. This strip is 0.075 in. thick, 8 ins. long, over all, and ⅞ in. wide. Midway on each edge is cut a square slot, *s*, 3-16 in. wide and deep; the ends of the strip are cut down to ½ in. in width for a distance of ⅞ in. from each extreme end, and two ⅛ in. holes are punched in the narrow part of the strip, as shown. The edges are folded along the dotted lines until the end view of the strip looks like *B*, Fig. 236.

Next three rubber forks, like *D*, Fig. 237, are provided for each plate (not each strip, but each group of twelve strips). Each fork is made

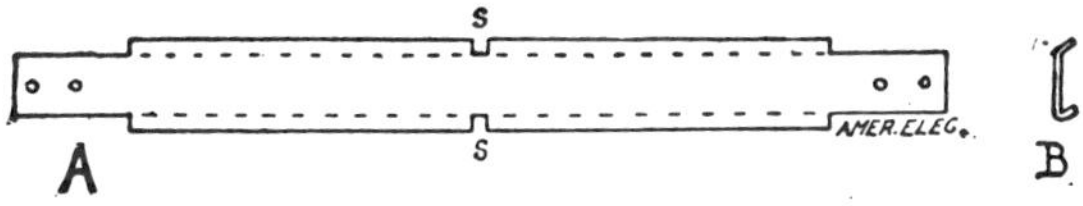

FIG. 236.—SHAPE OF STRIP.

from a strip of hard rubber, 3-16 in. thick, ⅝ in. wide, and 7½ ins. long; the slit down the center must be just wide enough to admit the thickness of the lead strips forming the plate (*A*, Fig. 236) with no "lost motion," and must stop exactly 1 in. from the lower end of the rubber. When these rubber forks are ready, a plate is built up in three of them as follows: Cut one of the lead strips in two longitudinally, exactly down its center, and assemble the strips on edge in the rubber fingers so that the end view of the lead part of the structure looks like *E*, Fig. 237. Fig. 238 shows the face view of the complete plate. The halves of the strip that was cut longitudinally go at the top and bottom of the plate, and are indicated by *a* and *b*, Figs. 237 and 238.

After the strips are assembled in the retaining forks, *D*, *D*, *D*, tie the upper ends of the fingers tightly with lead wire so that they clamp the plate; this tie-wire should go just above the top strip, at *z*, and it is imperative that no material other than lead be used, unless very strong, short rubber bands are obtainable, in which case they may be used. Then cut two strips of lead, *C*, *c*, each 5/8 in. wide, and of the same thickness as the lead strips composing the plate; the longer one, *C*, is 8½ ins. long, and the other one, *c*, 6½ ins. long. Rivet one of these to all the ends of the plate strips at one extremity of the plate and the other one to the ends at the opposite extremity, using lead rivets. The engraving shows holes punched in the connecting strips, *C*, *c*, to correspond with those in the

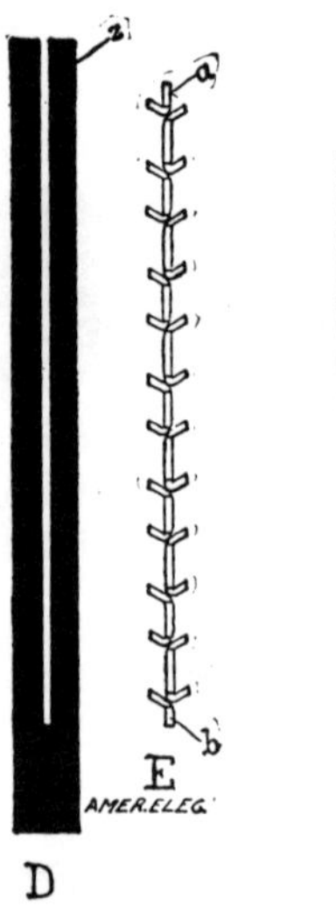

FIG. 237.—RUBBER FORK AND PLATE EDGE.

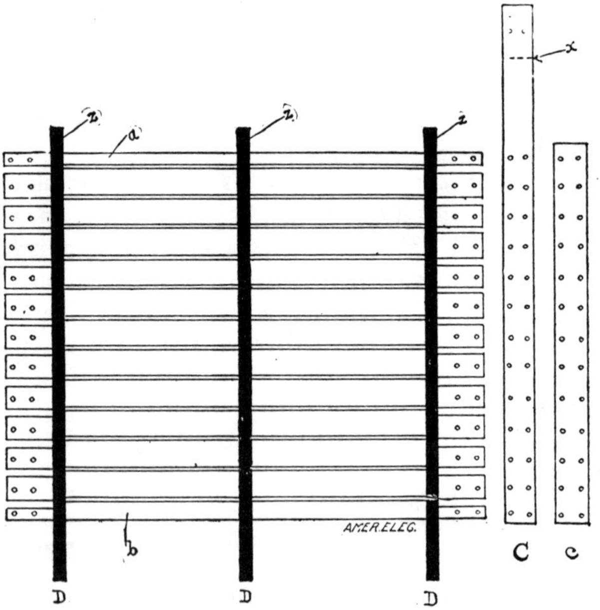

FIG. 238.—COMPLETED PLATE

ends of the plate strips, *A*. Bend the upper ends of the long connectors, *C*, about half an inch from the end, at right angles to the main body of the strip; the bending point is indicated by *x*, Fig. 238. The edges of the plate strips which were bent at an acute angle to the body strips (*B*, Fig. 236) are designed to serve as shelves to hold the paste or active material which is to be applied to each plate. The process of pasting will be described later on.

After the paste has been put on the plates and has hardened, nine plates (four positive and five negative) are assembled in a glass jar, as shown by the plan view, Fig. 239, where *A* is the lead portion of the outside plates; *P*, the paste, or active material; *D*, the rubber retaining fingers; *J*, the glass jar, and *C*, *c*, the vertical connecting strips. The strips, *C*, are shown coming straight upward instead of bent over, to avoid obscuring the view of the plate ends. The rubber fingers, *D*, in addition to holding the twelve strips composing each plate in their proper positions, serve to prevent the plates themselves from buckling and shifting in the jar. The edges of these rubber fingers are shown slightly separated, but as a matter of fact, they must abut each other so as to keep the plates firmly in position. The jar, *J*, is rectangular, and should be 8¼ ins. long, 5¾ ins. wide, and 7½ ins. deep, inside measurement. None of its dimensions can be smaller than specified; if the width or length be greater, a strip of wood may be used to bring it down to the figure desired. A

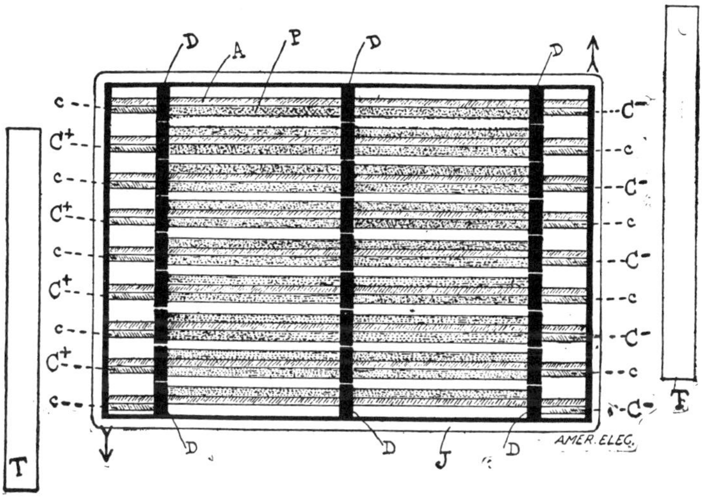

FIG. 239.—PLAN VIEW OF BATTERY CELL.

sheet of soft rubber must be laid against each end and side wall of the jar, and the plates must fit snugly against the rubber sheets so that they will not shift in the jar; soft rubber must be used for the wall sheets in order that the plates may expand without tending to buckle.

As above stated, each cell contains four positive plates and five negatives; the two outside plates are negative, and have paste on only one side each, as the sides next to the wall of the jar are not active. The terminal strips, *C* —, on the right hand end of the cell are all negative; those marked *C* + on the left hand end are the positives, as their signs indicate.

The short strips, *c*, serve simply to connect the parts of the plates to each other, and should not project above the upper edges of the plates.

All the positive ends, $C +$, at the left are riveted to a lead strip, *T*, which is laid along the top of the bent-over ends of the strips, $C +$. The negative connectors are similarly riveted to the other strip, *T* 2, and these two strips form the terminals of the complete cell. The arrows at the corners of the jar indicate the direction in which the ends of the terminals are led from the cells to connect to an adjoining cell or to leading wires, as the case may be.

Each cell of battery of the above dimensions, when properly pasted and "formed," will give an electro-motive force of about 2 volts during the greater part of its discharge, and it may be discharged at the rate of 18 to 20 amperes. To operate any of the small motors described by the writer on pages 1 to 14, inclusive, four of these cells will be required (The battery winding given in each case must, of course, be used on the motor.) After the cells have been "formed" as described below, they may be kept charged sufficiently for light, intermittent service by connecting up 10 cells of gravity battery in series with the four storage cells, the copper terminal of the blue-stone battery being connected to the positive terminal of the storage battery. This connection may be left permanently on, during the use of the storage cells as well as when they are idle, the only attention necessary being the replenishment of the gravity cells at comparatively long intervals.

The plates of the storage cells are pasted, the positives with a thick paste made of red lead and dilute sulphuric acid, and the negatives with a similar paste made of litharge and dilute acid. The acid should be one-tenth concentrated sulphuric acid and nine-tenths water, and the water should be distilled; the proportions of one and nine parts are by weight, not volume. In mixing, always pour the acid into the water, never the reverse. The pastes must be mixed with wooden spatulas in glass or earthenware vessels, and should be so thick (containing so little dilute acid) as to appear almost powdery. The pastes are applied to the sides of the plates and pressed firmly in with the spatulas until the surface of the paste is flush with the edges of the little shelves; the entire surfaces of the lead strips, except the edges of the shelves, must be covered evenly. The best procedure will be to take all the positive plates first; lay them flat on a board, and apply red lead paste to one side. Set them aside and mix the litharge paste (in a separate vessel and with a separate spatula), and then treat one side of all the negative plates. When the plates are all dry, turn them over and treat the other sides, being careful not to jar out the paste already on the under sides. It should be remembered,

100, that two of the negative plates in each cell are to be treated on one side only—the side which comes next to the neighboring positive plate.

When the plates are all pasted, assemble them in their cells, as described above, and then rivet the ends of the connectors, *C* +, and *C* —, to the horizontal terminal strips, *T* and *T2*. Connect the positive terminal of one cell to the negative terminal of its neighbor, and fill all the cells with a solution consisting of one part concentrated sulphuric acid and four parts distilled water, measuring by weight. Connect the series of cells in an arc light circuit, just as though they were arc lamps, and let the current pass through them from the positive to the negative terminal of the series until the paste on the negative plates has all turned color. The cells will then be "formed" and ready for service. They should not be allowed to remain charged long before being put into service, and it will be advisable, therefore, to have the apparatus for which they are to furnish current all ready to start up before putting the cells in circuit for formation. The arc light circuit on which the cells are "formed" may have any current value from 4 to 20, but as most of the circuits in this country carry either 6.8 amperes or 9.6 amperes, one of these values will doubtless be found in the charging circuit.

If the circuit is an intermittent one (does not run constantly, 24 hours a day), care must be observed to take the battery out of circuit as soon as the current is off at each shut-down, so that it cannot discharge in case the line is closed before current is restored.

CHAPTER XXVI.

CONSTRUCTION OF A CONSTANT-POTENTIAL ARC LAMP.

With a small screw-cutting lathe, a drill chuck and a few drills and other small tools, any mechanic of average ability, having a fair knowledge of electrical apparatus, can, by using the accompanying sketches as working drawings, make a reliable and efficient arc lamp for use on a 110-volt continuous-current circuit, in series with a resistance coil of 8 ohms, or a duplicate lamp and a resistance coil of 1½ ohms, preferably the latter. Fig. 240 shows the frame of the lamp, one-half in cross-section. *A* is the top-plate; *B* is the floor-plate; *C*, *C*, are short side-rods; *D*, *D*, are long side-rods; *E*, the yoke; *F*, the bottom carbon holder; *i*, *j* and *k*, are insulating washers of hard fibre. The under side of the top-plate, *A*, is shown by Fig. 241. It is of cast-iron or brass, and is provided with two lugs, *d*, *d*, ¾ in. in diameter and 1 in. long, drilled and tapped ⅞ in. deep to take ¼ in. gas pipe; a lug, *e*, ⅜ in. in diameter and 1 in. long, drilled and tapped to take a 5-32-in. machine-screw, and a flange around the outer edge, 1-16 in. thick and ½ in. deep. At diametrically opposite points, two pins, *x*, *x*, are set in the flange; these are of 1-16-in. steel wire, ½ in. long. The centers of the lugs, *d*, *d*, are 2⅜ ins. from the center of the plate, and the center of the lug, *e*, is 2½ ins. from the center of the plate. On the upper side of the plate is a neck (see Fig. 140) 1¼ ins. in diameter outside, and standing 1 in. above the upper surface of the plate. This neck is bored ¾-in. deep and tapped to fit a ¾-in. gas pipe; below this bore, a hole, 11-16 in. in diameter is drilled clear through the plate. A fibre washer, *i*, shown enlarged in Fig. 247, is fitted to this hole; the larger diameter of the washer, *i*, must be such as to allow it to slip down freely in the threaded neck, and the smaller diameter must fit snugly the 11-16-in. hole at the base of the neck; the bore at the top of the washer is ⅜ in., and the recess in the under side is ½ in. in diameter. The washer is ¼ in. thick, and the flange is ⅛ in. thick.

The short side-rods, *CC*, are pieces of ¼-in. gas pipe, 5¾ ins. long, threaded on the outside at the upper ends and on the *inside* at the lower ends; the long side-rods, *DD*, are similar pieces of gas pipe, 19½ ins. long, and the outside thread is an inch long. The short and long side-rods are held together by a steel plug, *l*, ¾ ins. long, threaded the whole length to correspond with the thread in the side-rods; the washers, *jj*, shown enlarged in Fig. 247, are interposed to insulate the floor-plate, *B*, from the side-rods and top-plate. These washers are ¾ in. in outer diameter and ½ in. diameter at the neck; the bore is such as to allow the threaded plug, *l*, to slip through without having to screw it; the flange is ⅛ in. thick, and the total thickness is ¼ in.

The floor-plate, *B* (Figs. 240 and 242), is of brass or iron, 5 15-16 ins. diameter, ⅛ in thick, and has a round lug, *c*, ⅜ in. in diameter, and

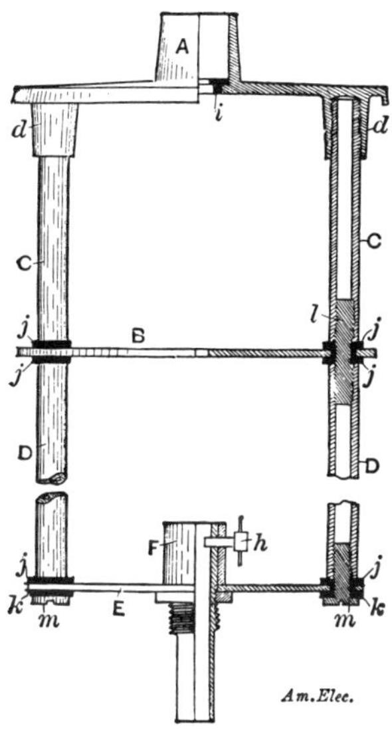

FIG. 240.—SEMI-SECTIONAL SKETCH OF FRAME.

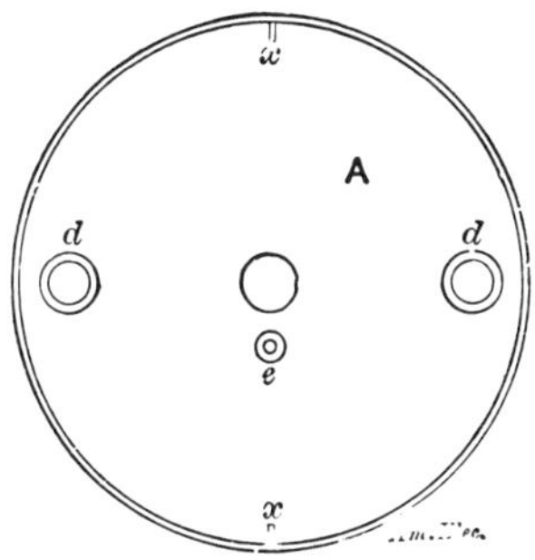

FIG. 241.—UNDERSIDE OF TOP-PLATE.

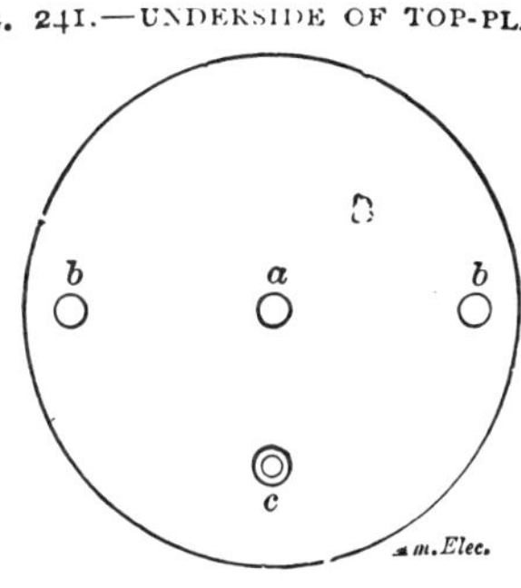

FIG. 242.—UPPER SIDE OF FLOOR-PLATE.

½ in. high. There are two ½-in. holes, *b*, *b*, a ⅜-in. hole, *a*, in the center, and a hole through the center of the lug, *c*, threaded for a 3-16-in. screw. The distances, center to center, are 2⅜ ins. from *a* to each *b* and 1¾ ins. from *a* to *c*; the lug, *c*, is 90 degs. from the holes, *b b*. If it is more convenient, the floor-plate may be cut from sheet brass and the lug, *c*, soldered on, or screwed in and riveted.

The yoke, *E* (Figs. 240 and 247), may be cut from a strip of brass 1¼ ins. wide and 3-16 in. thick; the center hole is ¾ in. and the others ½ in. in diameter; the distances are 2⅜ ins. each way, from center to center of the large hole and each of the smaller ones, corresponding to the location of the lugs on *A*, and the holes through *B*. The carbon holder, *F*, is made of two pieces of brass tubing, the long one ½ in. inside diameter and the short one of a size to fit snugly over the long tube; the two are sweated together and riveted as a precaution against the loosening of the solder by the heat from the arc when the carbons are almost burned out. The long piece of tubing is threaded from the end of the short piece ⅜ in. down, and turned down to ⅝ in. outside diameter the balance of its length to allow the nut, *g*, to slip up to the beginning of the thread, and to form a mandrel for the globe holder. A ⅛-in. pin, 1 in. long, must be driven horizontally through the shank of the carbon holder, ⅜ in. below the thread, after the nut is on; the pin holes should be drilled exactly across the center of the tube, so that when the pin is inserted the ends will project from diametrically opposite sides of the shank. This pin is to support the globe-holder, as will be explained further along. The carbon-holder is 3 ins. long over all, and the short piece of tubing,

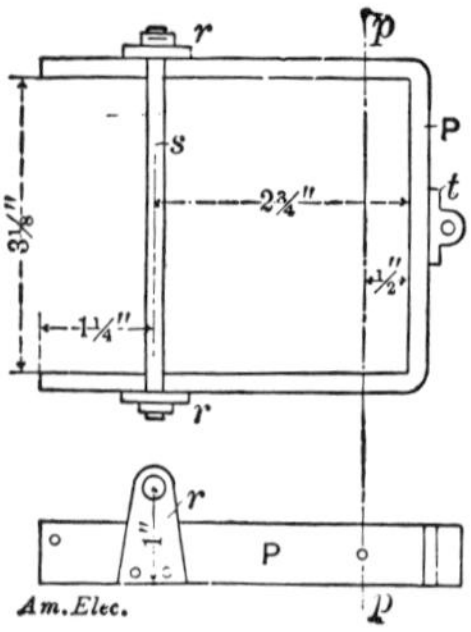

FIG. 243.—ELEVATION AND PLAN OF ARMATURE.

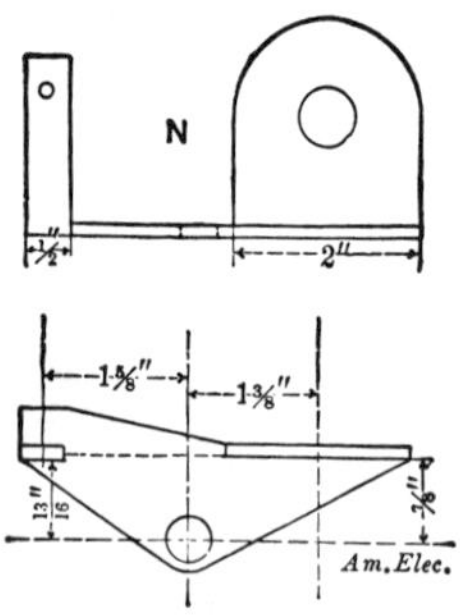

FIG. 244.—ELEVATION AND PLAN OF BRASS FRAME.

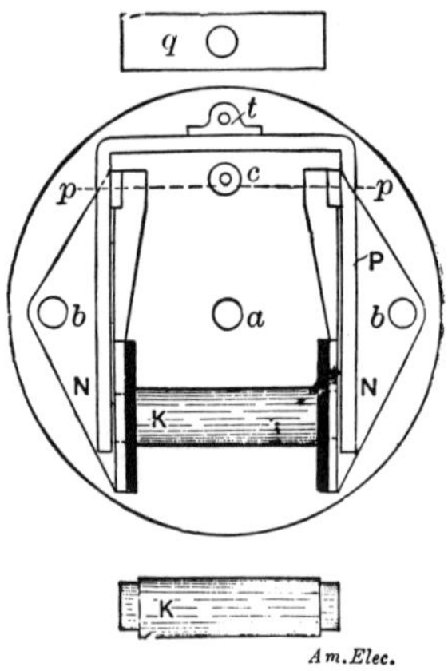

FIG. 245.—PLAN VIEW OF MECHANISM, MINUS CLUTCH.

which forms a sleeve over the holder-tube, is 1 in. long. A set-screw, *h*, serves to clamp the carbon in the holder. The yoke is insulated from the side-rods by washers, *j* and *k*; the former was described above, and the latter is a plain, flat fibre washer ⅛ in. thick, ¾ in. diameter, with a ½-in. hole; screws, *m*, *m*, hold the yoke to the side-rods.

The lamp is of the clutch type and the moving parts consist of a magnet-armature, a clutch and a carbon-carrying rod. The magnet is a straight, round bar, with a single coil; the core (*K* in Fig. 245) is ¾ in. in diameter and 3 ins. long, with a shoulder ¼ in. long at each end, the diameter there being 5/8 in. The core is provided with two insulating heads of fibre 1/8 in thick, the hole in which is a tight fit on the reduced ends of the core, and after it is wound it is mounted between two brass frames, *N*, *N* (Figs. 244 and 245). Each frame, *N*, has a 1/2 in. hole drilled in its base, the exact location of which may be found by reference to Fig. 244. These frames have standards to which is pivoted the armature, *P* (Figs. 243, 245 and 246). The thickness of the metal is 1/8 in. throughout. The back ends of the frames, *N*, *N*, are held down by a cross-bar, *q* (Fig. 245), which has a 3/8-in hole in the center to allow the lug, *c*, to come through. The lug is threaded on the outside to take a nut to hold down the cross-bar, *q*. The bar is 1/8 in. thick, 3/4 in. wide and a trifle over 2 3/4 ins. long so that the ends may be filed to fit exactly between the pivot standards of the frames, *N*, *N*.

The armature is a piece of flat Norway iron, 3-16 in. thick, 5/8 in. wide and 11 1/4 ins. long, bent into a U, as shown in Fig. 243, and provided with two ears, *r*, *r*, 1/8 in. thick, which carry a round rod, *s*, of steel,

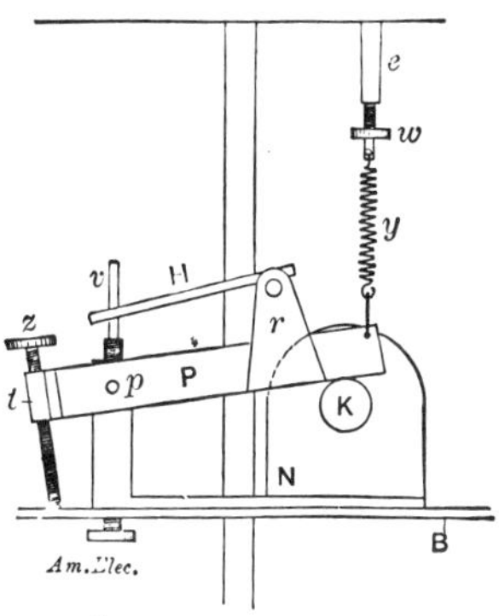

FIG. 246.—ELEVATION OF MECHANISM, COMPLETE.

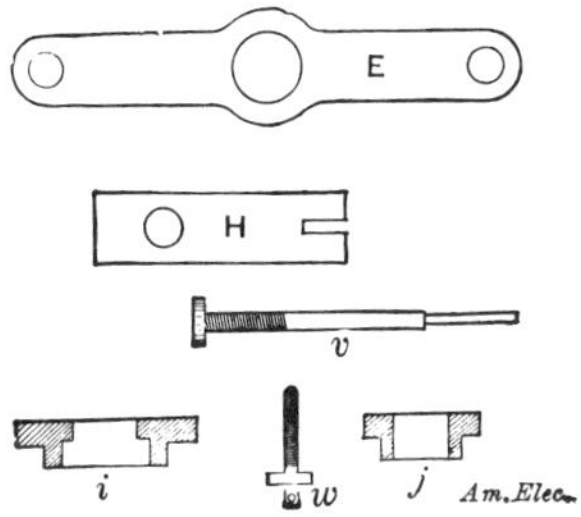

FIG. 247.—SMALL PARTS.

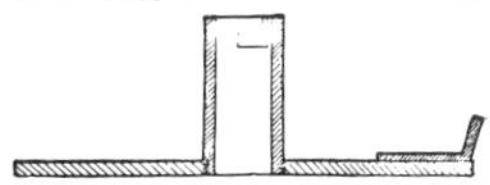

FIG. 248.—CROSS-SECTION OF GLOBE-HOLDER.

3-16 in. diameter. At the back end a clip, *t*, of brass, is riveted on; this piece has a hole through it, tapped for a 3-16-in. machine screw. Pivot-holes, whose centers coincide with the dotted line, *p*, *p*, are drilled in the sides of the armature. The dimensions specified in the drawing must be carefully observed.

The clutch *H* (Fig. 247) is a flat piece of brass 1/8 in. thick, 3/4 in. wide, and 2 3/4 ins. long, with a 7-16-in. hole drilled 3/4 in. from one end and a 1/8-in. slot, 1/2 in. deep, sawed in the other end. The edges of the hole must be very slightly rounded to prevent the clutch from cutting into the carbon rod; *v* is a regulating screw to trip the clutch; it is 3½ ins. long over all, 3-16 in. diameter at the threaded part, and ⅛ in. diameter beyond the shoulder. The shoulder is one inch from the end. The armature, frames, clutch and carbon rod are shown assembled in Fig. 246. The spring, *y*, which pulls the armature upward, is adjusted by means of the screw, *w*, which screws into the lug, *c*, on the top-plate. The screw is 5-32 in. diameter and the threaded end is an inch long. The play of the armature is limited by the back-screw, *z*, by means of which the length of arc first struck is adjusted; the screw, *v*, and the spring, *y*, regulate the length of arc while burning. The spring is attached to a stout brass wire strung from limb to limb of the armature.

The magnet is wound with No. 30 double cotton-covered magnet wire, 32 layers deep, and 125 turns long, the starting end being connected with the core and the outer end with the bottom carbon holder. Binding posts may be put on if desired, but the writer prefers to carry a piece of No. 12 rubber-covered and braided wire from the yoke up alongside one of the side-rods, making this the negative lamp terminal; the outer end of the magnet coil may be connected to this terminal by means of a piece of stout wire brought through the floor-plate, *B*, of the lamp, the hole being bushed with insulation, and the positive terminal may be a binding post screwed in the floor-plate (which is in electrical contact with the carbon rod).

The case of the lamp is a piece of thin sheet brass, 6 ins. × 19 ins., bent into a 6-in. tube and riveted at the lap; at one end of the tube thus formed and diametrically opposite each other, are bayonet slots which engage with two pins projecting inwardly from the flange of the top plate of the lamp. The carbon rod is a piece of brass tubing ⅜-in. diameter outside, 24 ins. long, with a 1-16-in. wall. The upper carbon holder can be purchased for a small sum and is not worth the trouble of making. The ball and shank must be made to fit the carbon-holder and the bore of the carbon rod; the shank should be an inch long, very slightlty tapered. Drill a 1-16 in. hole clear through the rod, ½ in. from the end, before inserting the shank; then tin the shank and the inside of the end of the carbon rod, drive the shank in, and solder through the holes.

The globe-holder (Fig. 248) consists of a disc of brass (or iron) 3-16 in. thick and 5 ins. in diameter, having a piece of brass tubing screwed into its center and three lips riveted at equidistant points around the edge. The tube in the center must fit snugly over the shank of the carbon-holder (Fig. 240) and it has two bayonet slots at the upper end which fit over the ends of the pin driven transversely through the carbon-holder shank. This tube must measure 1½ ins. long above the disc. The ears are simple brass strips each ⅛ in. thick, ½ in. wide and 2 ins. long, with ½ in. of its length bent up almost at right angles to the balance; 3-16 in. thumb-screws in the up-turned lips serve to hold the globe by its rim.

Two arc lamps such as the one above described will work together, in series with a resistance coil of 1½ ohms, on any 110-volt direct-current circuit.

CHAPTER XXVII.

AN EXPERIMENTAL NERNST LAMP.

This lamp, invented by the physicist, Nernst, of Göttingen, consists of a rod of dense magnesia with platinum terminals. This rod is connected in series with a dead resistance, and an e. m. f. (preferably alternating) of from 200 to 600 volts is applied to the arrangement. Upon heating the magnesia rod, by a blow pipe, for example, it becomes a conductor and passes sufficient current to raise its temperature to that of intense incandescence. In the more recent types Nernst uses a large proportion of thoria in the rod.

An increase of current in the lamp causes a rise in its temperature and a drop in its resistance and, at the temperature at which the lamp is used, this drop in resistance is so great that considerably less e. m. f. is required to push the increased current through the rod, so that the lamp is unstable, and without the dead resistance the lamp would be destroyed by the excessive current that would flow through it. The efficiency of the lamp, according to tests made abroad, is about 1.5 watts per candle-power, including the watts lost in the dead resistance. The lamp gives a beautiful and pleasant white light and its life is claimed to be very great.

A number of these lamps have been constructed at the physical laboratory in Bethlehem, Pa., by Prof. W. S. Franklin and Mr. R. B. Williamson. After many trials the following procedure was found to give good results: A mixture of calcined magnesium oxide (composition of mixture given below) is tamped as compactly as possible into a smooth bore brass tube lined with two or three thicknesses of stiff writing paper. This paper should be fixed in place with a little glue and baked dry. The tube full of magnesia is then slowly baked on a metal plate over a Bunsen burner until the paper is completely charred, when the magnesia rod may be pushed out. The rod is then calcined before a blow pipe, heating it slowly and uniformly to avoid cracking by unequal shrinkage. The rod is then broken to a length of about 2½ ins. and laid upon a bed of magnesia. Two ordinary arc carbons are brought up to the ends of the rod, one carbon being fixed by weights, the other being preferably held in the hand. Several hundred volts e. m. f. are applied to the carbons with dead resistance in circuit and the magnesia rod is heated by the blow pipe until the current starts. As the magnesia rod rises in temperature it shrinks greatly, and it must be subjected to very slight end pressure to prevent the formation of cross cracks; too much pressure will cause longitudinal cracks. The current is then increased until the magnesia rod becomes slightly soft, when it may be straightened if, as is likely, it has curled up in shrinking. The rod is then allowed to cool and ground on an emery wheel to the required shape, as described below.

The most convenient source of current for the purpose of this preliminary heating and for operating the finished lamp is a step-up transformer with a rheostat in the primary circuit; a secondary e. m. f. of 1,000 volts is satisfactory. This e. m. f., of course, falls off greatly when the current starts, because of the action of the primary rheostat.

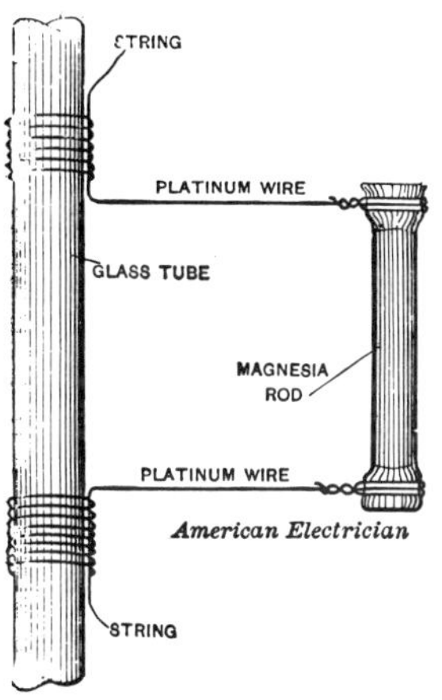

FIG. 249.—A NERNST LAMP.

The magnesia mixture may be pure calcined magnesia with a slight amount of magnesium chloride ground up with it to serve as a bond. A slight amount of soluble silicate of soda is also a good bond. The mixture should be only moist enough to pack like flour; it is better to have it perfectly dry than too moist. A lamp made of pure magnesia or of magnesia with 1 per cent or less of powdered silica, has a very high resistance and can scarcely be started with less than 1,000 volts, and then with difficulty. After it is once started, however, the resistance falls so that even a pure magnesia rod will operate with, say, 300 volts per inch of length. A lamp which is very much easier to start is made by mixing from 2 to 6 per cent of pounded glass with the powdered magnesia. Perhaps a lime glass would be best for this purpose.

The magnesia rod should be about 1 in. or 1½ ins. in length, and about ⅛-in. in diameter, with slightly enlarged grooved ends: platinum wire is wound two or three times around these ends and covered with a paste of magnesia, pounded glass, and water glass (or simply water). The lamp is conveniently mounted by binding the platinum wires to the side of a small glass tube. Fig. 249 shows the finished lamp full size.

A lamp made as above described, with about 1 per cent. of pounded glass and 1 per cent. of powdered silica, the rod being about 1¼ ins. long and ⅛-in. in diameter, operated on 250 volts (between platinum terminals), takes 0.8 ampere, and gives fully 175 candle-power, although the candle-power has not been measured at Bethlehem. It has been found that the silicates of sodium and potassium (or perhaps simply the sodium and potassium) are slowly expelled by the heat while the lamp is in use, causing the resistance to become slowly greater.

Commercial magnesia (calcined Grecian magnesite) makes good lamps without any admixture of silica, although its resistance is rather high unless it is mixed with powdered glass.

An attempt was made to fuse magnesia into a compact mass in an electric furnace (100 amperes at about 90 volts), but it was found that the boiling point of magnesia (at atmospheric pressure) is about the same as its melting point, so that the material vaporized about as rapidly as it was melted. The operation would, no doubt, succeed under pressure. During this work with the electric furnace it was necessary to keep a close watch of the action, and a small piece of heavily smoked glass was used to screen the eyes, leaving the forehead exposed, and a sever case of sunburn was produced, although the heat on the face was not excessive.

A most striking experiment is to mount a glass tube as a Nernst lamp. A large, thin walled tube gives the best effect. Wind copper wire terminals about 4 ins. apart on a thin walled glass tube ½-in. or ¾-in. in diameter. Connect to the secondary of a step-up transformer with a rheostat in the primary. Heat the tube along one side. The current starts along a narrow strip of the glass, heats it to bright redness, and this heated strip gradually widens until the whole tube is melted down This experiment was tried in Bethlehem with a 1,000-volt secondary, but it would certainly be possible to perform the experiment successfully with as low an e. m. f. as 100 volts, and direct current would answer as well as alternating. With low e. m. f. the distance between the copper terminals should be much less than 4 ins., and, of course, a rheostat should be included in the circuit.

CHAPTER XXVIII.

CONSTRUCTION OF AN INDUCTION COIL.

Since the advent of the Röntgen discovery the induction coil has risen to a much more prominent place as a scientific and practical instrument. It has very naturally been greatly improved in construction within the past year, but inasmuch as these improvements are not generally known and used, the writer has presumed to believe that a description of them may be interesting.

The basis of the discussion will be the construction of a 6-inch spark coil, but it may be profitably remembered that the average induction coil built in sections may be thus rebuilt, and oftentimes the length of spark it is capable of giving thereby trebled, even though thirty or forty per cent. of the secondary is removed in order to accomplish the construction.

Many modern coils are built on lines that make extensive internal leakage a great possibility. Some coils are made with as much as twenty-five pounds of wire in the secondary, and yet under the most favorable conditions the spark obtained is but six inches in length. The makers of such coils broadly claim that it is impossible to break down the insulation of their apparatus, but in view of the fact that a 6-inch coil can be made with a 5-pound secondary, it is easy to see that the coils just referred to are broken down already, and that it is a case of spoiling a bad egg—a manifest impossibility.

The principal leak in an induction coil is from the secondary to the primary, as is shown in Fig. 250. Between the points of leakage indicated the full difference of potential of the coil exists. The ⅛-in. of hard rubber and the almost negligible air gap usually provided can scarcely be expected to withstand the e. m. f. that will urge a discharge across a six-inch air gap.

A second source of leakage is shown in Fig. 250-A, and exists at the separator pieces between sections. The insulation between the primary and secondary is broken in its continuity by these pieces, and as it is impossible to make an electrically tight joint, such insulation as is provided is no more effective than an equivalent gap of air. The insulation between primary and secondary must be a continuous homogeneous mass, and sufficiently thick to withstand the maximum e. m. f. of the coil. Fig. 251 illustrates the method of insulating a secondary section. The spaces, *S S*, are to be filled with paraffine or some equivalent continuous insulator.

Covered wire for an induction coil is not necessary, and the use of silk wire is a most expensive construction, from which absolutely no ad-

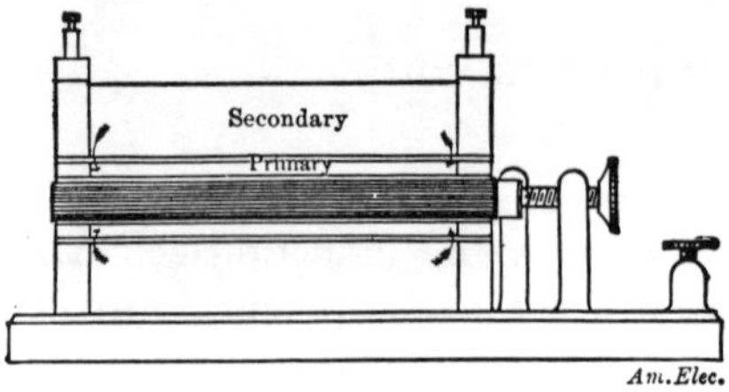

FIG. 250.—SHOWING LEAKAGE FROM SECONDARY.

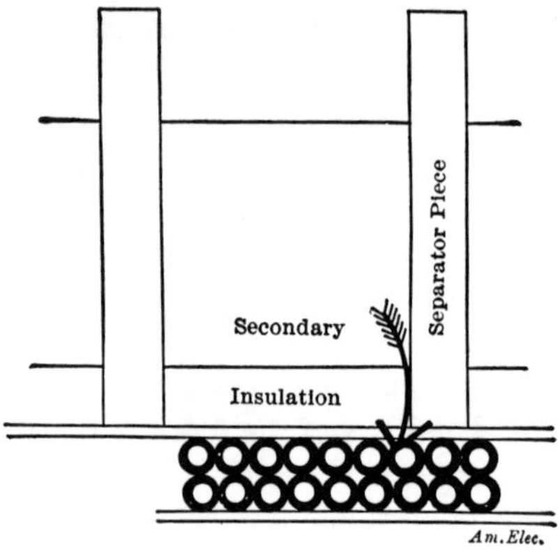

FIG. 250A.—LEAKAGE THROUGH JOINT.

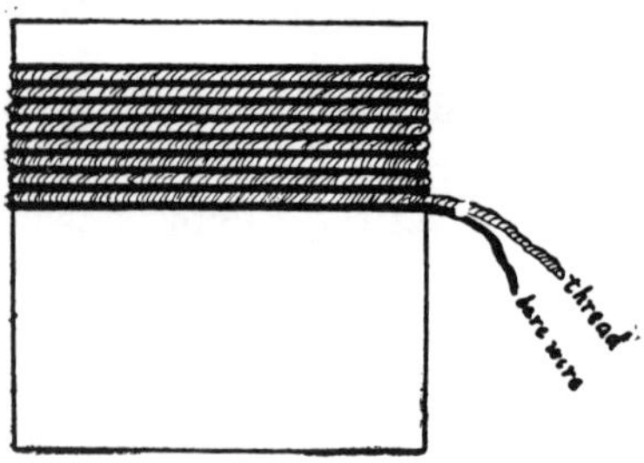

FIG. 252.

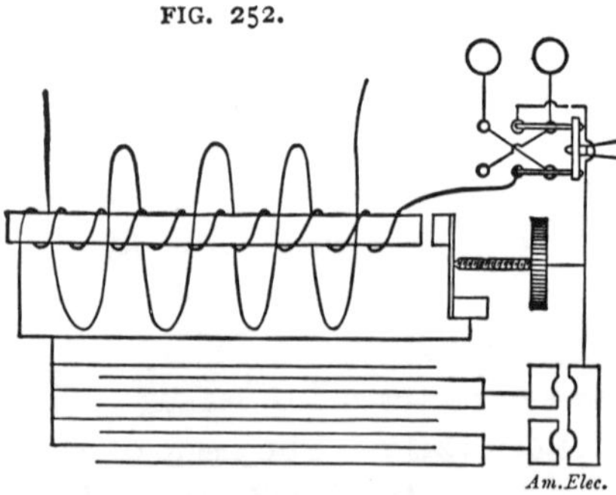

FIG. 253.—DIAGRAM OF COIL CONNECTIONS.

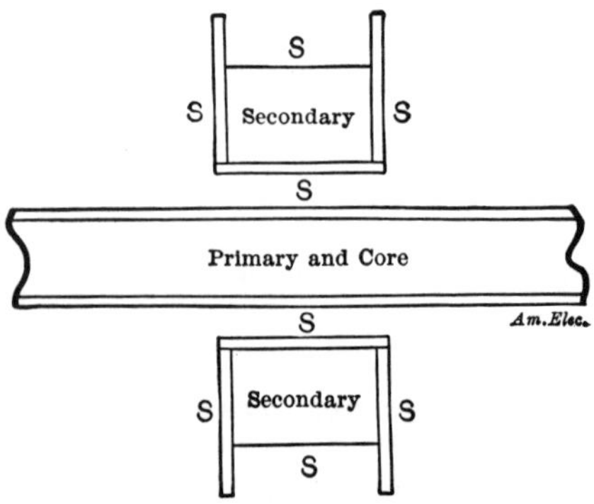

FIG. 251.—IDEAL INSULATION FOR SECONDARY.

vantage can be gained. One way is to use bare wire, winding a thread between adjacent turns, as shown in Fig. 252. Colored thread should be avoided. The space between the layers should be at least four or five times the thickness of the insulation between the turns. The insulation between the turns of an induction coil is about 5 mils (.005 in.) thick, and experience has shown that this is none too much. A space of 1-64-in. can be used between the layers to advantage. This space should be filled with absorbent paper that will readily soak up paraffine wax.

Fig. 254 shows a regular sectioned dimension drawing of the 6-in. spark coil already referred to. It would be idle to enter into a long dissertation on the various features of this coil that are common to every instrument of a similar nature, and only the novel ones will be discussed and the quantitative measurements given. The secondary coils are constructed of bare wire, absorbent paper and cotton thread, substantially as indicated heretofore. Care must be taken in the winding to keep away at least ¼-in. with the wire from the edge of the paper layer, partly for the added insulation between the layers and partly to prevent the annoyance of the end turn slipping out when handling the section. If an old coil is being rebuilt, it will not pay to thus rewind it. Sufficient wire from the inside of the secondary sections should be removed to admit of reassembling it as per drawing, a comparatively easy thing to do, and the results will be nearly as good as with the coil here described.

The great feature of the coil is the method of supporting and insulating its primary and secondary. A long box is constructed as per drawing, and from the geometrical center of the ends is supported the tube that forms the enclosing envelope for the primary coil and its core. The secondary coil is divided into six sections, each supported on a piece of hard rubber tube with end collars of glass or hard rubber. This hard rubber tube allows ½-in. in the clear between its interior surface and the primary envelope. The glass collars are square, and are of such a shape that they just fit the inside of the box, and in their lateral dimensions are a perfect measure of its interior section. The space between adjacent sections is ¼-in., and between the last coils and the end pieces, ½-in. is provided. The coil is wound to a diameter of 6 ins., the internal dimensions of the box surrounding it being 8 ins. square.

Before assembling, the coils are boiled for a long time in paraffine, and are removed therefrom only when the wax has cooled sufficiently to attain a mushy consistency. They are preferably assembled while in this state, for large soft clots of wax adhere to the coils and close in on the bobbin on which the coil is put, thus filling up objectionable air spaces. The assembling of the coil being complete, each secondary will be mounted on a tube in the box and will rest in a partition made on two sides of glass or hard rubber. Nowhere will any secondary section have any connection with any primary section except through paraffine wax in a continuous mass that cannot be broken down unless penetrated. The great

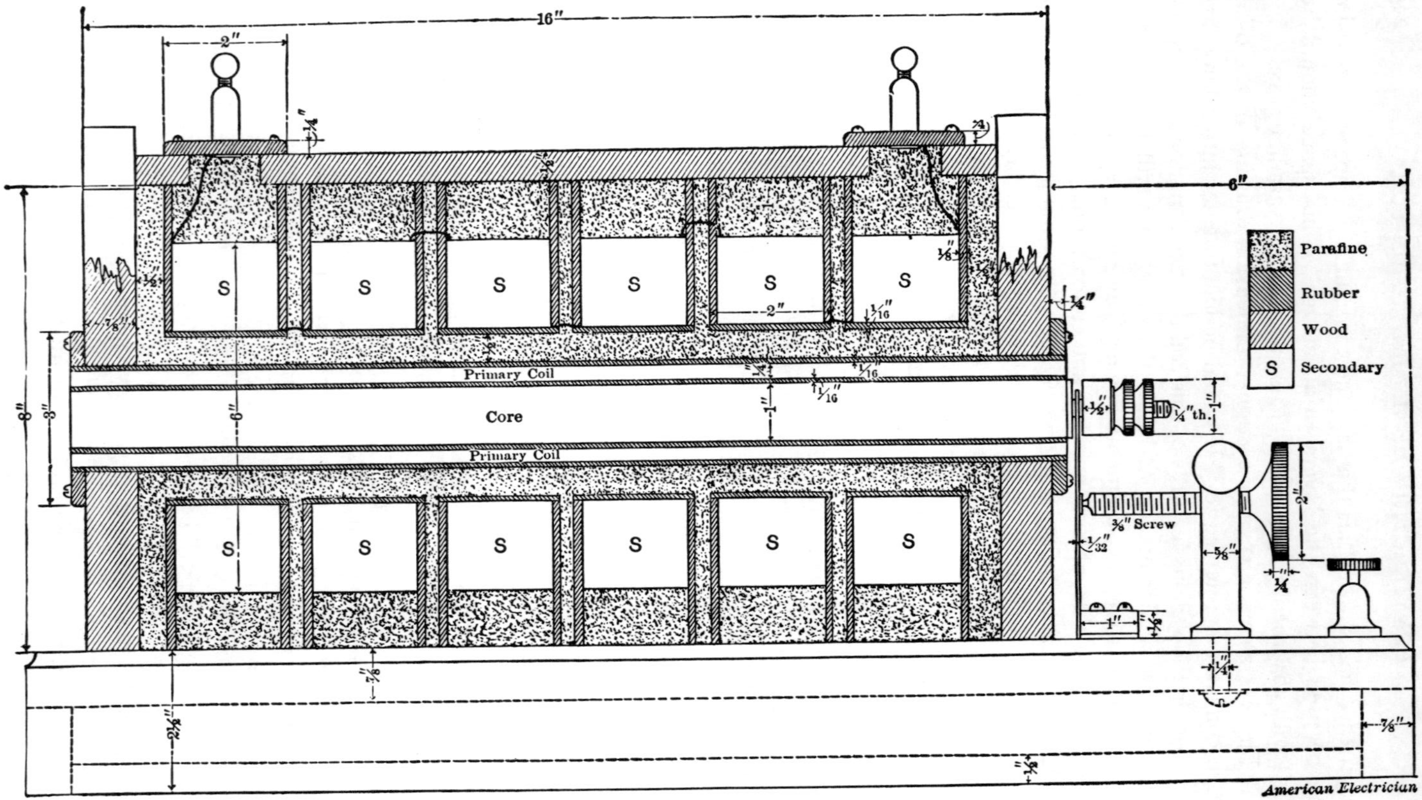

FIG. 254.—COMPLETED INDUCTION COIL.

merit is the continuity of the insulation and the entire absence of joints. To attain this result, the box must be filled with boiling paraffine at all partitions, thus filling up all the air spaces, of course, first making the proper connections. The top of the box is then put on and the paraffine is allowed to set. In setting is will shrink a certain amount, and this space must be filled with more paraffine.

The coil is to be mounted on a box containing the condenser in the usual way. It will be well to divide the condenser into sections, as shown in the diagrammatic connections of Fig. 253. If the coil is to excite a

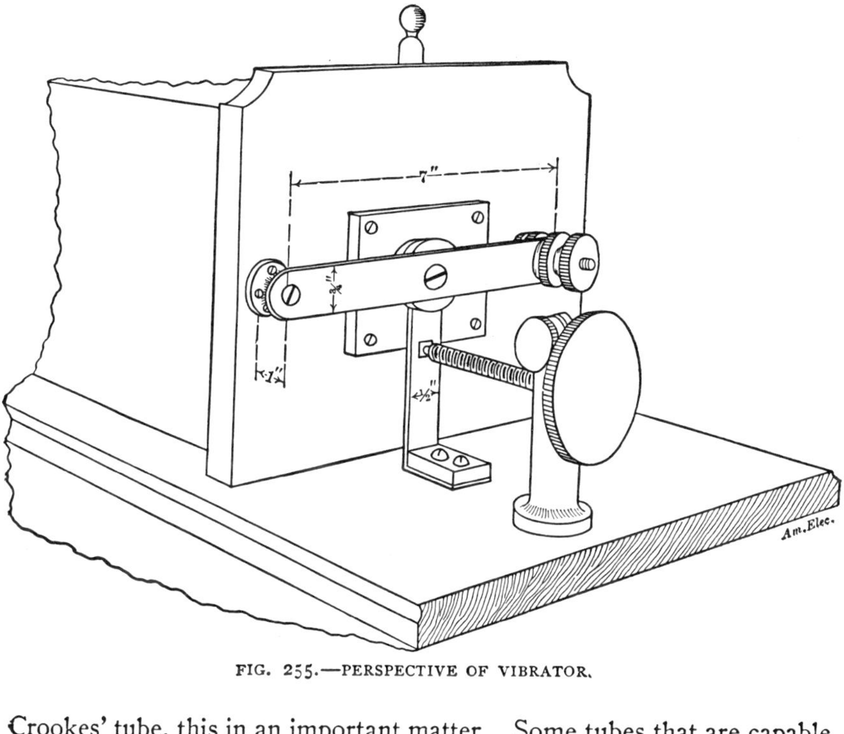

FIG. 255.—PERSPECTIVE OF VIBRATOR.

Crookes' tube, this in an important matter. Some tubes that are capable of giving admirable results often signally fail to do so on a coil of great capacity, but will operate perfectly on a smaller one. The reason of this is found in the fact that the large coil may not be in as close resonance with the tube as the smaller one. By the use of the variable condenser, the resonance of the coil can be varied in pitch and its range of excitation of tubes widened materially. The principal dimensions of the coil just described are as follows:

Primary coil. Two layers of No. 12 B. & S. wire, single cotton covered, wound on a fibre tube and surrounded with a hard rubber enveloping tube, as per drawing.

Secondary coil. Five pounds of No. 36 B. & S. bare wire, wound in six sections, as shown and described.

Support. A mahogany box supporting primary envelope, and glass partitions, as described and shown.

Condenser. Seventy-five sheets of tin-foil 7 ins. × 9 ins. alternated with sheets of paraffined paper 8 ins. × 10 ins.

A word about the secondary connections may not be out of place because of the confusion that has arisen among amateurs. It is customary to wind the secondary coils exactly alike with the outer lead on one flat face and the inner lead on the other. If such similar coils are slipped on the core in the *same way*, it will be necessary in order to connect them in series to join the inner end of one to the outer one of its next neighbor. This will require that the connecting wire must be brought up between sections and in this position it will be very difficult to insulate. Therefore the coils are slipped on in alternate reverse order. By this is meant that if the first coil is put on in one direction, the next is put on so that similar ends face each other. To connect the coils so placed in series, the *like ends* must be connected. A moment's inspection of this connection will show that the current travels about the core in the same direction through all bobbins, and that the arrangement does not connect the bobbins in opposition, as has been popularly supposed.

The circuit breaker or interrupter is one of the most important parts of the coil and little has been done to improve it. The ordinary vibrator is perhaps the most convenient automatic circuit breaker, but it is very defective in many respects. One of its chief faults is that it keeps the circuit open too long and closed for so short a period that the core does not have time to fully charge or the current to attain its full value. An interesting modification that tends to achieve this result is shown in Fig. 255 and is drawn in suitable form to apply to the coil just discussed. Its principle is as follows: The spring, *C*, presses tightly against its contact, *K*, at all times except when it is struck by the hammer of the vibrator, when contact is broken for an instant. Thus the break is instantaneous and the circuit is closed for a definite period of time. The other screws are to limit the motion and frequency of vibration of the hammer.

As indicated in the illustration, a double-pole switch and a means of varying the condenser are to be placed on this induction coil. A double-pole double-throw baby knife switch is the most suitable for the reversing device, and for the condenser a pair of plugs and plates will be found convenient. These are not shown on the drawing because they would tend to confuse the more important details of the vibrator. It is obvious that they should be placed in a convenient and symmetrical position and that further mention of them would be more perfunctory than interesting.

It will be noted that this coil is designed on lines that seem to directly defy all laws of magnetic efficiency with regard to the distance between primary and secondary. Many might hesitate before spending their time and money on such a construction. The reader is assured that the dimensions herein given are the result of a series of progressive experiments, and each coil in the series was constructed with the idea of improving the last. Not until the liberal insulation shown was adopted were maximum results obtained and even now the advisability of carrying the principle further is being considered. The smaller amount of wire and its inferior magnetic position are more than compensated by the absence of leakage and, moreover, the extremely low internal resistance of such a coil enables it to produce a much more highly calorific spark or as it is commonly termed, a fatter one, than if the older and more conventional construction were followed.

CHAPTER XXIX.

CONSTRUCTION OF A TESLA-THOMSON HIGH FREQUENCY COIL.

The following is a description of the construction of a Tesla-Thomson high frequency coil, large enough to give a five-inch spark and excite Röntgen ray tubes.

To excite the Tesla-Thomson coil, a high potential transformer of from 10,000 to 15,000 is necessary. The construction of this transformer will be first given. Fig. 256 gives a partial cross-section of the transformer, which is made as follows: A two-inch iron pipe, sixteen inches long, is slotted the whole length, either in a milling machine, planer or shaper. This slot need not be more than 1-16-in. in width. The pipe

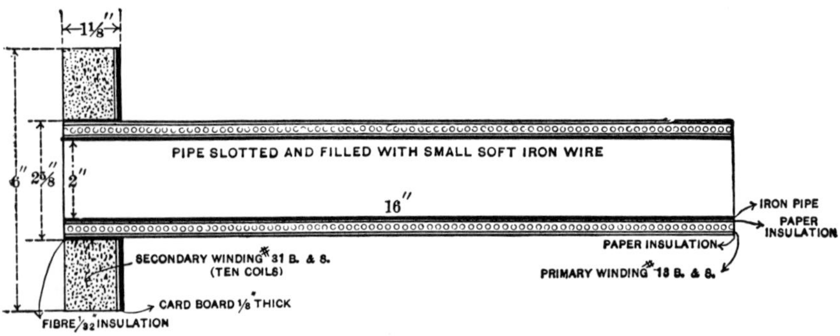

FIG. 256.—HIGH TENSION TRANSFORMER.

is then insulated with ordinary wrapping paper to an outside diameter of 2⅛ ins., shellac being freely used, and is then wound with No. 13 B. & S. double cotton-covered wire for its whole length (one layer). It is then covered with paper and shellacked until the outside diameter is 2⅝ ins.

The next step is to fill the pipe with soft iron wires, No. 16 B. & S., each wire being cut eighteen inches long. This completes the primary winding of the high tension transformer.

The secondary winding of this transformer consists of ten coils wound in a form and thoroughly taped and insulated. This form is shown in Fig. 257 and can be easily made of wood. The wire is wound in this form, shellacked, removed, taped and baked. These coils are then slipped over the primary winding, between each coil being placed a disc of cardboard ⅛-in. thick, care being taken to connect the coils so that none will be in opposition.

The spark gap (see Fig. 258) is made as shown in diagram. The

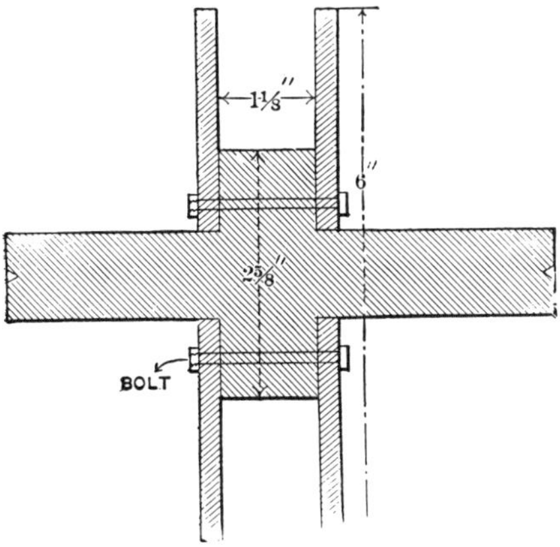

FIG. 257.—FORM FOR WINDING COILS.

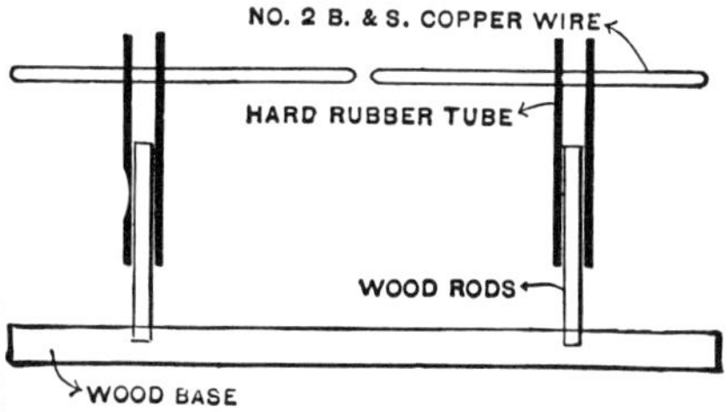

FIG. 258.—SPARK GAP.

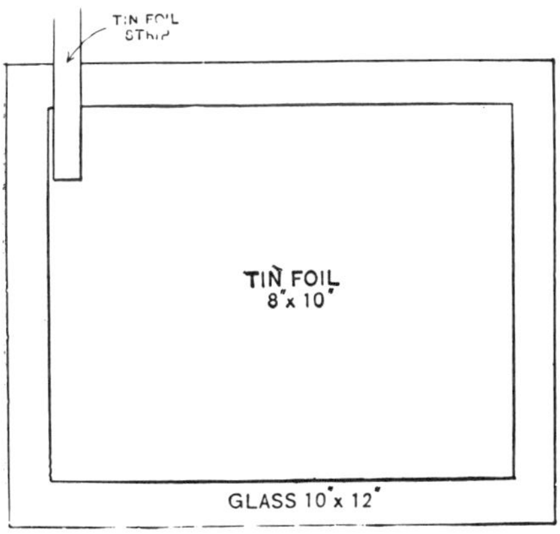

FIG. 259.—CONDENSER PLATE.

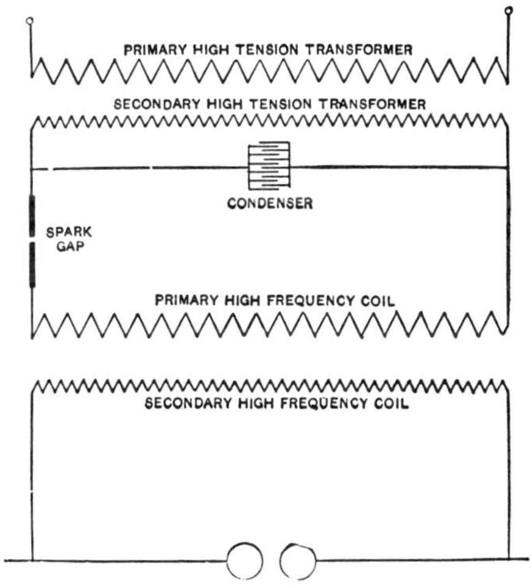

FIG. 261.—DIAGRAM OF CONNECTIONS.

copper wires fit rather close in the holes drilled through the hard rubber tubing, so that the length of gap can be adjusted with ease.

The condenser is made of ordinary 10-in. × 12-in. window glass. A sheet of tin foil 8 ins. × 10 ins. is pasted on one side of the glass with shellac, leaving a margin of one inch. (See Fig. 259.) A strip of tin foil two inches wide is placed across one corner, this strip being placed alternately on each side. For each side of this condenser there should be fifteen plates.

To build this condenser proceed as follows: Place on a smooth surface a condenser plate with the connecting strip projecting on the right. On top of this plate place another piece of glass, 10 ins. × 12 ins., that has no tin foil on it at all. Then place a condenser plate with the strip projecting on the left. Then a piece of glass without tin foil, and on top of this a condenser plate with the strip projecting on the right, and so on. This construction gives two thicknesses of glass between each sheet of tin foil, which is absolutely necessary.

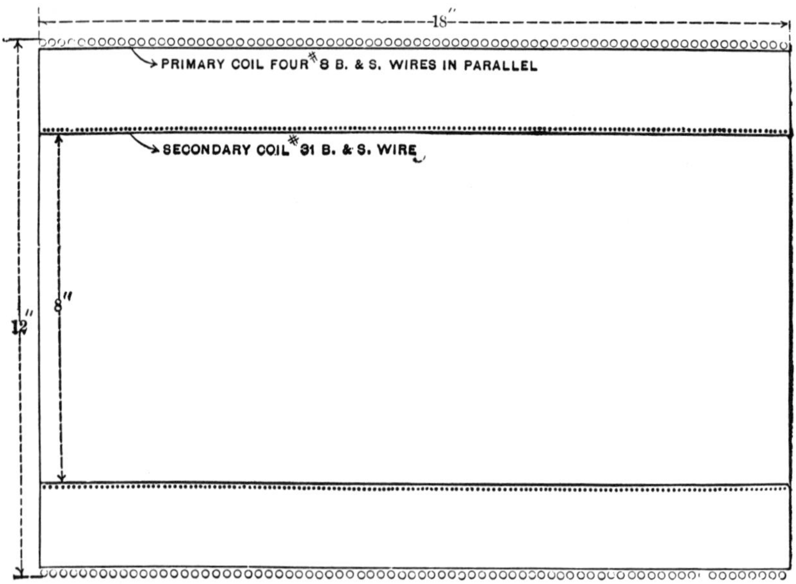

FIG. 260.—PRIMARY AND SECONDARY COILS OF HIGH FREQUENCY TRANSFORMER.

The high frequency coil is made as follows: Wind an 8-inch paper cylinder eighteen inches long with No. 31 B. & S. double cotton-covered wire (or larger), leaving a margin at each end of about one inch. This is the secondary winding. The primary winding is placed on a 12-inch paper cylinder eighteen inches long and consists of fourteen turns of four No. 8 B. & S. double cotton-covered wires in parallel. Each of these No. 8 wires is wound on separately, then the four ends at the beginning and ending are soldered together. Between wires of different polarity, as an extra precaution, two turns of cord are wound. The primary and secondary coils are then shellacked and baked. After being baked, the secondary coil is placed concentrically (see Fig. 260) inside the primary and the connections as shown in Fig. 261 then made.

The primary of the high tension transformer must be excited with an alternating current. With a frequency of 60 cycles per second, 50 volts will suffice, and for 125 cycles per second 100 volts. The length of the spark from the secondary of the high frequency coil will depend on the width of the "spark gap," consequently, in exciting a tube it is best to start with the "spark gap" very short, then gradually increase until the tube is properly excited. When the terminals of the secondary high frequency coil are separated farther than five inches, a spark will pass from the secondary to the primary of the high frequency coil. By the use of a good insulating oil a much longer spark can be obtained from the high frequency coil, but for exciting Röntgen ray tubes a five-inch spark will be sufficient.

CHAPTER XXX.

CONDENSER FOR EXTREMELY HIGH POTENTIALS.

A condenser for high potentials that is commercial has been a problem that has long defied complete solution, and the demand for one that will withstand the enormous potentials of so-called Tesla currents has been only partially met by the clumsy and ineffective Leyden jar. The writer's practical experience with the condenser herein described bears him out in offering it as a complete solution for Tesla currents as usually employed, and a partial solution for the problem of how to get a condenser for high voltage commercial currents.

This condenser has a capacity of about .02 microfarad according to the specific inductive capacity and thickness of the glass that is used. It will replace a battery of fifty or sixty quart Leyden jars and will only occupy the space of a couple of them. If it is stacked in banks of fifty or sixty, a capacity of one microfarad could be obtained, which is sufficient for experimentation on commercial circuits. This condenser can be made by the veriest amateur, at an expense not exceeding $2.50.

Procure of some good-natured photographer a supply of old negatives five by seven inches in lateral dimensions. About one hundred will be needed. Soak them in hot water till the gelatine film has dissolved, rinse them off, and when dry and clean they are ready for use.

A dealer in photographic supplies will sell ferrotype plates 14 ins. × 10 ins, for not more than four cents each. As each plate of this size will make four condenser plates, the total cost of the latter will not exceed $1. The plates should be laid out and cut as shown in Fig. 262. Through the center of each lug should be drilled a ⅛-in. hole. Procure an Edison-Lalande jar 5 ins. × 8 ins. in horizontal sectional dimension, this being a standard size. Select the jar with some care, being sure that the bottom and sides are perfectly flat, for otherwise the condenser plates will not pack in place nicely. Two pieces of ⅛-in. brass rod should now be obtained, together with a box of ⅛-in. copper rivet burrs, and some ⅛-in. standard tap brass nuts. The brass rods should be the length of the Edison-Lalande cell, and should be threaded for some distance on each end. Having obtained about one-half gallon of paraffine oil, the condenser is ready to be put together.

The first thing to do is to pack the jar full of glass and ferrotype plates, so adjusting their number that there will be one less ferrotype than glass plate. If the glass is not too thick, the jar will hold between ninety and one hundred plates, and it should have just enough that the

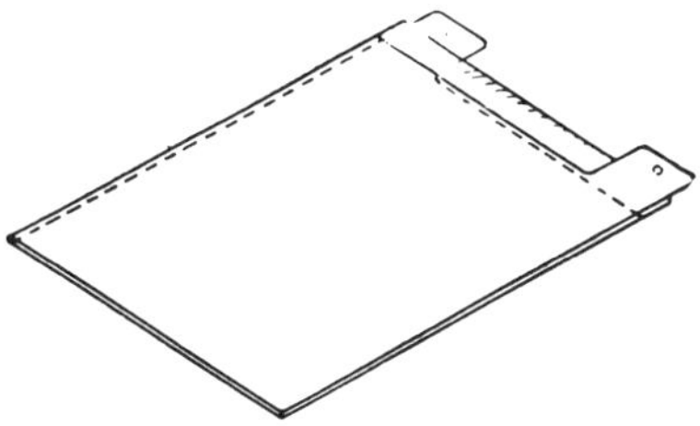

FIG. 263.—ARRANGEMENT OF PLATES.

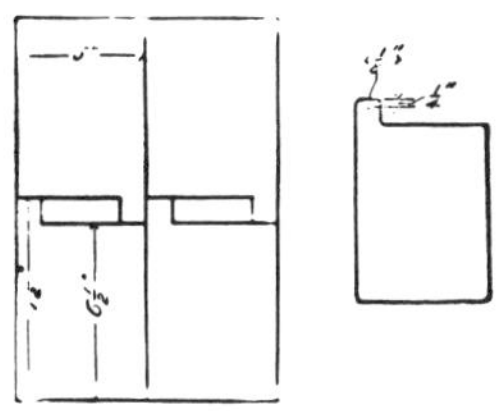

FIG. 262.—GLASS PLATES.

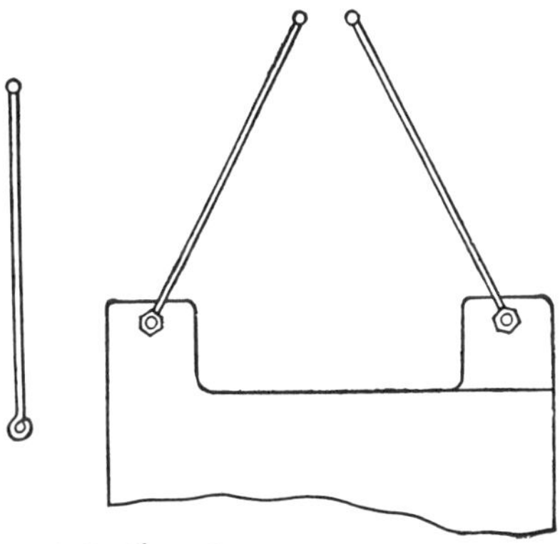

FIG. 264.—TERMINALS OF PLATES.

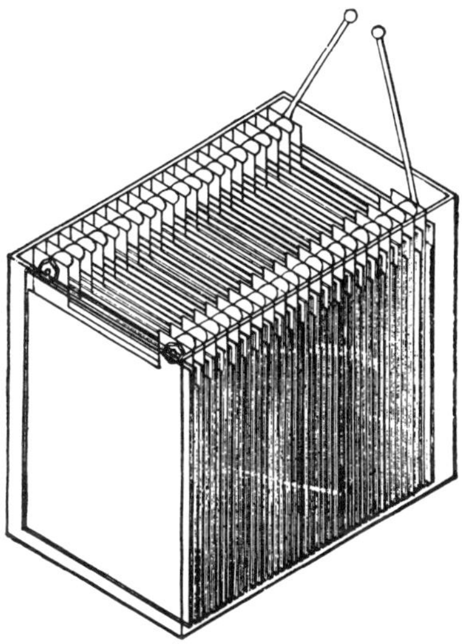

FIG. 265.—CONDENSER COMPLETED.

walls of the jar shall be effective in holding the plates together in a solid homogeneous mass.

The plates of glass and sheet iron should now be arranged alternately, as shown in Fig. 263. The lugs of each set of plates are to be threaded with the brass rods before mentioned, and rivet burrs interspersed so that when the nuts are set up as shown in the sketch of the complete condenser (Fig. 265) the tin plates will not bind the glass plates between them. The terminals may be simple wires, but preferably a ball and knob arrangement as shown in detail in Fig. 264 and in position on the condenser in Fig. 265. After the condenser is thus arranged, it remains to fill it up over the tops of the lugs with paraffine oil and it is complete.

As described, the condenser would be suitable for potentials of 10,000 volts or less. For higher potentials the plates between the conductors may be made thicker. This will reduce the capacity of the condenser both by increasing the thickness of the dielectric and reducing the number of plates that can be placed inside a jar of given dimensions.

The ball and knob arrangement is very simple. Some 1-16-in. brass rod is bent into a ⅛-in. eyelet at one end, while on the other is cast a round leaden bullet. These rods are bolted each to its system of plates on the rod holding the plates together. They will serve to separate the ten plates at the points where they are bolted in, instead of washers, and will bind a sufficient amount to hold them in any position that they may be placed; as they are placed opposite each other the discharge gap may be varied at pleasure.

For use with the higher potentials the jar had better be of hard rubber, for it is liable to be punctured, and if this happens the jar may crack and release the oil, to the great discomfiture of the experimenter.

CHAPTER XXXI.

CONSTRUCTION OF A WIMSHURST INFLUENCE MACHINE.

This machine is the easiest of all static machines to make, and one of the most satisfactory in its results. It is practically independent of the weather conditions. If made as described herein, the machine will be capable of giving a continuous stream of two-inch sparks, and will have sufficient power to excite a small Crookes tube, provided that the terminals of the tube are very near together.

The first and most difficult part of the work is to shape the glass discs. There are two of these and they are made exactly alike. They are to be twelve inches in diameter, and have a ¾-in. hole in the center. Inasmuch as many are not familiar with the cutting of glass into such a shape a few hints will be useful.

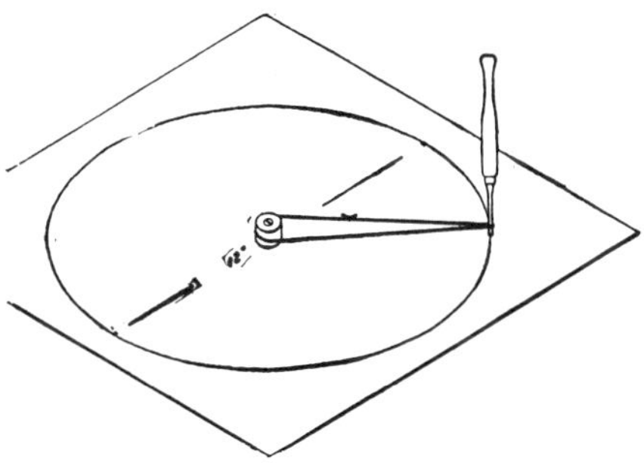

FIG. 266.—METHOD OF CUTTING GLASS DISCS.

Select a piece of window glass of the cheap green variety. Better grade glass contains lead and is less suitable. The hole in the center should be bored first. Prepare a solution of camphor in turpentine and use it to keep the boring tool moist. The boring tool may be made of a rat tail file. The end should be snapped off and the boring performed with a twisting motion of the hand, care being taken to keep the file moist. Patience is necessary, and when the hole gets so deep that it is nearly ready to break through, it is necessary to proceed with extreme caution. Once safely through, the hard part of the work is done. The hole must now be cautiously filed to size, still using the camphor and turpentine as a moistener. A mark to work by may be made by gluing a piece of cardboard carrying a hole of proper size onto the side of the glass.

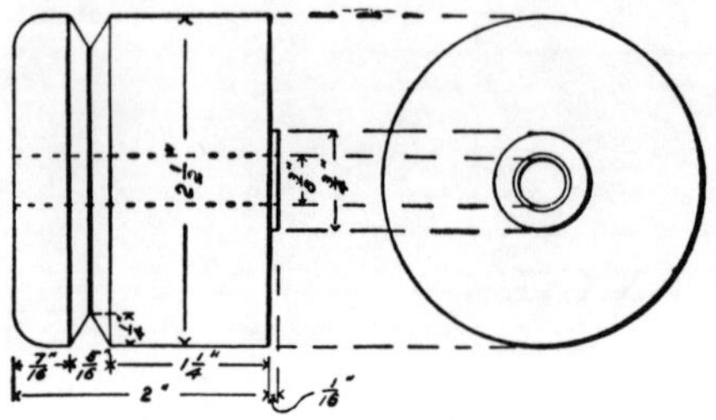

FIG. 267.—HUBS OF MACHINE.

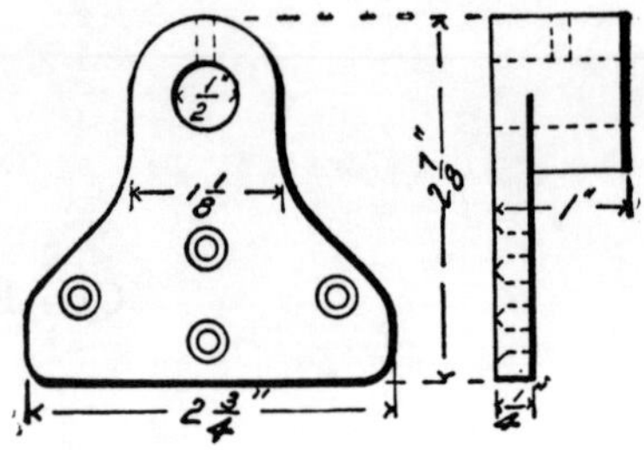

FIG. 268.—SIDE SUPPORTS.

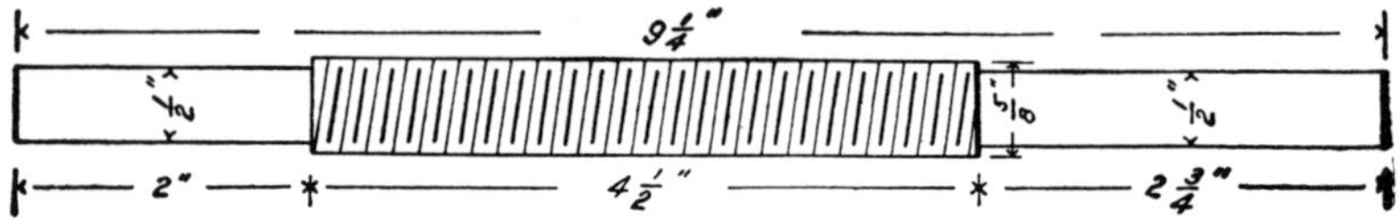

FIG. 269.—SHAFT.

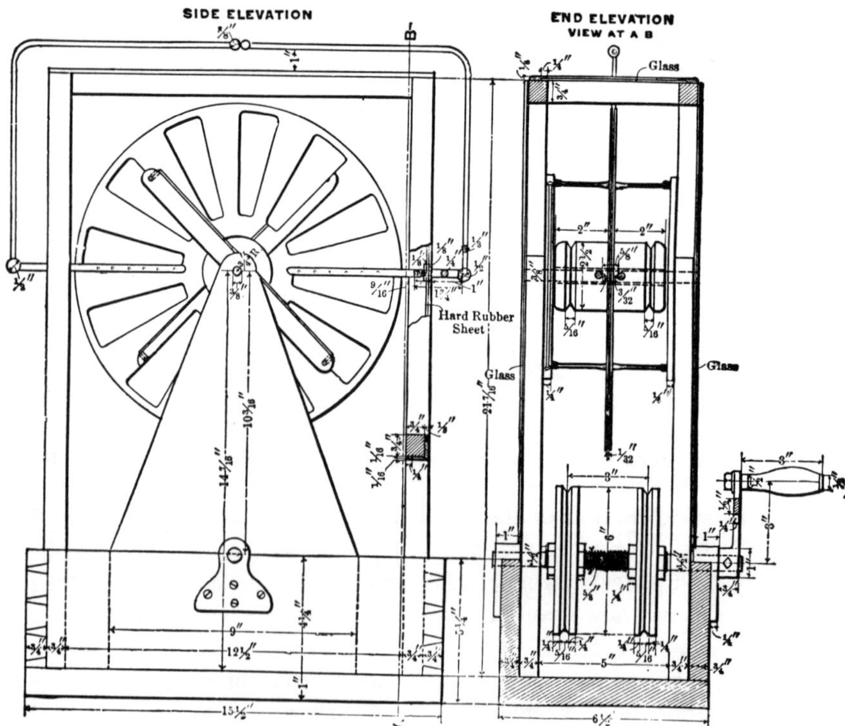

FIG. 270.—ELEVATION OF MACHINE.

Having completed the hole, it remains to trim the edge of the glass into circular form. This is a comparatively easy matter. Erect on a flat surface a little pillar of wood ¾-in. in diameter. Place the glass over this so that the pillar protrudes through the hole. Prepare a loop of string of such length that when it is looped around the pillar as in Fig. 266, the glazier's diamond will swing in a twelve-inch circle. Be sure to use a glazier's diamond, as the use of a cheap wheel glass cutter would be likely to spoil all the work in boring the holes. It may be better to have a glazier snap off the glass if the operator is not experienced in such work.

Prepare the wooden hubs as shown in the drawing (Fig. 267). Bush them with a brass tube ⅜-in. in internal diameter. These hubs are secured to the glass discs with cement. Major's cement or marine glue

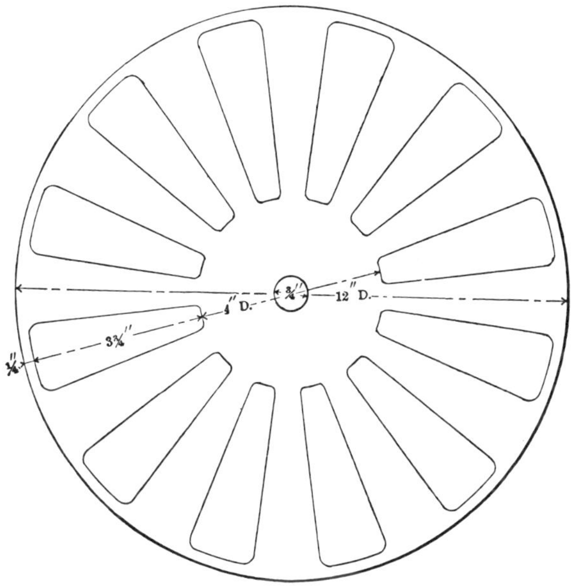

FIG. 271.

is excellent, and bicycle tire cement answers very well. After this is done the discs should be thoroughly shellacked with filtered shellac, and allowed to dry. In the meantime other parts may be prepared.

The side supports are of wood, and hard maple is preferable. They are finished to the size shown in the drawing (Fig. 268) and the holes in the uppper part are of such size as to tightly fit the ⅜-in. shaft they support. This shaft does not revolve. The hubs with their glass discs revolve upon it.

The shaft carrrying the two wheels is the only part that requires the services of a metal lathe; should this not be available, the metal parts can be made for a small sum by a machinist from the figured drawings in this article. In its largest diameter this shaft is ⅝-in., and all of this part is threaded. The ends are turned down to ½-in. journals, as shown in the drawing. One of these journals is sufficiently long to pass completely through its bearing and carry a small crank. The shaft is shown in Fig. 269, and the bearing in Fig. 270. This latter may be cast in brass from a wooden pattern.

The remainder of the wooden parts of the machine may be built and assembled as per drawing. They should be of hard, well-seasoned maple, and thoroughly varnished. The parts should be put together with glue. Nails and screws are to be avoided. They will be necessary to hold the main supports of the machine in place and in some other places where the strain is great, but they should be used sparingly. The whole should be given a coat of shellac varnish.

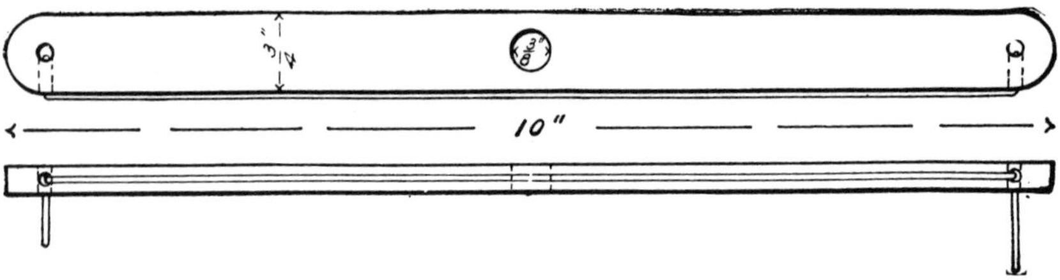

FIG. 272.—YOKE FOR CONNECTING OPPOSITE SECTIONS.

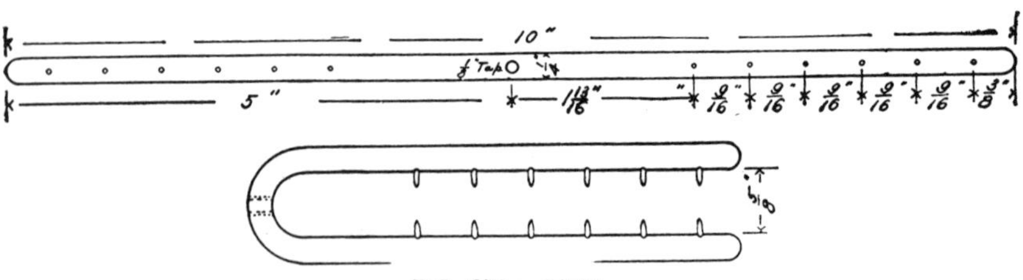

FIG. 273.—COMB.

When the discs are thoroughly dry they are ready to receive the tinfoil sectors (Fig. 271). There are twelve of these to each disc, and they are secured in place at equal angular intervals thereon. Follow the drawings closely and no mistakes can be made. Shellac is to be used as an adhesive, and the edges of the sectors are to be covered with varnish, overlapping at least 1-16-in., to prevent dissipation of charge. This completes the discs.

Mounted on the disc shaft with a tight driving fit are two pieces of hard rubber (Fig. 272). These carry stiff wires, on the ends of which are light brushes made of tinsel. Each rubber piece carries two brushes, one at each end, and the two brushes are electrically connected. They are adjusted so as to just touch the sectors, as the discs rotate and thereby put opposite sectors in contact. Their angular position can be easily adjusted to the position where the working of the machine is bound to be best.

Two U-shaped combs collect the output from the discs. They are conveniently made by drilling a ¼-in. brass rod with holes at suitable intervals and soldering pin points into the holes. The combs may then be bent to shape. In Fig. 273 is illustrated the method of forming the comb. The sides of the enclosure are of hard rubber and serve to support the combs. A small binding post may be threaded into a hole at the curvature of the U of the comb, and with the aid of a few washers the comb is neatly and securely held. See general view, Fig. 274.

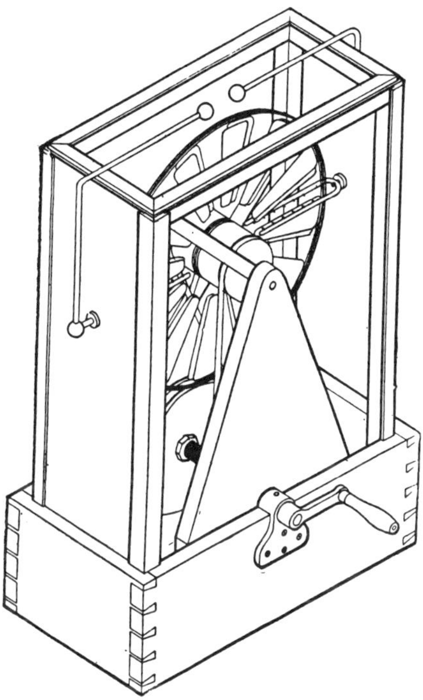

FIG. 274.—GENERAL VIEW OF MACHINE.

The other sides of the enclosure are of glass, both on account of its insulating quality and transparency. The plates are held in place by pieces of rabbeted moulding mitered on to the sides of the upright pillars. If the construction is followed out as shown in the cuts, the glass and rubber plates will lift like a window sash and render the machine completely accessible. The discs are driven in opposite directions by means of a straight and a crossed belt from the shaft below. In making the metal parts of the machine, all sharp corners are to be avoided with great care, for at every corner the charge disappears and leaks away.

The person building this machine must not be disappointed if at first trial it does not work at once. If the shellac is the least particle damp the machine will refuse to generate, but once dry it will generate without failure thereafter. The tinsel brushes must make positive contact with the sectors or the machine will not start. They must be so adjusted as to touch opposite sectors simultaneously. The best working angle for the tinsel brushes is 45° with the horizontal. The discs should rotate from the comb towards the nearest tinsel brush.

The entire cost of the machine, assuming that all of the metal working that requires the use of machine tools is hired out, should not exceed $5.

CHAPTER XXXII.

TELEPHONE TRANSMITTER AND RECEIVER.

The only thing that prevented Philipp Reis being honored the world over (as he is to-day in Germany) as the inventor of the telephone, was the fact that he could not—or those who have since tried cannot—make his first instruments talk. It is said that the difficulty now is to find a microphonic instrument of any kind—his kind included—that will *not* talk. And all the reason in the world is that we know how to adjust a single screw! The whole secret lies in keeping the electrodes together constantly. This is the only real difference between the Reis telephone and the Blake transmitter, which is in use all over the world and has proved the best all-around instrument on the market.

For talking, a Blake transmitter and a form of the standard Bell receiver will be found the best. The patents on both of these instruments have expired, and they can, therefore, be made and used by anyone at present.

The Blake transmitter is illustrated in Figs. 275 to 281, and the receiver in Fig. 282. The receiver is the easier to construct and will be described first. Procure a straight bar magnet of the best tool steel, hardened glass-hard and strongly magnetized (Tungsten steel is preferable). It should be long in proportion to its thickness, in order to maintain its magnetism well—say, ¼-in. thick by 7 ins. long.

Take three-sheet bristol board and cut out two discs, about 1½ ins. in diameter. Thin hard wood discs will also answer. Cut at the center of each a hole of the exact size of your magnet and slip the discs on one end of the magnet, ¼-in. apart. One-sixteenth of an inch of the end of the magnet should protrude. Wind a single winding of thin paper around the magnet, between the discs, for insulation. Then wind No. 36 silk-covered copper wire carefully between the discs until the space is nearly full. In winding, the best results can only be attained by the greatest care. There must be absolutely no kinks or twists in the wire and no breaks in the insulation. A little melted paraffine or a little shellac can be placed over the outside layer to hold the wires, and you should leave several inches of ends for connecting up.

Now take a sound piece of hard wood and turn up a case. In the illustration (Fig. 282) this is lettered *R*. It should be 1 in. in external diameter, except at the large end, which should spread out in a bulbous form, as shown, with a diameter of 2½ ins. outside at the edge. Turn up a cap of the same wood, in the shape shown. Its outside diameter is 3 ins., and it should fit neatly over the end of the case. The cap is lettered *A*. A depression should be turned in the outer face, as at *b*, and a similar, but shallower, depression, as shown in the inner face. In the center should be a hole, ½-in. in diameter. Drill small screw holes in the side flanges for the screws, *d*.

Through the center of your case bore out a straight smooth hole just

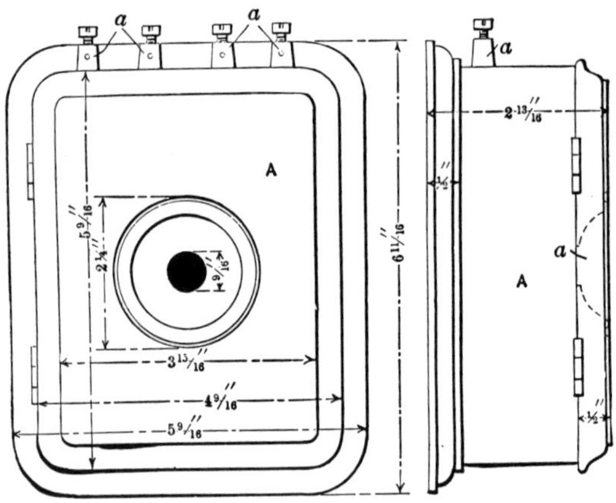

FIGS. 275 AND 276.

large enough to receive the magnet, and at the large end of the case turn the hole out to 2 ins. in width and ¾-in. deep, as shown.

Drill a screw hole for the set-screw, *S*, near the small end of the case. Now slip the long end of the magnet into place. If it proves too loose, wrap thin paper about it.

Take a pair of compasses and lay off on cardboard a circle with 1¼ in. radius—to fit exactly between the cap and case. With this as a template scratch a similar circle on a piece of photographer's ferrotype plate —"tintype" plate—which you can buy for a nickel. Cut out the disc with sharp scissors. Take care not to bend it, as this is the reason the compasses are not used directly on it. Any bend or buckle spoils it. This is the diaphragm. Now take two pieces of No. 16 wire, scrape the ends and make a kink in each. Solder the ends of your fine wire coil to them and pass them through the side of the case. They are shown at *w* in Fig. 282. The object of the kink or knot becomes apparent when you insert it. The knot comes against the side of the case, and prevents any pull coming on the thin wire to break it. Of course, if you want to take the trouble you can cut a channel each side of the magnet hole all the way back to the rear cap, *c*, and put binding posts there. Now put the diaphragm half over the large end of the case, and adjust the magnet until it nearly touches; 1-32-in. is the proper clearance. Then screw down the set-screw, *S*. Put on the diaphragm (shown at *c*, place the cap, *a*, over it, screw in the holding screws, *d*, and glue on a covering disc, *e*, for the rear end, and your instrument is complete.

It is advisable, in making the cap, *a*, to have the inner face over the

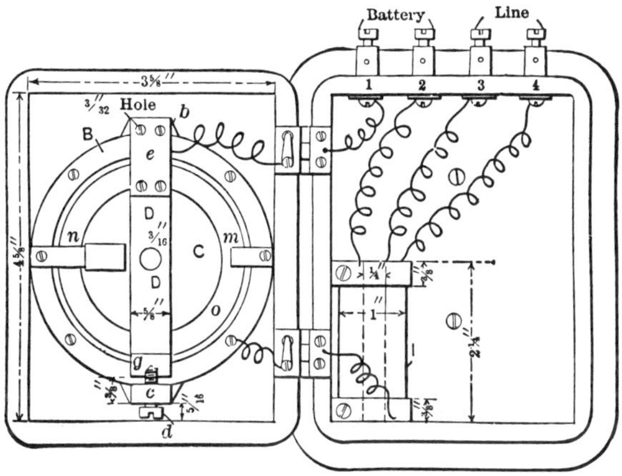

FIG. 277.

diaphragm with a clearance of not more than 1-16-in. If there is too much space between the cap and diaphragm the sounds received will be muffled. The cap should be firmly adjusted against the diaphragm at the edges so as to clamp it against rattling.

In making the transmitter the first thing is the case, *A* This is shown in Figs. 275 and 276. It should preferably be of hard wood, nicely smoothed, filled, rubbed and polished. The door of the case carries all the operative parts of the transmitter and is provided with brass hinges, which form part of the circuit, so that by simply opening the door you can get at the apparatus without breaking any connection. The mouthpiece, *a*, is simply a depression turned in the door face, as shown in dotted lines in Fig. 276.

The transmitter proper is all carried by the iron ring, *B*. This is best shown, with its dimensions, in Figs. 278 and 279. As seen, it is a flat ring having a circular rabbet or depression to receive the diaphragm, and an upper and a lower lug, *b*, and *c*. It is screwed directly on the inside of the door by screws, as shown in Fig. 277. It should be turned up smooth and true, and the extreme thickness of metal may be about ⅜ in. on the edge. The upper lug, *b*, is tapped in its face for two screws, and the lower lug, *c*, is vertically tapped for one, the adjusting screw, *d*. Carried on the upper lug, by the spring, *c*, as shown in Figs. 277 and 281, is the bar *D*, to which the electrodes are attached. This bar has an upper or head lug, *f*, and an inclined foot, *g*. The head lug carries the springs, *h* and *j*, carrying the electrodes, *E* and *p*. These springs are simply clamped to the head lug, *f*, the first one, *h*, resting directly

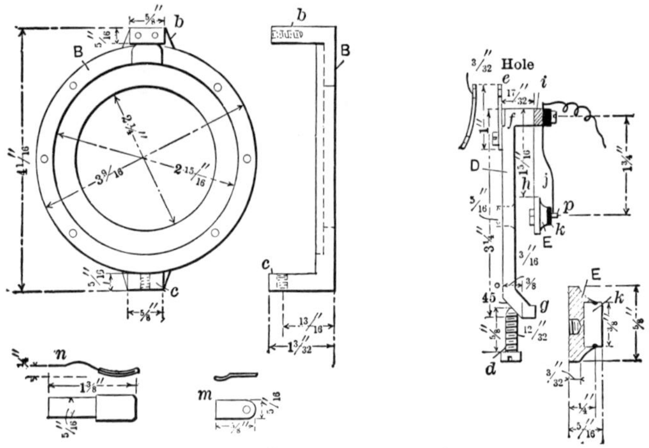

FIGS. 278, 279, 280 AND 281.

against the metal, and held on by the insulating block, *i*, against which rests the spring, *j*, clamped in turn by another little insulating block. The screws pass through both insulating blocks and are tapped into metal head lug, *f*. To the spring, *j*, *above* the insulation, is soldered a wire, forming one terminal of the transmitter.

In operation, the point, *p*, carried by spring, *j*, rests against the diaphragm, the spring, *c*, however, constantly tending to carry it away; so that by screwing up or down the screw, *d*, the pressure may be accurately adjusted.

The spring, *h*, is of fine spring steel, 1-100 in. thick and 9-64 in. wide. Secured to the end, either by clamping under a screw head, or by cutting a channel across and upsetting the edges over the spring, is a brass button, *E*, carrying the carbon electrode, *k*. The way to make this is as follows:

From a solid piece of brass turn up the button of the size indicated, leaving the edges very thin and sharp. Cut the carbon button, *k*, accurately and channel around it a shallow groove just where the edge comes. Then put the whole in the lathe and spin the edge around into the channel so that it tightly embraces the carbon.

The carbon button, *k*, must be pure homogeneous carbon free from grit, and highly polished on the contact surface. The way to get the

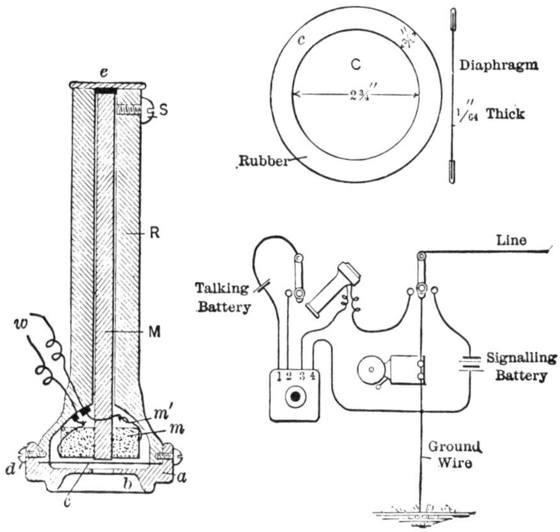

FIGS. 282, 283 AND 284.

best polish is to rub the button for a while on a smooth sheet of the same kind of carbon. As this is not usually available, however, you will probably find most convenient the old reliable emery paper or cloth. Take a piece of fine emery cloth about 6 ins. square and rub your button (which you must leave about ⅛ in. too high when mounting) on the emery, in a 3 in. circle. Keep it moving always in the same direction and after a while the carbon deposited on the emery will form a fine polishing surface, and give you a glass polish. Be sure, however, to use none but the finest emery.

Attached to the spring, *j* (which is of German silver, .005 in. thick, ⅛ in. wide), is the platinum point, *p*. This can best be secured by solder, and should be 5-64 in. across. A tiny end of platinum wire put through a corresponding hole in the end of spring, *j*, and soldered, is all that is required.

In the bar, *D*, is an opening, opposite the electrode to permit adjusting. The diaphragm, *C*, rests in the depression in the ring, *B*. Around its periphery is stretched a rubber band, *o*, to deaden or dampen the vibrations to some extent. It is held in place by two spring arms, *m* and *n*, made of flat spring steel, and shown best in Fig. 281. The arm, *n*, extends over beyond the rubber sleeve, covering the end that rests on the diaphragm. This produces a dampening effect that is very necessary because of the delicacy of the contacts in this form of instrument. The other arm, *m*, simply extends on to the rubber, and serves merely as a clamp. Both spring arms should press lightly on the diaphragm.

The diaphragm itself is to be made of sheet iron. Ferrotype iron, much heavier than that used for the receiver, is required.

One side of the circuit through the transmitter leads from the iron ring, *B* (to which the wire is soldered), to a spring on the lower hinge, *H*. This spring (one on each hinge) makes a scraping contact with the other leaf of the hinge when the door is shut, and so ensures a good contact there. The other side of the circuit leads from the spring, *j*, to the upper hinge. The current from the battery enters at binding post 1, Fig. 277, flows to the upper hinge, through the wire to the spring, *j*, platinum tip, *p*, carbon, *k*, brass button, *E*, steel spring, *h*, iron bar, *D*, screw, *d*, lug, *c*, ring, *B*, and wire, to the lower hinge; thence to the primary of the induction coil, *I*, to the second binding post 2, and so back to battery.

The secondary of the induction coil is connected to the binding posts 3 and 4. The induction coil itself may be made as follows:

Take a bundle of very soft and fine iron wires, 2¼ ins. long, and enough to measure ¼ in. or ⅜ in. through. Wrap a turn or two of thin tough paper about them, and fit on either end a square block of wood, ⅜ in. thick and 1⅛ ins. on a side. Wind between these blocks and on the paper, about 35 ft. of No. 24 silk-covered wire. The ends should be carried out through fine holes drilled in the wooden end pieces.

Over this primary winding lay on carefully about 600 ft. of No. 38 fine silk-covered wire, and carry the ends out at one end in a similar manner. Cover the coil with a wrapping of binder's paper gummed fast on the edges, and fasten the coil in the position shown in Fig. 277, by two long screws through and through the ends into the back of the case. Carry the ends of the secondary winding to binding post screws, 3 and 4, and solder them. Connect one end of the primary to binding post 1 and the other to the lower hinge, as shown.

In winding it is important to wind in regular layers from end to end, and to avoid the slightest kink or twist in the wire.

The connections of the instrument are clearly shown in Fig. 284. The switches cut off the bell and put on battery, when moved to the right, for calling, and cut in the telephone and close the local battery for talking when moved to the left.

If an outdoor line is used it is advisable to use some form of lightning arrester, which may be obtained from a dealer at small cost, *outside* the instrument.

In fitting up a telephone line for communication there are four elements necessary at each end—a transmitter, a receiver, a call-sending device, and a call-receiving device. For the purposes of this article I will presume that the telephone line is a short one, say less than a mile in length—perhaps 1000 yds., with a single wire, of iron, No. 12, galvanized. At each station you bring the line indoors to the instrument by connecting office wire at the window and leading it around the woodwork of the room. The joint outside the window must be soldered, the joint taped, and the wire bent down U-shaped before it comes in, to allow moisture to drip off. At your instrument another piece of office wire should be started and led off to the nearest water pipe. The end of the wire should be stripped for 12 ins., cleaned bright, the pipe likewise scraped bright, and the wire wound tightly around the pipe and soldered. If this is done carefully at both ends of the line, you have a good circuit completed over the iron wire from one station to the other and back by way of the pipes and the earth. It only remains to connect your instruments to the wire ends and you should be able to talk perfectly.

For such a line a push button and vibrator bell, with a battery, at each end, will furnish as good a call as may be. The arrangement of these is indicated in Fig. 284. They can be purchased of any supply dealer more cheaply than you can make them.

CHAPTER XXXIII.

CONSTRUCTION OF A DRY BATTERY CELL.

Dry batteries, so called, are only dry in the sense that there is no fluid spilled or slopped over when they are shaken or overturned. In every voltaic cell the current is derived from the chemical action which goes on within its substance, and no chemical action can take place between solids alone, but in all cases there must be present a liquid or a gas. Some few cells employ gaseous electrolytes, and some, fused salts, but the vast majority use aqueous solutions of their respective chemicals, and it is in this class that the ordinary dry cells are to be found. Even the old dry piles of Zamboni and others, which consisted of discs of paper coated with metals (gold and silver paper) laid up "dry," in reality contained a very small amount of moisture in the paper, and if the paper is really perfectly dry, the piles will not work. If the ordinary dry cell then requires moisture to make it work, and is in fact only a non-spilling wet cell, it is a natural inference that the wetter the cell, consistent with it not spilling or slopping over, the better. This inference is absolutely correct. The more fluid a dry cell contains the better, for many wet cells would be improved for having a larger amount of electrolyte than they do have.

It might seem in view of what has been said, that any wet cell, if well sealed up, would do for a dry cell, but such is not the case, for several reasons. Many cells will not stand sealing up tight, because they give off gases, and these must have free vent, and again it is not always practicable to seal up a cell so that it will not leak at all when inverted. Again, it is not worth while to seal up a cell, except one of the kind that will last for considerable time before it gives out or even needs replenishing. Another desideratum of a dry cell is that it should not be easily broken, as there are many places where cells are liable to fracture as well as upsetting, etc. For these reasons, it is customary to dispense with the glass jar, and to make the zinc serve the double purpose of containing jar and electrode, and further, to use an absorbent substance that will take up and hold the fluid electrolyte like a sponge, so that the seal is rather to prevent evaporation and creeping of salts, than spilling or slopping. The types of cells giving the best results on open circuit work, as wet cells, naturally do the best when put up in the dry form; consequently, as might be expected, the vast majority of dry cells on the market are some form or variety of the sal-ammoniac type. Several of the manufacturers of dry cells claim to have valuable secrets relating to their manufacture, but however true this may be in regard to the details, the main requirements are well understood.

In making a dry cell, the first thing requiring attention is the jar. This, as before remarked, is usually made of zinc. The cell is usually cylindrical, although sometimes square or oblong in section, but in any case a piece of moderately heavy sheet zinc is bent into the required form, the edges soldered together and a bottom soldered in. Any one who tries to solder zinc for the first time may be very much surprised and disgusted to find that it does not take kindly to soldering like tin plate, but balks and makes lots of trouble. However, by observing the proper precautions, zinc may be soldered with comparative facility. Thoroughly clean all the parts where it is intended that the solder should stick, by scraping; use clean chloride of zinc for a flux, and apply the solder as near the point where it is needed as possible, not trying to make it flow over the surface of the zinc as can be so readily done with tin plate, for the more the solder alloys with the zinc, the more intractable it becomes. Learn to make a joint quickly on the first application of the soldering bit, as the more you fuss and tinker with it the rougher, more unsightly and more uncertain it becomes. Another pleasant little habit of zinc is to strip the tinning off the soldering bit. You may have tinned your bit with the utmost care, but after using it a short time find it completely stripped. Some persons prefer an iron bit to the usual copper one, claiming that it holds the tinning better, although somewhat more difficult to tin in the first place.

Having made the jar, the next thing to attend to is the contents. One of the most important constituents of the contents is the absorbent. Several materials have been used for this purpose, among which are plaster of paris, gelatinous silica, gelatine, so called, which is really the starchy mass obtained from boiling Irish moss; gelatinous magnesium oxychloride and a material made from the granular portion of the rind of the cocoanut, called cofferdam. There are other materials, but these are the more important ones. As the process of filling the cell differs somewhat for each kind of filling, it is better to describe each one separately. The zinc usually has a brass binding post soldered to its rim on one side (there being no objection to this structure in a dry cell, because the electrolyte cannot possibly come in contact with the junction), and the other electrode, which is usually of carbon, has a binding post of the same kind, fastened in one of several different ways.

The seal is made of pitch or some similar material, which will form an air and water tight stopper, and also resist the tendency of the salts to creep, as does the paraffin coating on the upper part of the jar of an ordinary sal-ammoniac cell. It is simply melted, poured in on top of the charge and allowed to cool in most dry cells, but some manufacturers make a sort of safety valve or pressure regulator, by inserting a small piece of rattan so that it passes completely through the seal, its ends projecting slightly above and below the pitch. The natural porosity of the rattan is sufficient to relieve any pressure generated by the escape of gases, but it will not of course provide for the swelling of the more solid portion of the contents which sometimes takes place when the cell is subjected to too high a temperature, and cells are frequently destroyed by bursting when placed in boiler rooms and other situations where the temperature rises to an inordinate degree.

One of the important qualities of a dry cell is long life on open circuit, which means that the local action should be negligible, and it is in the prevention of local action that some manufacturers claim to have valuable secrets. Bi-sulphate of mercury is sometimes used to keep the zinc amalgamated, as in wet cells. This, of course, does some good, but it is not all. Anything that will tend to keep the chemical composition of the electrolyte uniform in all parts of the cell will help to prevent local action.

We will now consider some of the particular forms of dry cells. The Cox cell is formed by boiling Irish moss in sal-ammoniac solution until it is thoroughly gelatinized, and pouring it into the zinc jar, where the carbon electrode has been already placed. Bin-oxide of manganese in conjunction with the carbon as a depolarizer is, of course, used as in the wet form. The inventor also mixes a little bi-sulphate of mercury with the electrolyte. Some of the other sal-ammoniac cells as described do not use it, but there is no reason why they should not, and the reader should understand that he may use it or not in the other cells described. When the moss solution is cold it sets to a firm jelly, and is then ready to be sealed. Obach's cell is made by mixing plaster of paris with the sal-ammoniac solution and pouring it into the jar to harden. Mehner's cell is made by mixing the sal-ammoniac solution with chloride of calcium and calcined magnesia, forming a paste of about the consistency of cream, which is poured into the jar, and in two or three days forms a stiff jelly, owing to the formation of oxy-chloride of magnesium. In Gassner's cell the following composition is used: Oxide of zinc, 1 part; sal-ammoniac, 1 part; plaster of paris, 3 parts; chloride of zinc, 1 part, and water, 2 parts, all by weight. The oxide of zinc is intended to make the plaster more porous, giving this cell an advantage over the simple plaster cell before described.

Gelatinous silica is precipitated when any strong acid is added to a solution of silicate of soda, and several inventors have used silicate of soda to gelatinize the electrolyte in storage cells. This, of course, introduces sulphate of soda into the electrolyte, but this does no harm, and is even regarded as beneficial by some. The charge and discharge rate of the cell is much reduced, and it will not do to allow much gas to be generated into the jelly, neither can the cell be sealed up perfectly tight, but a vent must be left for the escape of gas. The spattering or spraying during charge, however, is cured, even if no cover is used.

Gelatinous silica may be used in any electrolyte in which the sodium salt, formed at the same time with the silica, is not detrimental. It is very difficult to wash the silica, and it does not pay to do it for such a purpose as this, so if it is intended to use the silica with sal-ammoniac it is better to precipitate it with hydrochloric acid, instead of sulphuric, as in that case the precipitated silica will contain chloride of sodium, instead of sulphate.

The Germain cell, which at one time attracted considerable attention, used the material known as cofferdam, previously mentioned in this article. The containing vessel of this cell is made of wood boiled in paraffine, the carbon plate is imbedded in lumps of peroxide of manganese and carbon, and the rest of the space is filled with cofferdam saturated with sal-ammoniac, the zinc, which is well amalgamated, is laid on top, and the cover (or rather the side) is screwed on, slightly compressing the contents.

It will be observed that some of the directions for the manufacture of dry cells are very particular about the amalgamation of the zinc, even when the electrolyte is sal-ammoniac, but, as a matter of fact, most of the dry cells on the market are made with unamalgamated zincs, which is a pretty good proof that the amalgamation is superfluous. There have been many other dry cells proposed, some quite elaborate in composition and others less so, but the different forms of the sal-ammoniac cell are at the present time used to the exclusion of everything else.

One practical difference between wet and dry cells is the different manner of using them. In the wet cell, we endeavor to use some of the parts, such as the containing jar, indefinitely, and the porous jar, if there is one, at least a very long time. We also endeavor to use up the zinc as completely as possible, excepting to renew the electrolyte several times before the zinc is all gone. With dry cells, however, it is the common practice to throw them away as soon as they fail from any cause to do their work. The zinc being also the containing jar, it manifestly cannot be all used up, but on the other hand must be discarded as soon as it is perforated, if indeed something else does not give out before this event, as is intended.

Dry cells are intended more particularly for a class of consumers who do not care to be bothered with the manipulation of wet cells. When they come from the factory they are ready for use without any preparation whatever, and when they are exhausted they may be thrown away without compunction, for they are made for such a low price that it does not pay to spend much time or trouble on them, even when they happen to be in such a condition that they may be restored, which is not usually the case.

The zinc electrode-jar must be insulated in some way, especially when several of the cells are used in series, as in this case any external conductor touching two of the cells would short circuit at least one of them. For this reason they are usually varnished, and often in addition placed in strawboard boxes.

Judging from the variations in the different cells on the market, and which work satisfactorily, it would appear that the exact proportions of the ingredients are not very important. In the nature of things the electrolyte is the first thing to give out, and there is no danger of getting too much of it, and the solution should be saturated. The carbon and manganese should not take up too much space, and as in the wet sal-ammoniac cells the manganese is usually thrown away before it is exhausted, from the fact that the inner part of each lump is unavailable, it is evident that a small amount of manganese will answer the purpose, if arranged so as to be available, for which purpose it is better to have it rather finely broken than in large lumps, and in as intimate contact as possible with the carbon. There being no danger in a dry cell that the manganese will become displaced after having been once fixed, there is no objection to its being in a state of powder.

CHAPTER XXXIV.

SOME HANDY COMMUTATOR TOOLS.

Direct current dynamos and motors have now come into very general use, exceeding in number, perhaps, steam, gas and oil engines combined. As is well known these electrical machines are subject to mechanical wear at only the bearings and the commutator, which have to be replaced from time to time. The very general demand for dynamo and motor repairs has been met to some extent by electric repair shops that have come into existence at many points through the country, but aside from these there is hardly a regular machine shop of any size that is not constantly called on for more or less work on the worn parts of dynamos and motors.

As bearings in electrical machines are usually fitted with bronze bushings, with standard reamed holes, well equipped machine shops are usually in position to make these parts, but in the matter of commutators there is not one regular shop in fifty with the simple tools necessary for their renewal, and the average machinist has but slight conception of how this work must be done to insure satisfactory results.

In electrical repair shops there are usually some tools for handling commutator work, but they are frequently of the crudest kind and such as to require too much labor, and even then lack certainty in results. Some even among dynamo and motor manufacturers lack the few and simple tools necessary to insure first-class commutator construction with a minimum of labor. As usually constructed, a commutator contains a number of copper "bars" or "segments," separated from each other and the clamping parts by mica strips and rings. These bars are parts of true circular sectors, though not of the exact circle of the commutator surface, and are held, with the intervening mica strips, by a sleeve and clamps.

Some of the main requirements of commutator construction are that each segment be insulated or free from metallic contact with any other or the sleeve and clamps, that each segment be so firmly held that the forces of expansion, due to heat when the machine is in use, shall not alter its relation to the other segments, and further that the relative position of segments and mica strips shall remain the same after the commutator has cooled.

Were all parts of commutators metal, the above requirements would make careful work necessary, but as each segment must be held entirely by contact with mica, the problem is much more difficult, in fact, it has

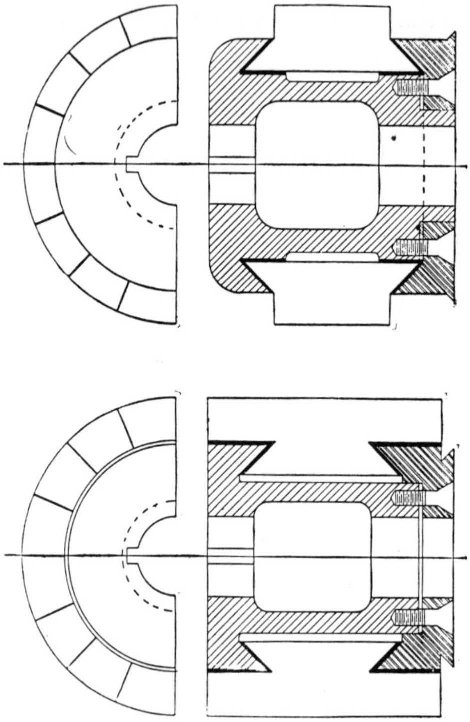

FIGS. 285 AND 286.

required more study and experiment than any other mechanical question that electrical manufacturers have had to meet.

Mica has come into very general use for commutator insulation because of its high insulating properties, non-injury by heat and power to sustain great pressures with but small compression. On dynamos and motors of moderate capacity in common use the number of commutator segments varies from six to eight to about two hundred according to the purpose and capacity of the machine, the most common numbers of segments being from twenty-four to one hundred.

As for every segment there must be a strip of mica, in addition to the mica rings, the number of separate parts that must be held in their exact position in a commutator will vary from fifty to about four hundred in machines of moderate capacity and common use. In order to hold so many pieces of materials, to a large extent contrary as to their qualities, rigidly together, it has been found necessary to assemble them with great pressure, and then set the permanent clamps as tightly as possible before the external pressure is removed.

The two most common and successful methods of clamping commutators are shown in Figs. 285 and 286, the solid black lines in each case representing the mica strips and rings. The better class of segments are forged or of drawn stock, so that no labor is required on the sides before assembly in the way shown. As soon as assembled it is necessary to compress and hold the segments securely so that the surfaces which come in contact with the mica rings may be machined.

A method to compress and hold the segments, common in many electric repair shops and with some manufacturers, employs a solid forged ring turned on the inside to the diameter which the segments are estimated to have when compressed, and tapered slightly at one edge so as to start easily over the segments. In the correct use of this ring it should be forced over the segments with a pressure of some tons, as this source of pressure is the only one to bring the segments and mica strips solidly together. To do a good job with this solid ring a hydraulic or large screw press is necessary, and in many cases machine and repair shops are without either of these presses, so that the ring can only be forced on and off the segments with a hammer, a very unsatisfactory method. A serious objection to this solid ring method is that it is very hard to estimate the exact diameter to which the ring should be turned in order to properly compress the segments, and the trials to see how hard the ring crowds on all take time. Again a solid forged ring must be had for every size of commutator, even though they vary by only a small fraction of an inch in diameter, and there is great temptation when a ring goes on too easy to let it go as "good enough." To do away with the necessity for presses, also forged rings for every commutator, save time and insure means for the desired compression in every case, the tool shown in Fig. 287 is now much used.

This tool consists of two rings, the outer a solid forging and the inner an iron casting, split along one side so that its diameter may be slightly changed.

The outer forged ring is fitted with six, eight or more radial set screws, which bear upon the inner split ring. The commutator segments and mica strips having been assembled in circular form, the split cast-iron ring, having been turned as near as possible to the correct diameter, is pushed over them and pressure applied by means of the large set screws in the forged ring. Any desired pressure can be obtained by the combined action of the heavy set screws, the split ring readily conforming to the slightly reduced diameter of the segments. The work can also be done much more quickly than when a solid ring and press are used. It is not necessary to have one of the above tools for each size of commuta-

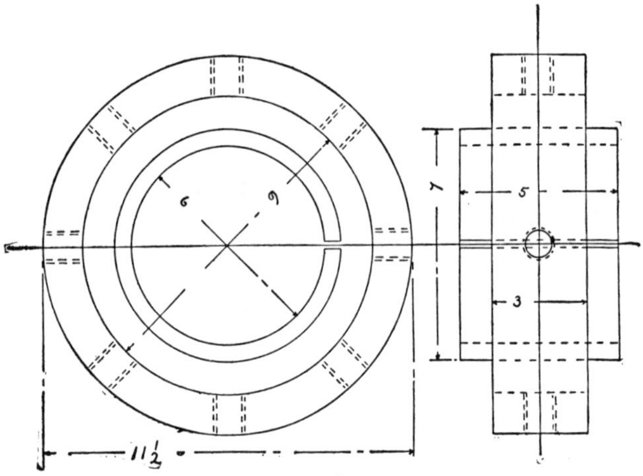

FIG. 287.

tor, as only the inner cast ring need be changed until the commutator diameters vary by as much as three inches, so that the forged ring becomes either too small or so large as to be unhandy. In this way two or three forged rings will cover a large line of commutators, while the cast ring for each commutator is cheaply and quickly made.

When the segments are securely clamped in the ring the next step is to turn up the surfaces to which the mica rings transmit the pressure; the lathe chucks are the only means for holding the clamped commutator segments while the surfaces at each end are machined, and here comes a waste of time and inaccurate results due to the effort, after one end of the segments has been turned up to reverse them in the chuck, so as to machine the other end in line with the first. The tendency with work done in this way is to force the permanent clamps slightly out of line with each other, and this may result in loose segments at some point in the commutator. Proper expansion mandrels, as shown in Figs. 288 and 289, not only enable the finished surfaces at the two ends of the segments to be brought practically into line, but also save much time on the work.

Fig. 288 shows a mandrel adapted for use with segments of the type in Fig. 285, where there is no undercut work to be done, while the mandrel of Fig. 289 is more convenient for undercut segments. The mandrel of Fig. 288 mounts in the usual way on lathe centers, has a taper of one in twenty-four, is fitted with cast-iron expansion sleeve and a screw collar at each end, to force the sleeve on and off. The cast-iron sleeve should be cut entirely through once along its entire length and nearly through, say, to within one-fourth or three-eighths inch at three or more other points, that it may expand as evenly as possibly when forced onto the mandrel.

A sleeve for this mandrel should be turned outside to correspond with the inside diameter of the commutator segments it is intended to mount, and if very accurate work is desired the inside of segments should be turned out before mounting them on the expansion sleeve, though some makers think this unnecessary. When the segments are mounted on the sleeve this latter is expanded by forcing it on the mandrel with the screw collar, and the ends of segments in Fig. 285 can be turned up without changing their position.

The segments shown in Fig. 286 can also be turned on the above mandrel, but the work on the undercut is done at a disadvantage and much time can be saved by the use of the mandrel of Fig. 289. This mandrel, like the other, is on a taper and fitted with expansion sleeve, but one end is forged into a flange, adapted to bolt to a face plate and allow free access to one end of the commutator segments, so that the undercut can be made quickly. One end of the segments being finished, they are forced off the mandrel with the sleeve by the screw collar, and then put on again reversed and the other end finished. There is, of course, no reason to take the mandrel from the face plate until both ends of the segments are finished, so that time is saved and the finished surfaces brought very nearly in line. Quite a number of expansion sleeves can be used on the same mandrel for different commutators, so that two or three mandrels will be enough for a factory turning out a fair line of machines.

Having turned up the ends of the segments, it is next necessary to mount them on the insulating rings and permanent clamps. As segments are held only at the finished surfaces, it is necessary that each

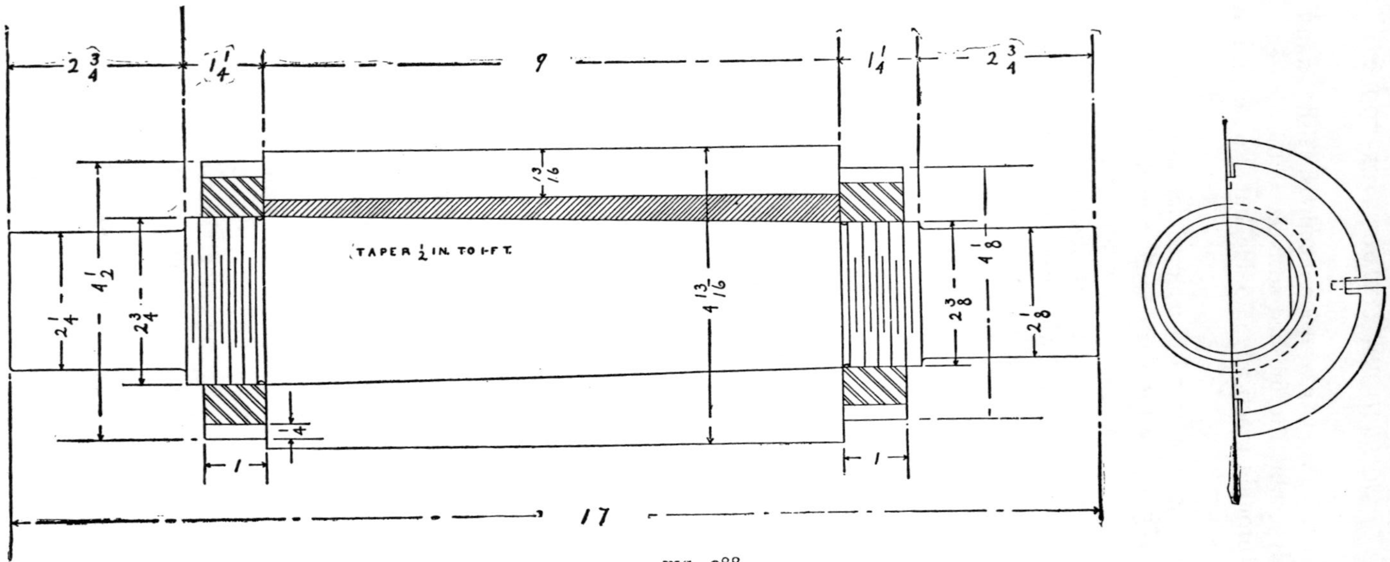
2 3/4
1 1/4
9
1 1/4
2 3/4
13/16
TAPER 1/2 IN. TO 1 FT.
2 1/4
4 1/2
2 3/4
4 13/16
1/4
1
1
2 3/8
4 1/8
2 1/8
17

FIG. 288.

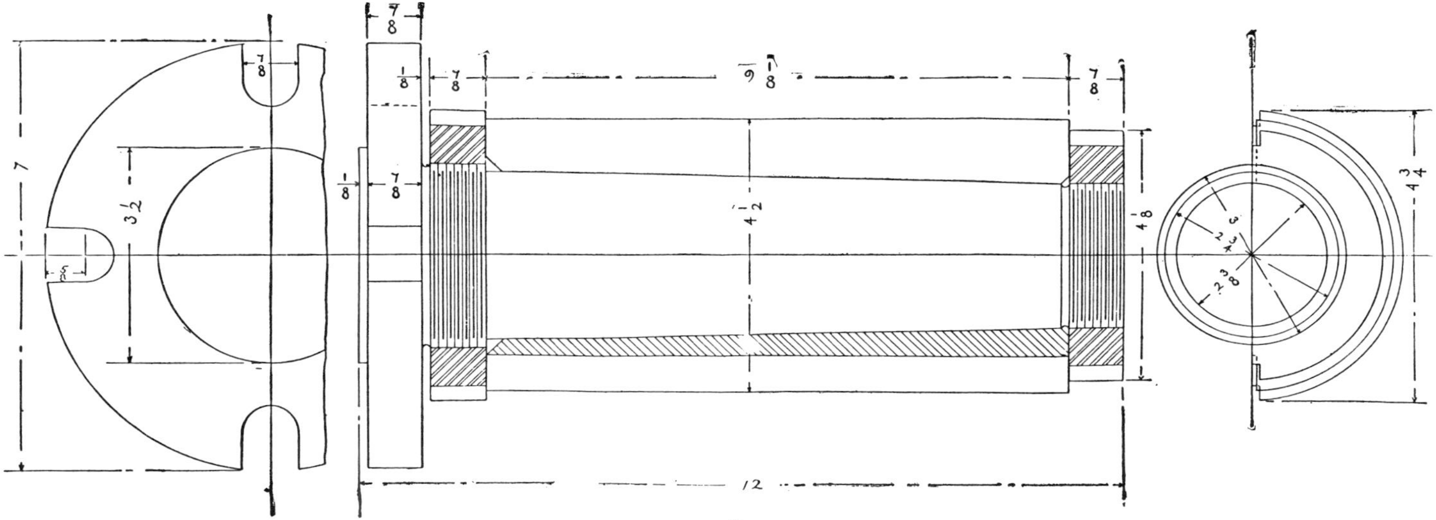

FIG. 289.

make solid contact with all of the mica rings, as the side pressure of the other segments is not sufficient to prevent motion either up or down. The double ring clamps are of special value when the segments are to be brought to a firm bearing on the mica rings, since, when necessary. the set screws can be let up a little in order to allow the segments to slide over the horizontal mica rings, and then set up until each segment beds firmly on the mica. The permanent clamps once in place, the outside rings are removed and the surface of the segments finished up, first by turning with a diamond-pointed tool and later with a single cut file or the finest grade of sand paper.

It may be well to add the oft repeated warning that emery cloth should never be used on a commutator, as the emery sticks in the copper.

The above simple and inexpensive tools will multiply several times the amount of commutator work a man can turn out daily with the devices now common in some shops devoted to the repair and even the manufacture of electrical machinery.